The Insects and Arachnids of Canada

Part 1. Collecting, Preparing, and Preserving Insects, Mites, and Spiders, compiled by J. E. H. Martin, Biosystematics Research Institute, Ottawa, 1977.

Part 2. The Bark Beetles of Canada and Alaska (Coleoptera: Scolytidae), by D. E. Bright, Jr., Biosystematics Research Institute, Ottawa, 1976.

Part 3. The Aradidae of Canada (Hemiptera: Aradidae), by R. Matsuda, Biosystematics Research Institute, Ottawa, 1977.

Part 4. The Anthocoridae of Canada and Alaska (Heteroptera: Anthocoridae), by L. A. Kelton, Biosystematics Research Institute, Ottawa, 1978.

Part 5. The Crab Spiders of Canada and Alaska (Araneae: Philodromidae and Thomisidae), by C. D. Dondale and J. H. Redner, Biosystematics Research Institute, Ottawa, 1978.

Part 6. The Mosquitoes of Canada (Diptera: Culicidae), by D. M. Wood, P. T. Dang, and R. A. Ellis, Biosystematics Research Institute, Ottawa, 1979.

Partie 7. Genera des Trichoptères du Canada et des États adjacents, par F. Schmid, Institut de recherches biosystématiques, Ottawa, 1980.

Part 8. The Plant Bugs of the Prairie Provinces of Canada, by L. A. Kelton, Biosystematics Research Institute, Ottawa, 1980.

Part 9. The Sac Spiders of Canada and Alaska, by C. D. Dondale and J. H. Redner, Biosystematics Research Institute, Ottawa, 1982.

Part 10. The Spittlebugs of Canada, by K. G. A. Hamilton, Biosystematics Research Institute, Ottawa, 1982.

# THE INSECTS AND ARACHNIDS OF CANADA

## PART 11

## The Genera of Larval Midges of Canada

## Diptera: Chironomidae

Donald R. Oliver and Mary E. Roussel

Biosystematics Research Institute
Ottawa, Ontario

Research Branch
Agriculture Canada

Publication 1746        1983

© Minister of Supply and Services Canada 1983

Available in Canada through

Authorized Bookstore Agents
and other bookstores

or by mail from

Canadian Government Publishing Centre
Supply and Services Canada
Ottawa, Canada K1A 0S9

Catalogue No. A 42-42-1983-11E         Canada: $11.95
ISBN 0-660-11385-6                     Other countries: $14.35

Price subject to change without notice

**Canadian Cataloguing in Publication Data**

Oliver, Donald R.

The genera of larval midges of Canada

(The Insects and arachnids of Canada,
ISSN 0706-7313 ; pt. 11)
(Publication ; 1746)

Includes bibliographical references and index.

1. Chironomidae. 2. Insects — Larvae. 3. Insects — Canada. I. Roussel, Mary E. II. Canada. Agriculture Canada. Research Branch. III. Title. IV. Series. V. Series: Publication (Canada. Agriculture Canada). English ; 1746.

QL537.C45604     595.77'1     C83-097201-3

# Contents

| | |
|---|---|
| Introduction | 7 |
| Classification | 8 |
| Systematic position of Canadian genera of Chironomidae | 9 |
| Anatomy | 14 |
| Geographic distribution | 18 |
| General biology | 19 |
| Collecting, preserving, and preparing | 21 |
| Acknowledgments | 23 |
| Key to subfamilies of larval Chironomidae | 24 |
| Subfamily Tanypodinae | 26 |
|     Key to the tribes of Tanypodinae | 27 |
|     Tribe Coelotanypodini | 28 |
|         Key to the genera of Coelotanypodini | 29 |
|     Tribe Tanypodini | 30 |
|     Tribe Macropelopiini | 32 |
|         Key to the genera of Macropelopiini | 33 |
|     Tribe Pentaneurini | 38 |
|         Key to the genera of Pentaneurini | 38 |
| Subfamily Podonominae | 49 |
|     Key to the tribes of Podonominae | 50 |
|     Tribe Podonomini | 50 |
|     Tribe Boreochlini | 51 |
|         Key to the genera of Boreochlini | 51 |
| Subfamily Chironominae | 54 |
|     Key to the tribes of Chironominae | 55 |
|     Tribe Chironomini | 56 |
|         Key to the genera of Chironomini | 57 |
|     Tribe Pseudochironomini | 91 |
|     Tribe Tanytarsini | 92 |
|         Key to the genera of Tanytarsini | 93 |
| Subfamily Orthocladiinae | 100 |
|     Key to the genera of Orthocladiinae | 102 |
| Subfamily Prodiamesinae | 148 |
|     Key to the genera of Prodiamesinae | 149 |
| Subfamily Diamesinae | 151 |
|     Key to the tribes of Diamesinae | 153 |
|     Tribe Boreoheptagyini | 153 |
|     Tribe Diamesini | 154 |
|         Key to the genera of Diamesini | 155 |
|     Tribe Protanypodini | 158 |
| Subfamily Telmatogetoninae | 159 |
| Figures | 160 |
| Glossary | 239 |
| References | 242 |
| Index | 259 |

# Introduction

Larvae of the family Chironomidae are abundant in most freshwater and semiaquatic habitats, and in some terrestrial, brackish, and saline habitats. The red larvae found in many aquatic habitats, commonly called bloodworms, belong to this group of insects. The number of species in an aquatic habitat may comprise over 50% of the total number of species of the larger invertebrates. Larval densities over $30\ 000/m^2$ are not unusual. The total number of species occurring in Canada is probably over 2000.

Despite the common occurrence of chironomid larvae in the freshwaters of Canada their taxonomy has been little studied, although recently there has been an increase in taxonomic literature about larvae (e.g., Sæther 1969–80; Oliver 1971*b*, 1976, 1977; Oliver et al. 1978; Cannings 1975; and Soponis 1977). The neglect is due partly to the small size of many species, superficial similarity in appearance within many groups, and the large number of species in a particular habitat. But it is also due to the fact that microscope slide preparation and microscopic examination at high power is necessary to reveal the structural characters used in naming them. This presents the non-taxonomist with a bewildering array of structural variation, which is not easily handled because of the lack of taxonomic literature. Therefore an extensive section on anatomy is given to assist the non-taxonomist in identifying larvae. General sections on classification, biology, and collecting, preserving, and preparing are also provided.

This publication gives keys for the identification of the last larval stage of 123 of the 133 genera of chironomids known to occur in Canada. In addition, each genus has been provided with a brief description and notes on its geographic distribution and habitat. Where possible the keys and descriptions are based on specimens associated with adults, as the taxonomy of chironomids is largely based on the adult male.

Specimens on which this publication are based, come primarily from central Canada, the Canadian northwest, and the islands of the Arctic Archepelago. Limited larval material was available from huge tracts of Canada such as the Atlantic Provinces, the Prairie Provinces, British Columbia, and southern Yukon. Therefore this publication is only an introduction to the study of Canadian chironomid larvae, as many genera and species remain to be discovered. The manuscript for this publication was completed in early 1980, but some generic name changes published since then have been included.

Bilingual keys are provided and Figs. 1–15 appear twice in order to show the labeling in English and French.

# Classification

The classification of chironomids at the familial and subfamilial level is confusing to the non-taxonomist. Two family names, Chironomidae and Tendipedidae, have been used. Similarly several subfamily names have been repeatedly interchanged; Tanypodinae with Pelopiinae; Chironominae with Tendipedinae; and Orthocladiinae with Hydrobaeninae. This instability was due primarily to two papers published by Meigen (1800, 1803). Each paper proposed a different set of names for the same insects. The earlier paper was unknown for about 100 years and during this period the names proposed in the 1803 paper gained acceptance. There has been much discussion about the use of the 1800 names versus the 1803 names (see Stone 1941; Oldroyd 1966) and it is sufficient to note that nomenclature used in this publication is in accordance with the ruling of the International Commission on Zoological Nomenclature (see Fittkau 1966), which suppressed the use of the Meigen 1800 names in favor of the 1803 names. Thus *Chironomus* Meigen, 1803 and Chironominae are used in place of *Tendipes* Meigen, 1800 and Tendipedinae, and *Tanypus* and Tanypodinae in place of *Pelopia* and Pelopiinae. Consequently the family name is Chironomidae, not Tendipedidae. Although not covered by a ruling of the International Commission on Zoological Nomenclature it is generally accepted that Orthocladiinae be used, not Hydrobaeninae.

The subfamilial classification used here follows that of Brundin (1966) as modified by Sæther (1977*b*). Seven subfamilies are recognized: Tanypodinae, Podonominae, Chironominae, Orthocladiinae, Prodiamesinae, Diamesinae, and Telmatogetoninae.[1] Most of the subfamilies have tribal divisions—see following section on Systematic position of Canadian genera of Chironomidae. Although a number of tribal names have been used in the Orthocladiinae (see Brundin 1966), none are used here. Sæther (1977*b*) has briefly reviewed the group relationships within the Orthocladiinae and suggested that until more is known about the immature and female stages the subfamily should remain undivided. It should be noted that the tribe Calopsectrini (Townes 1945), widely used in North American literature, is equivalent to Tanytarsini.

The generic level classification generally follows that of Hamilton et al. (1969). It has been modified to incorporate new information published between 1969 and early 1980 (Hansen and Cook 1976; Hirvenoja 1973; Roback 1971; Sæther 1969–80; and Soponis 1977). The classification of Hamilton et al. (1969) was based on a number of major revisions, including those of Brundin (1956, 1966), Fittkau (1962), Pagast (1947), Townes (1945), and Wirth (1949). This classification adapted the European concept of small genera (Beck and Beck 1968) to the Nearctic fauna, thus differing considerably from the broader generic concepts used by North American workers (e.g., Sublette and Sublette 1965).

---

[1]The subfamilies Aphroteniinae and Buchonomyiinae do not occur in North America.

In developing a classification of insects with several distinct life stages the characteristics of all life stages should be considered. This presents difficulties in the Chironomidae as the ecological requirements of each stage sometimes result in greater diversity, both ecologically and anatomically, in one stage than in the others. The adults of the Chironomidae are generally more uniform in structure than are the larvae (Fittkau 1962). In certain groups of chironomids their adults appear to have close taxonomic affinities, whereas their larval stages may be quite diverse. The opposite also occurs. The adult stage is somewhat ephemeral, completing the reproductive aspects of the life cycle in a comparatively short period of time. By contrast, the longest part of the life cycle is spent in the larval stage and the range of habitats occupied is perhaps unparalleled among other insect families. This ecological diversity between the life stages was reflected in the two systems of classification that evolved; the larval and pupal system of Thienemann and his associates and the imago system of Edwards and Goetghebuer (see Brundin 1956). The generic limits in the system based on the immature stages generally were narrower than those based on adults. Fortunately many recent studies have been based on all stages (e.g., Brundin 1956, 1966; Fittkau 1962; Hirvenoja 1973; Sæther 1969–80; and Soponis 1977) and a stable system of classification is beginning to develop.

## Systematic position of Canadian genera of Chironomidae

The following is a systematic list of the genera of chironomids known to occur in Canada. Genera with no described larvae are marked with an asterisk. The number of Canadian species reported in the literature prior to 1980 is given after each genus. A plus sign following this number indicates that unreported or new species also occur.

### Subfamily Tanypodinae

Tribe Coelotanypodini
    Genus *Clinotanypus* Kieffer 1
    Genus *Coelotanypus* Kieffer 1
Tribe Tanypodini
    Genus *Tanypus* Meigen 3
        Subgenus *Apelopia* Roback
        Subgenus *Tanypus* Meigen
Tribe Macropelopiini
    Genus *Apsectrotanypus* Fittkau 3
    Genus *Derotanypus* Roback 1 +
        Subgenus *Merotanypus* Roback
    Genus *Djalmabatista* Fittkau 1

　　　　Genus *Macropelopia* Thienemann 1
　　　　Genus *Procladius* Skuse 18+
　　　　　　Subgenus *Procladius* Skuse
　　　　　　Subgenus *Psilotanypus* Kieffer
　　　　Genus *Psectrotanypus* Kieffer 2+
Tribe Pentaneurini
　　　　Genus *Ablabesmyia* Johannsen 9+
　　　　　　Subgenus *Ablabesmyia* Johannsen
　　　　　　Subgenus *Karelia* Roback
　　　　*Genus *Arctopelopia* Fittkau 2
　　　　　　Subgenus *Arctopelopia* Fittkau
　　　　　　Subgenus *Meropelopia* Roback
　　　　*Genus *Cantopelopia* Roback 1
　　　　*Genus *Conchapelopia* Fittkau 8+
　　　　　　Subgenus *Conchapelopia* Fittkau
　　　　　　Subgenus *Helopelopia* Roback
　　　　　　Subgenus *Mesopelopia* Roback
　　　　Genus *Guttipelopia* Fittkau 1+
　　　　Genus *Krenopelopia* Fittkau 1
　　　　Genus *Labrundinia* Fittkau 2
　　　　Genus *Larsia* Fittkau 2
　　　　Genus *Monopelopia* Fittkau 1
　　　　Genus *Natarsia* Fittkau 1
　　　　Genus *Nilotanypus* Kieffer 1
　　　　Genus *Paramerina* Fittkau 1
　　　　Genus *Pentaneura* Philippi 1
　　　　*Genus *Thienemannimyia* Fittkau 5+
　　　　　　Subgenus *Rheopelopia* Fittkau
　　　　　　Subgenus *Thienemannimyia* Fittkau
　　　　Genus *Trissopelopia* Kieffer 1+
　　　　Genus *Zavrelimyia* Fittkau 3

## Subfamily Podonominae

Tribe Podonomini
　　　　Genus *Parochlus* Enderlein 1
Tribe Boreochlini
　　　　Genus *Boreochlus* Edwards 2
　　　　Genus *Lasiodiamesa* Kieffer 4
　　　　Genus *Trichotanypus* Kieffer 1

## Subfamily Chironominae

Tribe Chironomini
　　　　Genus *Acalcarella* Shilova 1
　　　　Genus *Beckidia* Sæther 1

    Genus *Chernovskiia* Sæther 1
    Genus *Chironomus* Meigen 29+
        Subgenus *Chaetolabis* Townes
        Subgenus *Chironomus* Meigen
    Genus *Cladopelma* Kieffer 4
    Genus *Cryptochironomus* Kieffer 4+
    Genus *Cryptotendipes* Lenz 3
    Genus *Cyphomella* Sæther 0+
    Genus *Demicryptochironomus* Lenz 2
    Genus *Dicrotendipes* Kieffer 6+
    Genus *Einfeldia* Kieffer 3+
    Genus *Endochironomus* Kieffer 2
    Genus *Glyptotendipes* Kieffer 7+
        Subgenus *Glyptotendipes* Kieffer
        Subgenus *Phytotendipes* Goetghebuer
   *Genus *Graceus* Goetghebuer 0+
    Genus *Harnischia* Kieffer 1
    Genus *Kiefferulus* Goetghebuer 1
    Genus *Lauterborniella* Bause 3
    Genus *Microtendipes* Kieffer 1+
    Genus *Nilothauma* Kieffer 1
    Genus *Omisus* Townes 1
    Genus *Pagastiella* Brundin 1+
    Genus *Parachironomus* Lenz 7+
    Genus *Paracladopelma* Harnisch 4
    Genus *Paralauterborniella* Lenz 1
    Genus *Paratendipes* Kieffer 2
    Genus *Phaenopsectra* Kieffer 8+
    Genus *Polypedilum* Kieffer 25+
        Subgenus *Pentapedilum* Kieffer
        Subgenus *Polypedilum* Kieffer
        Subgenus *Tripodura* Townes
    Genus *Robackia* Sæther 1+
    Genus *Saetheria* Jackson 1
    Genus *Stenochironomus* Kieffer 3+
    Genus *Stictochironomus* Kieffer 6+
    Genus *Xenochironomus* Kieffer 2+
        Subgenus *Axarus* Roback
        Subgenus *Xenochironomus* Kieffer
Tribe Pseudochironomini
    Genus *Pseudochironomus* Malloch 8
Tribe Tanytarsini
    (Larvae of several genera of Tanytarsini are known to occur in Canada but no species have been reported.)
    Genus *Cladotanytarsus* Kieffer 1+
    Genus *Constempellina* Brundin 0+
    Genus *Corynocera* Zetterstedt 1+
    Genus *Lauterbornia* Kieffer 2

Genus *Micropsectra* Kieffer 2+
Genus *Paratanytarsus* Bause 1+
Genus *Rheotanytarsus* Bause 1+
Genus *Stempellina* Bause 0+
Genus *Stempellinella* Brundin 0+
Genus *Tanytarsus* Wulp 2+
Genus *Zavrelia* Kieffer 1+

## Subfamily Orthocladiinae

Genus *Abiskomyia* Edwards 1+
Genus *Acricotopus* Kieffer 2+
Genus *Baeoctenus* Sæther 1
Genus *Brillia* Kieffer 5
Genus *Bryophaenocladius* Thienemann 2+
Genus *Camptocladius* Wulp 1
Genus *Cardiocladius* Kieffer 3
Genus *Chaetocladius* Kieffer 5+
*Genus *Chasmatonotus* Loew 8+
Genus *Corynoneura* Winnertz 2+
Genus *Cricotopus* Wulp 23+
    Subgenus *Cricotopus*
    Subgenus *Isocladius*
Genus *Diplocladius* Kieffer 1
Genus *Epoicocladius* Zavřel 1
Genus *Eukiefferiella* Thienemann 4+
Genus *Euryhapsis* Oliver 1
Genus *Gymnometriocnemus* Goetghebuer 2
Genus *Halocladius* Hirvenoja 1
Genus *Heleniella* Gowin 2
Genus *Heterotanytarsus* Spärck 2
Genus *Heterotrissocladius* Spärck 5
Genus *Hydrobaenus* Fries 9
Genus *Krenosmittia* Thienemann & Kruger 2
Genus *Limnophyes* Eaton 6+
*Genus *Mesocricotopus* Brundin 0+
Genus *Metriocnemus* Wulp 4+
Genus *Nanocladius* Kieffer 9
    Subgenus *Nanocladius* Kieffer
    Subgenus *Plecopteracoluthus* Steffan
Genus *Oliveridia* Sæther 1
*Genus *Oreadomyia* Kevan & Cutten-Ali-Khan 1
Genus *Orthocladius* Wulp 25+
    Subgenus *Eudactylocladius* Thienemann
    Subgenus *Euorthocladius* Thienemann
    Subgenus *Orthocladius* Wulp
    Subgenus *Pogonocladius* Brundin

*Genus *Parachaetocladius* Wülker 1
Genus *Paracladius* Hirvenoja 2
Genus *Paracricotopus* Thienemann & Harnisch 0+
Genus *Parakiefferiella* Thienemann 2+
Genus *Parametriocnemus* Goetghebuer 4+
Genus *Paraphaenocladius* Thienemann 2+
Genus *Paratrichocladius* Santos Abreu 1
Genus *Phycoidella* Sæther 1
Genus *Psectrocladius* Kieffer 10+
Genus *Pseudorthocladius* Goetghebuer 1+
Genus *Pseudosmittia* Goetghebuer 1+
Genus *Rheocricotopus* Thienemann & Harnisch 4
Genus *Synorthocladius* Thienemann 1
Genus *Smittia* Holmgren 6+
Genus *Symbiocladius* Kieffer 1
Genus *Thalassosmittia* Strenzke & Remmert 3
Genus *Thienemanniella* Kieffer 1+
Genus *Zalutschia* Lipina 7

## Subfamily Prodiamesinae

Genus *Monodiamesa* Kieffer 4
Genus *Odontomesa* Pagast 1
Genus *Prodiamesa* Kieffer 1

## Subfamily Diamesinae

Tribe Boreoheptagyini
    Genus *Boreoheptagyia* Brundin 1
Tribe Diamesini
    Genus *Diamesa* Meigen 16+
    *Genus *Pagastia* Oliver 1+
    Genus *Potthastia* Kieffer 1+
    Genus *Pseudodiamesa* Goetghebuer 3
    Genus *Pseudokiefferiella* Zavřel 1
    Genus *Sympotthastia* Pagast 1+
Tribe Protanypodini
    Genus *Protanypus* Kieffer 2

## Subfamily Telmatogetoninae

Genus *Paraclunio* Kieffer 1

# Anatomy

A major problem confronting users of taxonomic literature on chironomid larvae is the different terminology used by various authors. The same larval part has been given different names. Generally the terminology used here follows that of Sæther (1971–77).[2]

The following account is of fourth (last) instar larvae; however, as the basic structure of a species remains more or less the same throughout the last three instars it applies generally to second and third instar larvae. Changes that occur between these instars are primarily related to size increase, e.g., the first antennal segment becomes longer relative to the combined length of the terminal segments in each successive instar and the head capsule enlarges. Instars of a species can be differentiated by head capsule width or length (McCauley 1974). There is usually no overlap in these measurements of the last three instars. However, head capsule width or length will not always separate the first two instars of some species (Rosenberg et al. 1977).

Marked structural changes may occur between the first and second instars of a species: e.g., number of teeth and shape of the hypostoma (Kalugina 1959; McCauley 1974); development of ventral tubules (Danks 1971b); loss of egg burster (Danks 1971a); increase in number of antennal segments (personal observation); and increase in number of branches of SI (personal observation).

**General appearance** (Figs. 1–3). Chironomid larvae are elongate, usually cylindrical, and slender. Mature larvae range in length from about 1 to 30 mm, although few are longer than 10 mm. The strongly sclerotized, non-retractile *head capsule*, bearing the antennae and mouthparts, extends forward in the same plane as the longitudinal axis of the body. The body has 12 segments; 3 thoracic and 9 abdominal. There is no marked differentiation between the thorax and abdomen except that prior to pupation the thoracic segments enlarge. Paired *anterior parapods* bearing apical claws extend beneath the head capsule from the first thoracic segment. The anal end usually bears paired *posterior parapods* bearing apical claws, paired *procerci* with tufts of *anal setae*, paired supraanal setae, and one to three pairs of *anal tubules* (Figs. 1, 2). Sometimes the anal end is without one or more of the above structures (Fig. 3), and the anterior parapods may be fused. Color is variable, ranging from white or translucent to some shade of yellow, brown, red, green, or purple; the head capsule is generally darker than the body.

---

[2]Since preparation of the manuscript, Sæther (1980a) has published an extensive glossary of anatomical terms for chironomids. The main differences between the terminology used here and that of Sæther are the use of hypostoma and paralabial plate for mentum and ventromental plate, respectively. In the following account of the external anatomy of chironomid larvae information is freely drawn from the studies published by Hirvenoja (1973), Mozley (1970, 1971), Sæther (1971b), and Zavřel (1942).

**Head capsule** (Figs. 4–9). Generally the head capsule is longer than wide and narrower anteriorly than posteriorly. The relationship of length to width of head capsule is expressed as a head ratio (length/width). It is composed of two ocular sclerites fused ventrally and separated dorsally by a narrow frontoclypeal apotome or a frontal apotome and clypeus. One to three *eye spots* (Figs. 1–9) occur laterally on either side of the anterior part of the head capsule. The number and shape of the eye spots are useful taxonomic characters, although prior to pupation they may become diffused.

**Antenna.** The antenna, except in the Tanypodinae, is articulated directly to the anterodorsal wall of the head capsule or is borne on an *antennal tubercle* (Fig. 7). In the Tanypodinae it is retractile within the head capsule (Figs. 4, 5). It consists of four to eight segments, with five being the common number. Segments past the first or basal segment are collectively called the *terminal segments*. In some Orthocladiinae with six- or seven-segmented antennae the terminal segment is 'hairlike' and often difficult to see. Usually the segments are evenly sclerotized, but in some Diamesinae and Podonominae the third segment has alternating bands of weak and strong sclerotization or a spiral band of weak sclerotization. Both types of differential sclerotization are called *annulate* (Fig. 10). The position of the *ring organ* (Figs. 6, 8) on the basal segment is given by dividing the segment into halves or thirds, i.e., on basal half, on middle third, etc. The length of the *blade* (Figs. 6, 8, 10) is given in relation to the combined length of the terminal segments. *Lauterborn's organs* either arise directly from the antennal segments (Figs. 6, 8) or are borne on *petioles* (Fig. 7). They are either opposite to each other on the apex of the second segment or alternate on the apices of the second and third segments. These are the usual positions of the Lauterborn's organs, but in a few genera one may arise from the side of the second or third segments or both may be on the third segment. Sometimes the Lauterborn's organs are not visible and are presumed absent or vestigial and occasionally only one organ is present.

**Mouthparts.** The mouthparts consist of a ventral *hypostoma* and paired *paralabial plates*, lateral paired *mandibles*, a dorsal *labrum* and *palatum* bearing paired *premandibles*, a *prementohypopharyngeal* complex dorsal to the hypostoma, and paired *maxillae* between bases of the mandibles and the hypostoma (Figs. 4–9).

**Hypostoma.** Sæther (1971*b*) should be consulted for a discussion of the name for this structure. He proposed mentum, but the neutral and topographic name, hypostoma, is used here. The hypostoma is broad, strongly sclerotized, and toothed in all subfamilies except the Tanypodinae in which it apparently consists of the strongly sclerotized area of the anteromedial margin of the head capsule. In most genera there are one or two *median teeth* flanked on either side by a row of *lateral teeth* (Figs. 6–8). Rarely teeth are absent or restricted to the lateral part of the hypostoma (Fig. 9). The lateral teeth are designated by number from the median tooth or teeth laterally, i.e., the first lateral tooth is adjacent to the median tooth or teeth. Sometimes it is difficult to decide, especially in the Orthocladiinae, whether a median tooth is tripartite or if the first lateral teeth are closely appressed to the median tooth. The decision taken is sometimes arbitrary and for this reason a figure of a representative hypostoma for each genus is given.

**Paralabial plate.** In the Tanypodinae paralabial plates are usually membraneous, but in a few genera they are either sclerotized bearing teeth (Figs. 4, 22) or membraneous bearing teeth (Fig. 18). They are probably always present in the remaining subfamilies, although they are small and difficult to see in some Orthocladiinae and Diamesinae, and are termed vestigial. In the Chironominae they are distinct and almost always striated (Figs. 6, 7). In some Orthocladiinae and Prodiamesinae setae arise from the ventral surface of the plate (Figs. 11, 12).

**Mandible.** The basic structure is similar in all subfamilies. The apical half is usually attenuated bearing an *apical tooth* (Figs. 5, 6, 8), *inner teeth* (Figs. 6, 8) along the *inner margin* and a *seta subdentalis* (Figs. 6, 8) posterior to the inner teeth. The inner teeth are usually arranged in a row (Figs. 6, 8), but in most Tanypodinae there are two adjacent teeth; an *inner tooth* mesal to an *accessory tooth* (Fig. 5). The area posterior to the last inner tooth is often dark and the same color as the teeth and may be mistaken for a tooth. Chironominae, except most members of the *Harnischia* complex, have a row of setae (*pecten mandibularis*) across the mandible about the level of the inner teeth and one or two *apicodorsal* teeth on the outer margin posterior and somewhat dorsal to the apical tooth (Figs. 6, 17). Except in the Tanypodinae, and some Orthocladiinae, Diamesinae, and Chironominae, a *seta interna* arises from the base of the mandible posterior to the inner teeth (Figs. 6–8). It consists of a single or branched seta, a tuft of setae, or a row of setae.

**Premandible.** Premandibles are absent in the subfamilies Tanypodinae and Podonominae. In the remaining subfamilies they are nearly always present, and consist of apically inwardly curved and movable structures articulating with the *torma* (Figs. 6, 8). One to 14 teeth occur on the apical half and in some groups one or more low projections are present on the inner margin proximal to the teeth on the apex (Fig. 7). These are not counted as teeth and the term *apical teeth* is used to denote those projections clearly differentiated as teeth. When the differentiation is obscure or if there are no low inner projections then only the term teeth is used. Premandibles with no apical subdivisions are called simple.

**Labrum.** This structure forms the roof of the mouth cavity extending anteroventrally from the dorsal part of the head capsule. Two transverse sclerotized bars, the *tormae*, form the anteroventral margin of the labrum. In the Tanypodinae the labrum, except for the tormae, is weakly sclerotized and bears vesicles and modified setae (see Roback 1974a, 1976); (Figs. 4, 5) that are difficult to discern in normal slide mounts. By contrast, in the remaining subfamilies this area is strongly sclerotized and bears a variety of useful taxonomic structures (Figs. 6–8, 12). In making slide mounts care should be exercised to arrange this area so that the various structures may be seen. Centrally the anterodorsal part of the labrum bears four pairs of setae. Only the two pairs, *SI* and *SII*, nearest the anterior edge of the labrum are used here (Figs. 6–8, 12). The SI has five basic shapes; simple, tapering to a point (Figs. 122, 147, 268, 291); bifid (Figs. 8, 12); pectinate, with more or less equal branches along one side (Figs. 6, 7); plumose, multibranched with branches on both sides (Figs. 114, 206, 322); and palmate, branches arranged like tynes of a fork (Fig. 392). The distinction between palmate, pectinate, and plumose is not always clear as the various types intergrade. However, the term plumose is used for those not clearly one of the other types and within this category there is great variation. The SI and SII in the Podonominae and the SII in the Tanytarsini are mounted on long tubular extensions of the setal socket (*socles*) (Fig. 7). Paired or single *labral lamellae* can be present between the bases of the SI and the tormae or between the bases of the SI. It consists of a transverse comblike structure or 1–5 lobes (Figs. 6, 329). Generally they are pectinate but may be smooth or irregularly toothed. The *pecten epipharyngis* lies between the tormae and a U-shaped ungula bearing a variety of modified setae. Usually the pecten epipharyngis consists of 3 scales, which may be smooth, subdivided, or toothed (Figs. 7, 8, 201). Sometimes more than three scales are present, or it is replaced by a transverse toothed bar (Fig. 6); occasionally the scales are fused into a single plate (Fig. 288).

**Prementohypopharyngeal complex.** Except in the Tanypodinae this complex consists of a single structure that is frequently obscured by the hypostoma in normal slide mounts (Fig. 6). It bears a variety of structures that are potentially useful in taxonomic studies (Mozley 1971). It is not used here in subfamilies other than in the Tanypodinae except to note if it bears long setae or scales (Figs. 14, 15). In the Tanypodinae the complex has three sclerotized parts: *ligula*, paired *paraglossae*, and a *pecten hypopharyngis* (Figs. 4, 5). The ligula has four to eight, usually five, teeth. In a five-toothed ligula the teeth are designated (medial to lateral) median tooth, first lateral tooth, and outer lateral tooth. This complex is hinged and can move dorsally through 180 degrees, and hence in some microscope slide mounts it may point posteriorly.

**Maxilla.** The maxilla has a variety of characters that are potentially useful in taxonomic studies (Mozley 1971). However, it is often obscured in normal slide mounts and only the *maxillary palpus* (Figs. 4–8) is used here. This palpus, borne on the lateral part of the maxilla, usually consists of a single basal segment with a variety of apical sensory structures (see Mozley 1971, Figs. 1–4; Fig. 67). In a few Tanypodinae the *basal segment* is divided into two to six parts (Figs. 5, 50–51, 70); all are referred to as basal segments. The relationship of the length to width is a useful taxonomic character. The length includes only the sclerotized part of the basal segment and the width is measured at the middle of this segment.

**Body.** The *anterior parapods* are usually paired structures with an apical crown of claws (Figs. 1, 2), but sometimes they are fused, with reduced claws (Fig. 3). As outlined in "General appearance" the posterior end normally appears, as is shown in Figs. 1 and 2. Usually it is directed backward in the same plane as the longitudinal axis of the body, but in a few Orthocladiinae it is directed ventrally at a right angle to the longitudinal axis (Fig. 376). Modification of the posterior end occurs by loss or reduction of one or all of the following structures: *procerci, anal setae, anal tubules,* and *posterior parapods* (Fig. 3). The body usually bears setae in addition to those on the posterior end. Generally these are short and simple and they are sometimes difficult to see, but they may be modified in a variety of ways: long, darkly pigmented, bifid, plumose, or arranged in groups or in longitudinally lateral rows.

## Geographic distribution

Distribution of Chironomidae is worldwide. Two species occur in Antarctica (Wirth and Gressitt 1967). In the northern hemisphere chironomids extend to the northern limits of land and make up from one-fifth to one-half the total number of species of insects living in the Canadian Arctic Archipelago (Oliver 1968). Between these geographical extremes they have radiated into almost every habitat that is aquatic or wet.

The distribution of each subfamily, within its geographical range, is primarily governed by the availability of water ecologically suited to the requirements of the larvae. However, in extreme environments, such as the arctic region, the ecological requirements of the adults may be the limiting factor in some cases. The Aphroteniinae, the smallest subfamily, is more or less confined to swift mountain streams in the southern hemisphere; it does not occur in the northern hemisphere. The Podonominae, commoner in the southern hemisphere, generally occur in montane regions. The Diamesinae and Prodiamesinae occupy colder parts of circumpolar land and mountain ranges throughout the world. All the above subfamilies are absent in tropical lowlands (Brundin 1966). The Telmatogetoninae are marine living along the margins of tropical and temperate oceans.

Most of the species in the family belong to one of the subfamilies Tanypodinae, Chironominae, or Orthocladiinae. The Tanypodinae and Chironominae are abundant in the warmer regions of the world and decrease in numbers of species with increasing latitude or its climatic equivalent. By contrast the Orthocladiinae is the dominant subfamily in cold regions such as the Canadian arctic, and decreases in numbers of species in warmer regions.

Most genera recorded from North America occur in Canada. Only a few, such as *Goeldichironomus* and *Nilodorum*, which occur in southern

United States, are not expected to be found in Canada. Knowledge of the distribution of genera within Canada is limited, because most taxonomic studies have been based on specimens from central Canada and areas in the northwest and the Arctic; huge areas to the east and west are poorly investigated. However, it is expected that most genera will be transcontinental and only a few will be restricted to the western, eastern, or northern parts of Canada. The number of genera decreases from north to south, but over 35 genera occur north of the treeline. At least 24 genera occur at Hazen Camp in Ellesmere Island near the Canadian northern limit of land.

## General biology

The literature on the biology of chironomid larvae is extensive, comprising many hundreds of papers, although there is a great lack of studies of complete life histories of single species. Information published prior to 1950 was thoroughly reviewed by Thienemann (1954) and since then several publications with extensive bibliographies have appeared: life history (Oliver 1971b); overwintering (Danks 1971a); and larval dispersal (Davies 1976). Useful lists of habitats and ecological conditions for North American species are given by Curry (1965), Roback (1974a), and Beck (1977).

Freshwater is the primary habitat of chironomid larvae. To a lesser extent, but still common in some areas, they also occur in saline and brackish waters and in semiaquatic to terrestrial habitats including wetlands, wet areas adjacent to water bodies, wet leaf litter, pitcher plants, and cow dung. In bodies of freshwater they occur in almost every conceivable habitat and undoubtedly they are the most ubiquitous of all aquatic insects. They live in or on the substrate at all depths of deep lakes and rivers to shallow pools, springs and streams, on submerged objects such as stones, wood, and plants, and in many specialized habitats—boring in decomposing wood (Kalugina 1959), mining in leaves and stems of aquatic plants (Berg 1950), in *Nostoc* (Wirth 1957; Brock 1960), or in algal mats (Roback 1974a). Some live in close association, sometimes parasitic, with mayflies, stoneflies, sponges, bryozoans, and molluscs (Thienemann 1954; Steffan 1968).

There are four larval stages (instars). Newly hatched larvae (first instar larvae) of still water species are planktonic and positively attracted to light (Lellák 1968; Davies 1976). The planktonic behavior permits dispersal to suitable habitats and compensates for the lack of selection of egg-laying sites by females. Later instars become more or less sedentary, at least among the case (or cocoon, tube) building species: Davies (1976) reviews the evidence that they are not permanently sedentary. Larvae living in running water apparently are not planktonic and tend to react negatively to moving water; if caught in the water current they return with difficulty to the substrate (Davies 1976).

The more or less sedentary larvae construct cases of fine particles cemented together with a salivary gland secretion (silk) (Walshe 1951; Edgar and Meadows 1969). The free-living forms do not build cases. The cases range in structure from tunnels within a loose mass of particles cemented together to well defined cylindrical tubes. Some cases resemble those built by hydrophilid caddisflies. Tunnels of leaf miners substitute for cases.

Generally chironomid larvae feed on small plants or animals and on detritus; most Tanypodinae and a few other free-living species in other subfamilies are predaceous, feeding on larger aquatic invertebrates such as chironomid larvae, oligochaetes, and copepods. Diatoms are a common food source throughout the family (Brundin 1966). Except for the free-living species, most probably seek out and ingest food from their immediate surroundings. Food may be engulfed directly or trapped by the sticky salivary secretion. Walshe (1951) described several types of feeding using salivary secretion.

In uniform environments the potential for uninterrupted growth and development exists. In Canada, due to the cooler winter months environments are not uniform. Chironomids overwinter as larvae, and cold-hardness or adaptation is a common characteristic of most Canadian chironomids. Some build winter cocoons, which differ in structure from summer tubes (Danks 1971a). Freezing tolerance is found in every major group, except the Tanypodinae (Danks 1971a). In bodies of water that do not freeze to the bottom some species may continue to develop during the winter months (Hamilton 1965; Mozley 1970), but generally all activity ceases during the winter months. In the opposite extreme, e.g., habitats that periodically dry up, some larvae are able to withstand desiccation (Hinton 1960; Grodhaus 1976). This adaptation has not been observed in Canadian chironomids, but temporary ponds are inhabited and continuity of species obtains through successive dry periods. In permanent bodies of water, cessation or reduction in growth or feeding activity has been recorded for some species during the summer months (Hamilton 1965; Jonasson 1965).

With increasing latitude or its climatic equivalent the duration of the larval period becomes longer. In the arctic region all species require 1 year or more to complete one generation (Oliver 1968; Danks and Oliver 1972; Welch 1976). Farther south many species complete more than one generation per year, particularly those which emerge during the summer, although those inhabiting cool water, e.g., springs and deeper parts of large lakes, and those emerging soon after ice-off, tend to have only one generation per year. Species that have an extensive north–south distribution usually have fewer generations per year in cooler waters (Rempel 1936; Rosenberg et al. 1977). Temperature is an important factor in determining the length of the larval period, and others such as oxygen concentration and availability of food are also important (Jonasson 1965).

The broad environmental requirements of the larvae of the various subfamilies are relatively well known (Oliver 1971b). The Podonominae, Diamesinae, and Prodiamesinae are primarily cold-adapted, living in cool flowing water, but members of each subfamily occur in still water. The Tanypodinae and Chironominae are essentially warmwater forms adapted to living in still water, but occur in flowing water and cool habitats. Members of these two subfamilies are able to withstand conditions of severe oxygen depletion. The free-living Tanypodinae swim to higher strata with more oxygen and most Chironominae possess hemoglobin (Brundin 1966). The larvae of Orthocladiinae are primarily cold-adapted, but are not uncommon in warm habitats. The Telmatogetoninae are usually associated with rocky peripheral areas of the oceans that receive freshwater runoff. Certain groups of chironomids are characteristic of flowing or still water, oxygen-poor or oxygen-rich water, cold or warm water, etc. Thus the species composition of chironomid larvae has been frequently used as an indicator for different types of lakes; (see Brundin 1949 and Brinkhurst 1974 for a discussion of lake classification). However, as pointed out by Brinkhurst (1974) the knowledge of immature stages, both taxonomically and ecologically, is sadly deficient in North America and their value as ecological indicators remains to be tested.

## Collecting, preserving, and preparing

**Collecting.** Methods suitable for collecting chironomid larvae differ according to different habitats and whether qualitative or quantitative results are desired. Quantitative techniques may be found in Brinkhurst (1974) for still water, and in Cummings (1962) and Hynes (1970) for flowing water. General collecting techniques plus illustrations of some apparatus are found in Needham and Needham (1941), Pennak (1953), and Martin (1977). More detail is given by Welch (1948), Macan (1958, 1970), and Edmondson and Winberg (1971).

Dip nets, hand picking, nets pushed through the substrate, or a net capturing dislodged larvae by disturbing the substrate upstream from the net (kick-method) are all suitable in habitats that can be approached from shore or boat. Brundin (1966) half immersed a net in streams, which captured larvae drifting with the current over a period of time. In deeper water where the substrate cannot be directly approached by the collector, except by scuba diving, grabs (dredges) and corers attached to a line are used. These enclose and retain a portion of the substrate as the sampler is pulled to the water surface. Many types have been devised (Welch 1948; Edmondson and Winberg 1971), but the Ekman dredge is commonly used for sampling soft substrate. Artificial substrates are used where substrate type (e.g., rocky), water depth, and current velocities make conventional sampling difficult. Made of wire baskets or trays filled with stones, concrete blocks, webbing, plates of various material, etc., (Anderson and Mason 1968; Hilsenhoff 1969; Simmons and Winfield 1971; Rosenberg and Wiens 1976), they are placed in the habitat and left for a period of time to allow colonization.

**Sorting.** Separation from substrate is one of the most time-consuming aspects of collecting chironomid larvae. Samples can be concentrated by sieving them through a net or screen. Coarse material, such as rocks, can be rinsed or scrubbed with a soft brush in a net suspended in water. The common mesh size used is 0.6 mm gauge, but many chironomid larvae less than 10 mm long will pass through this mesh size (Jonasson 1958). Except for early instars, 0.2 mm gauge will retain most larvae. After concentration the sample is placed in a white tray with water and the larvae picked out with forceps or a pipette. Examination of the sample bit by bit with a binocular microscope will ensure total removal, but this is very time-consuming. Several techniques aid in separating larvae from fine material but will not work well when large amounts of plant material are present; flotation in salt or sugar solutions (Anderson 1959; Kajak et al. 1968) or staining with dyes (Hamilton 1969).

**Killing.** Larvae may be killed directly in a preservative but killing in 95% ethanol or in water slowly heated to near boiling relaxes the mouthparts.

**Preserving and storing.** Larvae are best preserved in 70–80% ethanol; formalin or lacto-phenol can be used but these are unpleasant reagents. Preservative in which larvae are killed should be replaced within 1 or 2 days. Larvae may be stored in any airtight container, but three dram vials with neoprene stoppers (Martin 1977) or straight-sided vials stoppered with cotton wool plugs and stored in museum jars are among the best containers available. The latter take up less space but the specimens are not as accessible as those stored in three dram vials set in racks (Martin 1977). Many plastic containers now available with screw-on or snap-on tops allow evaporation and are unsuitable for long-term storage.

**Rearing.** This term is used to denote raising larvae to the adult stage. Positive association of immature stages with corresponding adults is achieved by individual rearing of larvae. Each prepupal larva (i.e., with enlarged thorax) is placed in a 20–25 × 70–100 mm shell vial with 2–5 mm of distilled water. Roback (1976) suggested that more water may be used but we have had best results when the depth of the water is about three to four times the diameter of the larva. The vials, stoppered with a cork or cotton wool plug, are kept at 1°–5°C above habitat temperature and examined daily for adult emergence. Upon emergence, the adult is transferred to a second vial with a strip of moist filter paper and stored in a cool dark place for 24–48 hours to harden. Allowing the adults to harden results in less distortion during slide preparation and mature color development. After hardening the adult is killed in ethanol and stored with the larval and pupal exuvia (skins).

Mass rearing or culturing can be used to bring larvae to the prepupal stage for individual rearing or to obtain adults from a particular habitat. The larvae are placed in shallow trays with about 1 cm of substrate from their habitat, plus a small amount of activated charcoal and 3–5 cm of water (distilled or from habitat). The tray is placed in a cage, aerated, and kept at a temperature somewhat above that of the habitat. If necessary a small amount of food can be added once or twice a week. If representative larvae are preserved from the sample they may be directly associated with the emerging adults, although the association is positive if only one species per genus is present. Periodically the trays are examined for prepupal larvae, which are transferred to individual rearing vials. Literature pertaining to rearing chironomid larvae has been compiled by Campbell et al. (1973).

**Microscope slide preparation.** Specimens used in this study were mounted in Canada balsam. Other mounting media, e.g., Euparal, Permount, Hoyer's, CMC 10, ACS Mountant, etc., are easier to use as specimens may be mounted directly from ethanol or water, but the permanency of slides made with Canada balsam is well established.

The head capsule is severed from the body and both, if clearing is needed, are placed in cold 10% KOH for 8–12 hours. The amount of clearing required depends upon the darkness of the specimens and can only be judged through practice. After clearing the larva is run through the following series: distilled water (5 min), glacial acetic acid (10 min), 2-propanol (15 min), and equal parts of 2-propanol and cedarwood oil (15 min), and is mounted directly from the last mixture into Canada balsam. The head is mounted ventral side up and the body is mounted on its side under separate cover slips. Reared specimens (adult plus larval and pupal exuviae) are mounted on the same slide as a permanent record of the association. The larval exuvium is placed under one cover slip, the pupa under another, and the dissected adult under five cover slips arranged as shown by Soponis (1977, Fig. 1).

The importance of good slides cannot be overemphasized. Mouthparts must not overlie and obscure each other. It is especially important that the SI and SII and pecten epipharyngis be visible and that the mandibles do not overlie the antennae. Slight pressure on the cover slip will often spread out the mouthparts or they may be teased apart, in the Canada balsam, with fine needles.

# Acknowledgments

Dr. A. L. Hamilton, as a Postdoctoral Fellow working with the senior author, compiled manuscript keys for the genera in the subfamilies Tanypodinae and Chironominae. These keys provided a very useful working basis for preparing the keys for these two groups. The drawings (Figs.

1–15) were made by Mr. R. Idema, Biosystematics Research Institute, Ottawa. We wish to thank Drs. D. M. Wood and L. LeSage for reading and criticizing the manuscript.

## Key to subfamilies of larval Chironomidae

1. Antenna retractile. Separate ligula and paraglossae present. Hypostoma never a toothed plate (Figs. 18, 22) .......... **Tanypodinae** (p. 26)
   Antenna nonretractile. Separate ligula and paraglossae absent. Hypostoma usually a toothed plate, rarely without teeth, sometimes pale in color (Figs. 34, 46) ................................................... 2
2. Procercus long, at least 5 times as long as wide (Fig. 87). Premandible absent. SI set on elongate socles (Fig. 83). .......... **Podonominae** (p. 49)
   Procercus shorter, or absent; if present, less than 5 times as long as wide. Premandible almost always present. SI not set on elongate socles (Figs. 6, 7, 8) ............................................... 3
3. Paralabial plate distinct, striated (except *Stenochironomus*, and never with setae (Figs. 6, 7). Eye spots usually separate and lying in a vertical plane (Fig. 2), or with dorsal eye spot lying anterior to ventral eye spot ....
   ................................................ **Chironominae** (p. 54)
   Paralabial plate present or absent, if present and distinct, never striated; with or without setae (Figs. 8, 11, 12). Eye spots usually single, or if separate, then dorsal eye spot lying posterior to ventral eye spot (Fig. 8) .... 4
4. Procercus present, bearing at least one anal seta ..................... 5
   Procercus absent (Fig. 3); anal setae, if present, arising directly from body wall.................................................................. 8
5. Third antennal segment annulate (Fig. 10). Prementohypopharyngeal complex with long setae in three groups (Fig. 441) ....................
   ............................................. **Diamesinae** (in part) (p. 151)
   Third antennal segment not annulate (Fig. 8). Prementohypopharyngeal complex with scales or modified setae (Fig. 15), rarely with long setae, if long setae present, then not in three groups (Fig. 14) ............ 6
6. Occipital margin of head capsule with a distinct posteriorly directed ventrolateral projection on each side (Fig. 9); head capsule with numerous setae (Fig. 9). Central three-quarters of hypostoma straight to concave, without teeth (Fig. 446) .................... **Diamesinae** (in part) (p. 151)
   Occipital margin of head capsule without or with small posteriorly directed ventrolateral projection on each side; head capsule with fewer setae, or rarely, if numerous, then hypostoma with teeth on central part. Hypostoma variable, usually convex and central three-quarters of plate with teeth (Fig. 8) ............................................................ 7
7. Paralabial plate large, with setae (Fig. 12). Antenna 4-segmented ........
   ............................................. **Prodiamesinae** (p. 148)
   Paralabial plate present or absent and with or without setae, if large and with setae (Figs. 11, 294), then antenna with 5 or more segments ........
   ............................................. **Orthocladiinae** (in part) (p. 100)
8. Third antennal segment annulate ........ **Diamesinae** (in part) (p. 151)
   Third antennal segment non-annulate ................................. 9
9. Hypostoma with 1 median and 5 or more pairs of lateral teeth (Fig. 461). SI simple .................................. **Telmatogetoninae** (p. 159)
   Hypostoma with 1 or 2 median and 4 pairs of lateral teeth or if with 1 median and 5 pairs of lateral teeth, then SI bifid ........................
   ............................................. **Orthocladiinae** (in part) (p. 100)

# Clé des sous-familles des larves de Chironomidæ

1. Antenne rétractile. Ligule séparée et paraglosses présentes. Hypostome jamais une plaque dentée (fig. 18, 22) ........ **Tanypodinæ** (p. 26)
   Antenne non rétractile. Ligule séparée et paraglosses absentes. Hypostome habituellement une plaque dentée, rarement sans dents, parfois de couleur pâle (fig. 34, 46) ........................................ 2
2. Procerque long, au moins 5 fois plus long que large (fig. 87). Prémandibule absente. SI fixées sur des socles allongés (fig. 83) ................
   ........................................ **Podonominæ** (p. 49)
   Procerque plus court ou absent; si présent, moins de 5 fois plus long que large. Prémandibule presque toujours présente. SI non fixées sur des socles allongés (fig. 6, 7, 8) ........................................ 3
3. Plaque paralabiale distincte, striée, (sauf *Stenochironomus*) et jamais avec des soies (fig. 6, 7). Taches ocellaires habituellement séparées et disposées sur un plan vertical (fig. 2) ou avec stemmates dorsaux antérieurs aux stemmates ventraux ....................... **Chironominæ** (p. 54)
   Plaque paralabiale présente ou absente; si présente et distincte, jamais striée; avec ou sans soies (fig. 8, 11, 12). Stemmates habituellement simples ou, si séparés, stemmates dorsaux postérieurs aux stemmates ventraux (fig. 8) ........................................ 4
4. Procerque présent, portant au moins une soie anale ................ 5
   Procerque absent (fig. 3); soie anale, si présente, s'élevant directement de la paroi du corps ............................................... 8
5. Troisième article antennaire annelé (fig. 10). Structure prémentohypopharyngienne avec de longues soies disposées en trois groupes (fig. 441)
   ............................. **Diamesinæ** (en partie) (p. 151)
   Troisième article antennaire non annelé (fig. 8). Structure prémentohypopharyngienne avec écailles ou soies modifiées (fig. 15), rarement avec de longues soies, mais le cas échéant, soies non disposées en trois groupes (fig. 14) ........................................ 6
6. Bord occipital de la capsule céphalique avec, sur chaque côté, un appendice ventrolatéral distinct dirigé vers l'arrière (fig. 9); capsule céphalique avec nombreuses soies (fig. 9). Trois quarts centraux de l'hypostome de droits à concaves, sans dents (fig. 446) ..... **Diamesinæ** (en partie) (p. 151)
   Bord occipital de la capsule céphalique avec ou non de chaque côté un petit appendice ventrolatéral dirigé vers l'arrière; capsule céphalique avec moins de soies ou, cas rare, si soies nombreuses, hypostome avec alors des dents au centre. Hypostome de forme variable, habituellement convexe et trois quarts centraux de la plaque avec des dents (fig. 8) ....... 7
7. Plaque paralabiale large, avec soies (fig. 12). Antenne à 4 articles ........
   ............................. **Prodiamesinæ** (p. 148)
   Plaque paralabiale présente ou absente, avec ou sans soies; si grosse et avec soies (fig. 11, 294), alors antenne à 5 articles ou plus ...............
   ............................. **Orthocladiinæ** (en partie) (p. 100)
8. Troisième article antennaire annelé ...... **Diamesinæ** (en partie) (p. 151)
   Troisième article antennaire non annelé ........................... 9
9. Hypostome avec 1 dent médiane ou 5 paires ou plus de dents latérales (fig. 461). SI simple .................... **Telmatogetoninæ** (p. 159)
   Hypostome avec 1 ou 2 dents médianes et 4 paires de dents latérales ou si avec 1 dent médiane et 5 paires de dents latérales, alors SI bifide ........
   ............................. **Orthocladiinæ** (en partie) (p. 100)

# Subfamily Tanypodinae

Figs. 1, 4–5, 16–78

Color variable, usually white, yellow, or brown, sometimes pink, red, or green. Eye spot single, kidney-shaped to semicircular. Antenna 4-segmented, retractile into head capsule; first segment with ring organ, and blade on apex; Lauterborn's organs on apex of second segment, often minute. Hypostoma apparently consisting of sclerotized area of anteroventral margin of head capsule. Paralabial teeth present or absent. Prementohypopharyngeal complex consisting of a median 4- to 8-toothed ligula, paired paraglossae, and an arched toothed pecten hypopharyngis. Mandible hooked to sickle-shaped, usually with 1 inner tooth and 1 accessory tooth, rarely with row of inner teeth; seta interna, pecten mandibularis, and apicodorsal tooth absent; seta subdentalis present (always?), often difficult to see; base rarely enlarged. Maxillary palpus elongate, with 1 (usually) to 6 basal segments. Premandible absent. Labrum weakly sclerotized, generally vesicular, bearing simple setae or appendages. Abdomen with scattered setae and with or without a row of lateral setae. Procercus elongate, bearing 7–20 anal setae. Paired anterior and posterior parapods elongate, with apical claws. Two, occasionally 3, pairs of anal tubules present.

**Remarks.** Larvae of this subfamily are recognized by their generally narrow and elongate appearance. The anterior region is distinctive as the head and anterior parapods appear to extend further from the body than they do in the other subfamilies. The structure of the head is unique with retractile antennae, no toothed hypostoma, and the prementohypopharyngeal complex consisting of three sclerotized parts—ligula, paired paraglossae, and pecten hypopharyngis.

The classification used here follows that of Fittkau (1962) and Roback (1971). They recognized five tribes; Coelotanypodini, Tanypodini, Anatopyniini, Macropelopiini, and Pentaneurini. Members of the tribe Anatopyniini do not occur in North America despite the frequent usage of the name of its only genus, *Anatopynia*, in the literature. Fittkau (1962) based his classification on both the immature and adult stages, attributing greater importance to the characteristics of the immature stages, whereas Roback (1971) considered the characteristics of the adults to be more important. As a result Roback transferred *Natarsia* from the Pentaneurini to the Macropelopiini and consigned several genera recognized by Fittkau to subgeneric status. Except for retaining *Natarsia* in the Pentaneurini and returning *Apsectrotanypus* to full generic status (Roback placed it as a subgenus of *Psectrotanypus*), Roback's modification of Fittkau's classification is followed.[3]

---

[3] Roback and Moss (1978) and Roback (1978) have proposed a further modification of the classification of the Tanypodinae: the subtribes Macropelopiina and Procladiina are given tribal status; *Natarsia* is placed in a new tribe, Natarsiini; *Alotanypus* and *Apsectrotanypus* are placed as subgenera of *Macropelopia*; and *Derotanypus* is placed as a subgenus of *Psectrotanypus*. Although parts of this classification have merit we have elected to use a classification based on Fittkau (1962) and Roback (1971) as outlined above.

The larvae of several North American genera, *Parapelopia* Roback, *Telopelopia* Roback, and *Cantopelopia* Roback, are unknown. Only *Cantopelopia* has been recorded from Canada (Roback 1971).

**Distribution.** The subfamily Tanypodinae is worldwide, occurring in all major geographical regions except Antarctica. Four of the five tribes, namely, Coelotanypodini, Tanypodini, Macropelopiini, and Pentaneurini, occur in Canada.

Members of the Tanypodinae occur throughout Canada, although there is a decrease in the number of species from north to south and from warm to cool habitats. Only three genera, *Arctopelopia*, *Derotanypus*, and *Procladius* occur in the arctic region.

**Biology.** Larvae of Tanypodinae are generally warm-adapted, living in shallow still water, but occur in flowing water and deeper parts of lakes. Some species live in cool habitats; none are adapted to completing development below 4°C. No case is built and there is no report of salivary gland secretion. The salivary glands are small compared to those of the Chironominae (Kurazhkovskaya 1966). Most are strong swimmers and probably range widely within the confines of their habitat, resting and feeding on substrate or submerged aquatic plants, although some species penetrate into the substrate (Shiozawa and Barnes 1977).

Most species attack and consume aquatic invertebrates including chironomid larvae, crustaceans, and oligochaetes. Few species are strictly carnivorous (Roback 1969; Oliver 1971*b*). Most also ingest algae and phytoplankton, and one species, *Larsia acrocinata*, feeds exclusively on diatoms and detritus (Hamilton 1965). Little is known about the food of early instars but predaceous behavior appears to increase with succeeding instars. Small prey are swallowed whole and larger ones are seized by the rear end and engulfed to the level of the head, which is bitten off (Kurazhkovskaya 1966). Larvae also feed by piercing the prey and sucking out body fluids (Morgan 1949). Food selection is apparently precise as detritus is rarely ingested.

## Key to the tribes of Tanypodinae

1. Paralabial teeth present (Figs. 4, 18, 22, 31). Abdominal segments with lateral setal fringe. Head ratio less than 1.5 .......................... 2
   Paralabial teeth absent (Fig. 5). Abdominal segments without lateral setal fringe. Head ratio usually more than 1.5 ........................................... Tribe **Pentaneurini** (p. 38)
2. Conical projection between bases of procerci present (Fig. 20). Ligula usually with 6–8 teeth, rarely with 5 ....... Tribe **Coelotanypodini** (p. 28)
   Conical projection between bases of procerci absent. Ligula with 4 or 5 teeth ........................................................... 3

3. Basal two-thirds of mandible enlarged (Fig. 28). Toothed margin of ligula usually convex, rarely straight (Figs. 24, 25) ........................
.................... Tribe **Tanypodini**;*Tanypus* Meigen (p. 30)
Basal two-thirds of mandible not enlarged (Figs. 32, 35). Toothed margin of ligula concave or straight (Figs. 29, 33, 37, 41, 45) .................
................................ Tribe **Macropelopiini** (p. 32)

## Clé des tribus de Tanypodinæ

1. Dents paralabiales présentes (fig. 4, 18, 22, 31). Segments abdominaux avec frange sétale latérale. Rapport céphalique inférieur à 1,5 ......... 2
Dents paralabiales absentes (fig. 5). Segments abdominaux sans frange sétale latérale. Rapport céphalique habituellement supérieur à 1,5 ........
................................ Tribu des **Pentaneurini** (p. 38)
2. Appendice conique entre les bases des procerques présent (fig. 20). Ligule avec habituellement de 6 à 8 dents, rarement 5 ....................
............................. Tribu des **Cœlotanypodini** (p. 28)
Appendice conique entre les bases des procerques absent. Ligule avec 4 ou 5 dents ................................................... 3
3. Deux tiers proximaux de la mandibule élargis (fig. 28). Bord denté de la ligule habituellement convexe, rarement droit (fig. 24, 25) ...............
.................. Tribu des **Tanypodini**; *Tanypus* Meigen (p. 30)
Deux tiers proximaux de la mandibule non élargis (fig. 32, 35). Bord denté de la ligule concave ou droit (fig. 29, 33, 37, 41, 45) ..................
................................ Tribu des **Macropelopiini** (p. 32)

## Tribe Coelotanypodini

### Figs. 16–23

Color pinkish to pale red or orange; head light brown. Head ratio less than 1.5; strongly tapering anteriorly. Ligula with 6–8 teeth; toothed margin concave. Paraglossa simple, with lateral and/or mesal spines. Pecten hypopharyngis with 20–40 teeth per side. Paralabial teeth arising from weakly sclerotized or membraneous area, without distinct paralabial plates; 5–10 teeth present. Mandible strongly hooked to moderately curved, with base not enlarged. Maxillary palpus with 2 basal segments; proximal segment sclerotized, distal segment weakly sclerotized to membraneous. Abdominal segments with lateral setal fringe. Procercus about 1.5–4.0 times as long as wide, bearing 14–20 anal setae. Conical projection arising from between bases of procerci. Two pairs of anal tubules present, shorter than posterior parapods, usually triangular. Posterior parapods with yellow or yellow brown claws.

**Remarks.** The six- to eight-toothed ligula, the presence of paralabial teeth but no distinct paralabial plates, and the presence of a conical projection between the bases of the procerci distinguish the larvae of

Coelotanypodini. The conical projection is also present in the Tanypodini, but the mandible of the larvae of this tribe has an enlarged base. Very rarely some larvae have a five-toothed ligula but these may be distinguished by the above characters.

## Key to the genera of Coelotanypodini

1. Mandible strongly hooked, with 1 large and several smaller inner teeth (Fig. 19). A.R. greater than 10.0. Ligula usually with 6 teeth .............
   ........................................... *Clinotanypus* (p. 29)
   Mandible not hooked, but moderately curved, with no distinct inner teeth (Fig. 23). A.R. less than 8.0. Ligula usually with 7 or 8 teeth ........
   ........................................... *Coelotanypus* (p. 30)

## Clé des genres de Cœlotanypodini

1. Mandibule fortement crochue, avec 1 grosse dent et plusieurs petites dents internes (fig. 19). R.A. supérieur à 10,0. Ligule avec habituellement 6 dents ........................................ *Clinotanypus* (p. 29)
   Mandibule non crochue, mais modérément incurvée, sans dents internes distinctes (fig. 23). R.A. inférieur à 8,0. Ligule avec habituellement 7 ou 8 dents ........................................ *Cœlotanypus* (p. 30)

## Genus *Clinotanypus* Kieffer

Figs. 16–20

*Clinotanypus* Kieffer, 1913a:157; Roback 1971:16; 1976:188.

Antenna more than one-half as long as head; A.R. 10.9–16.2. Ligula usually with 6 teeth, rarely with 5 or 7 teeth; outer 2 teeth separate. Pecten hypopharyngis with 23–38 teeth per side. Six to 8 small pointed paralabial teeth. Mandible strongly hooked, with 1 large inner tooth flanked by 2–4 smaller inner teeth and 1 large pointed accessory tooth.

**Remarks.** Generally, larvae of *Clinotanypus* may be distinguished from those of *Coelotanypus* by the number of teeth of the ligula; 6 in *Clinotanypus* and 7 or 8 in *Coelotanypus*. Otherwise the characters given in the key may be used.

Roback (1971) erected two subgenera, *Clinotanypus* (*Clinotanypus*) and *Clinotanypus* (*Aponteus*), based on adult characters. Larvae are described by Boesel (1974) and Roback (1976). *Clinotanypus* (*Aponteus*) is recorded only from Texas and the larvae are unknown.

**Distribution.** Southern British Columbia; northern Manitoba; southern Ontario and Quebec.

**Habitat.** Larvae of *Clinotanypus* live in bottom mud of warm shallow regions (usually less than 3 m) of both still and flowing water. They prefer soft mud, occasionally mixed with sand or silt in clean water, but may occur in organically enriched habitats (Boesel 1974; Paine and Gaufin 1956; Roback 1971, 1976).

## Genus *Coelotanypus* Kieffer

Figs. 21–23

*Coelotanypus* Kieffer, 1913a:154; Roback 1971:26; 1974b:9.

Antenna less than one-half as long as head; A.R. 6.2–7.5. Ligula usually with 7 teeth, rarely with 6 or 8 teeth; outer teeth appressed to each other. Pecten hypopharyngis with 18–26 teeth per side. Five to 10, broad, apically rounded paralabial teeth present. Mandible moderately curved, not hooked, without distinct inner or accessory teeth.

**Remarks.** See remarks under *Clinotanypus* for discussion of characters that distinguish the larvae of *Coelotanypus* and *Clinotanypus*. Larvae are described by Boesel (1974) and Roback (1974b).

**Distribution.** Southern Ontario and Quebec.

**Habitat.** Larvae of *Coelotanypus* inhabit medium to large bodies of still water but may occur in quieter reaches of flowing water. They usually live in soft mud but may occur in harder bottoms composed of sand or silt. Water depths range from shallow inshore regions down to 40 m (Boesel 1974; Roback 1971, 1974b).

## Tribe Tanypodini

Figs. 24–28

This tribe consists of the single genus *Tanypus* and the description given for the genus will be sufficient to define the characters of the tribe.

## Genus *Tanypus* Meigen

Figs. 24–28

*Tanypus* Meigen, 1803:261; Roback 1971:52; 1977:55.
*Pelopia* Meigen, 1800:18.
*Protenthes* Johannsen, 1907:400.

Head ratio less than 1.5. Antenna less than half as long as head; A.R. 6.0–8.5. Ligula yellow brown, with 5 teeth; toothed margin convex, rarely straight. Paraglossa with 2–9 long apical and lateral branches, sometimes with smaller lateral spinelike projections. Pecten hypopharyngis with 6–8 teeth per side. Paralabial plate straight to concave with 5 or 6 teeth; outer tooth often bifid. Mandible base enlarged, with 2 or 3 small inner teeth. Maxillary palpus with 1 basal segment about 2.5–3 times as long as wide. Body with lateral setal fringe. Procercus about 3–6 times as long as wide, bearing 13–20 anal setae. Two or 3 pairs of anal tubules present; tubules pointed, shorter than posterior parapods. Posterior parapod claws yellow brown.

**Remarks.** Larvae of *Tanypus* (and of Tanypodini) resemble those of Coelotanypodini and Macropelopiini in the presence of paralabial teeth and lateral abdominal setal fringe, but differ in that the mandibular base is enlarged and the toothed margin of the ligula is usually convex.

The subgenera *Tanypus* (*Tanypus*) and *Tanypus* (*Apelopia*) may be distinguished by the characters given in the following key. Larvae are described in Roback (1977).

## Key to the subgenera of *Tanypus*

1. Toothed margin of ligula evenly convex (Fig. 24) ..................... ............................................. ***Tanypus (Tanypus)***
   Toothed margin of ligula with median and first lateral teeth subequal in size and longer than outer lateral teeth (Fig. 25) ...................... ............................................ ***Tanypus (Apelopia)***

## Clé des sous-genres de *Tanypus*

1. Bord denté de la ligule uniformément convexe (fig. 24) ............... ........................................... ***Tanypus (Tanypus)***
   Bord denté de la ligule avec dent médiane et premières dents latérales de taille à peu près égale et plus longues que les dents latérales externes (fig. 25) ........................................ ***Tanypus (Apelodia)***

**Distribution.** Southern British Columbia to southeastern Ontario; District of Mackenzie.

**Habitat.** The larvae of *Tanypus* generally inhabit shallow warm still water, but some species occur in flowing water and in deeper parts of lakes. They prefer soft mud but are occasionally associated with sand or silt substrates. Some species can tolerate a wide range of environmental conditions including organic pollution (Buckley and Sublette 1964; Grodhaus 1967; Roback 1969, 1977).

# Tribe Macropelopiini

Figs. 5, 29–47

Head ratio less than 1.5. Antenna less than one-half length of head. Ligula with 4 or 5 teeth, with toothed margin straight to concave. Paraglossa single and spined to multibranched. Pecten hypopharyngis with 4 to over 20 teeth per side. Paralabial plate with 4–8 teeth. Mandible usually with 1 inner tooth and 1 accessory tooth, sometimes with row of inner teeth; base not enlarged. Maxillary palpus with 1 basal segment 2.0–3.5 times as long as wide. Body with lateral setal fringe. Procercus 4–10 times as long as wide, bearing 12–20 anal setae. Two pairs of anal tubules present; tubules shorter than posterior parapods, usually apically pointed. Posterior parapod claws yellow brown.

**Remarks.** The combination of strongly sclerotized paralabial plates bearing a row of teeth, mandible without enlarged base, and a four- or five-toothed ligula distinguishes the larvae of Macropelopiini.

Fittkau (1962) recognized five genera, *Macropelopia*, *Psectrotanypus*, *Apsectrotanypus*, *Procladius*, and *Psilotanypus*, in the Macropelopiini. Roback (1971) described three new genera (*Parapelopia*, *Derotanypus*, and *Alotanypus*), placed *Apsectrotanypus* as a subgenus of *Psectrotanypus* and *Psilotanypus* as a subgenus of *Procladius*, and transferred *Natarsia* from the Pentaneurini to the Macropelopiini. Roback (1971) also erected two subtribes, Macropelopiina and Procladiina, primarily based on adult characteristics.

Larvae of Macropelopiina may be divided into two groups, *Psectrotanypus* plus *Derotanypus* and *Apsectrotanypus* plus *Macropelopia*. Larvae of *Psectrotanypus* and *Derotanypus* have a four-toothed ligula with straight toothed margin, multibranched paraglossa, mandible with row of inner teeth, whereas *Apsectrotanypus* and *Macropelopia* have a five-toothed ligula with concave toothed margin, single paraglossa with one or two mesal spines, and mandible with only two adjacent teeth. The adults of these genera can not readily be separated into two groups; for example, the males of *Psectrotanypus* and *Apsectrotanypus* are distinguished only by the presence or absence of setae on the ninth tergite. This problem of lack of concordance between larval and adult groupings cannot be resolved until more larvae of North American Macropelopiina are described. Therefore Fittkau's (1962) basic classification, which reflects larval differences, is followed here; *Apsectrotanypus* and *Psectrotanypus* are treated as separate genera, *Natarsia* is retained in Pentaneurini, but *Psilotanypus* is accepted as a subgenus of *Procladius*, and *Derotanypus* though similar to *Psectrotanypus* is treated as a separate genus.[4]

---

[4] See footnote 3, p. 26

## Key to the genera of Macropelopiini

1. Ligula with 4 teeth; toothed margin straight (Figs. 33, 45). Mandible with row of 3 or more inner teeth (Figs. 35, 47) .......................... 2
Ligula with 5 teeth; toothed margin usually concave, rarely straight (Figs. 29, 37, 41). Mandible without row of inner teeth, but with 1 inner tooth and usually 1 accessory tooth (Figs. 32, 40, 44) ...................... 4
2. Teeth and adjacent area of ligula black. Mandible with row of 3 or more inner teeth ............................. *Djalmabatista* **Fittkau** (p. 35)
Teeth yellowish, unicolorous with rest of ligula. Mandible without row of inner teeth, but with 1 inner tooth and 1 accessory tooth ......... 3
3. Anterior margin of paralabial teeth concave; outer lateral tooth enlarged (Fig. 34). Lateral branches of paraglossa subdivided (Fig. 33) .......
.................................. *Derotanypus* **Roback** (p. 35)
Anterior margin of paralabial teeth usually arching posterolaterally, rarely straight, outer lateral tooth smaller than next mesal tooth (Fig. 46). Lateral branches of paraglossa usually simple (Fig. 45) ...........
.......................................... *Psectrotanypus* **Kieffer** (p. 37)
4. Teeth and adjacent area of ligula black. Paraglossa broad with long apical point and with lateral and mesal branches (Fig. 42) ...............
....................................... *Procladius* **Skuse** (p. 36)
Teeth yellowish, unicolorous with rest of ligula. Paraglossa narrower, simple with weak mesal branch or branches, or weakly bifid (Figs. 30, 38) . 5
5. Median and first lateral teeth of ligula about equal in length, sometimes median tooth slightly shorter. Maxillary palpus less than four times as long as wide (Fig. 29) .............. *Apsectrotanypus* **Fittkau** (p. 34)
Median tooth of ligula distinctly shorter than first lateral teeth. Maxillary palpus greater than four times as long as wide (Fig. 37) ............
............................... *Macropelopia* **Thienemann** (p. 36)

## Clé des genres de Macropelopiini

1. Ligule avec 4 dents; bord denté droit (fig. 33, 45). Mandibule avec une rangée de 3 dents internes ou plus (fig. 35, 47) ........................ 2
Ligule avec 5 dents; bord denté habituellement concave, rarement droit (fig. 29, 37, 41). Mandibule sans rangée de dents internes, mais avec 1 dent interne et habituellement 1 dent accessoire (fig. 32, 40, 44) ....... 4
2. Dents et partie adjacente de la ligule noires. Mandibule avec une rangée de 3 dents internes ou plus ............. *Djalmabatista* **Fittkau** (p. 35)
Dents jaunâtres, de même couleur que le reste de la ligule. Mandibule sans rangée de dents internes, mais avec 1 dent interne et 1 dent accessoire
.................................................................. 3
3. Bord antérieur des dents paralabiales concave; dent latérale externe élargie (fig. 34). Ramifications latérales des paraglosses subdivisées (fig. 33) .
.................................. *Derotanypus* **Roback** (p. 35)
Bord antérieur des dents paralabiales s'arquant habituellement postérolatéralement, rarement droit; dent latérale externe plus petite que la dent mésale suivante (fig. 46). Ramifications latérales des paraglosses habituellement simples (fig. 45) ...... *Psectrotanypus* **Kieffer** (p. 37)

4. Dents et partie adjacente de la ligule noires. Paraglosses larges, avec une longue pointe apicale et des ramifications latérales et mésales (fig. 42) ........................................ *Procladius* **Skuse** (p. 36)
   Dent jaunâtre, de même couleur que le reste de la ligule. Paraglosses plus étroites, simples avec une ou des ramifications mésales faibles, ou faiblement bifides (fig. 30, 38) ........................................ 5
5. Dent médiane et premières dents latérales de la ligule à peu près de la même longueur, dent médiane parfois légèrement plus courte. Palpe maxillaire moins de quatre fois plus long que large (fig. 29) .............
   ............................ *Apsectrotanypus* **Fittkau** (p. 34)
   Dent médiane de la ligule nettement plus courte que les premières dents latérales. Palpe maxillaire plus de quatre fois plus long que large (fig. 37)
   ............................ *Macropelopia* **Thienemann** (p. 36)

## Genus *Apsectrotanypus* Fittkau

Figs. 29–32

*Apsectrotanypus* Fittkau, 1962:141.
*Psectrotanypus* (*Apsectrotanypus*); Roback 1971:103.

Color dark brown with yellow marbling; head capsule dark brown. A.R. 6.5–8.0. Ligula yellow brown, with 5 teeth, and with median and first lateral teeth subequal and shorter than outer lateral teeth, sometimes median slightly shorter. Paraglossa bifid, with inner branch shorter than apical branch, sometimes with 2 mesal branches. Paralabial plate weakly concave, with 4–6 teeth, with inner pair of teeth truncated and notched, and with outer pair of teeth small. Mandible with 1 inner tooth and 1 accessory tooth. Procercus about 3.5–5.0 times as long as wide.

**Remarks.** Larvae of *Apsectrotanypus* and *Macropelopia* are similar, but are distinguished primarily by length of median tooth of ligula. It is about the same length as the first lateral teeth in *Apsectrotanypus* and distinctly shorter in *Macropelopia*. Also the inner spine (or spines) on the paraglossa of *Apsectrotanypus* is distinct, whereas in some *Macropelopia* it is almost absent. Shape and size of the mesal tooth of pecten hypopharyngis relative to the next tooth may aid in separation. In specimens seen by us this tooth in *Apsectrotanypus* is larger and apically rounded, whereas in *Macropelopia* it is apically pointed and about the same size as the adjacent tooth. Thus anatomical differences are slight and difficulty may be expected in assigning an individual larva to one of the two genera.

**Distribution.** Alberta; southern Ontario and Quebec.

**Habitat.** Larvae of *Apsectrotanypus* inhabit small flowing water (including cold springs) and shallow areas of still waters (Fittkau 1962).

## Genus *Derotanypus* Roback

Figs. 33–36

*Derotanypus* Roback, 1971:91.

Color reddish; head capsule brown. A.R. about 9.0. Ligula yellow brown with 4 teeth; toothed margin straight. Paraglossa weakly sclerotized, broad, with about 8 main branches; most branches subdivided. Paralabial plate strongly concave with 6 or 7 teeth; outer pair of teeth enlarged. Mandible with row of 5 or 6 inner teeth; accessory tooth absent; small spines on dorsal surface at level of inner teeth. Procercus about 5–10 times as long as wide.

**Remarks.** Larvae of *Derotanypus* may be distinguished from those of *Psectrotanypus* by concave paralabial plates, paraglossa with most branches subdivided, and a higher A.R. Otherwise larvae of the two genera are similar and, as a group, distinct from the *Apsectrotanypus* and *Macropelopia* (see remarks under Macropelopiini).

Roback (1971) divided *Derotanypus* into two subgenera, *Derotanypus* and *Merotanypus*, based upon the presence or absence of setae on the ninth tergite of the male. Larvae of *Derotanypus (Derotanypus)* are unknown and adults have not been recorded from Canada. Larvae are described by Roback (1978).

**Distribution.** British Columbia and Yukon Territory to northern Ontario.

**Habitat.** Larvae of *Derotanypus* inhabit ponds, shallow regions of still water, and slow flowing water. They occur in a variety of ponds ranging from prairie sloughs to tundra ponds and are apparently able to tolerate high salinities.

## Genus *Djalmabatista* Fittkau

*Djalmabatista* Fittkau, 1968:328; Roback and Tennessen 1978:11.
*Procladius (Calotanypus)*; Roback, 1971:152.

Body color unknown; head capsule pale brown. A.R. 3.4–4.5. Ligula yellow brown, with 4 black teeth; toothed margin concave with middle 2 teeth distinctly shorter than lateral teeth. Paraglossa broad, with short apical tooth, and with 1–2 inner and 1–3 outer branches. Mandible with 1 inner tooth and 1 accessory tooth. Paralabial plate straight, with 7 or 8 teeth; outer tooth smaller than adjacent tooth. Procercus about 4 times as long as wide.

**Remarks.** The ligula with four black teeth will distinguish the larvae of *Djalmabatista* from those of other genera of Macropelopiini.

**Distribution.**   Eastern and southern Ontario and southern Quebec.

**Habitat.**   The larvae of *Djalmabatista* occurs in shallow regions of small bodies of still water (Roback and Tennessen 1978).

## Genus *Macropelopia* Thienemann

### Figs. 37–40

*Macropelopia* Thienemann *in* Thienemann and Kieffer, 1916:497; Fittkau 1962:102; Roback 1971:87.

Color red; head capsule brownish yellow. A.R. 6.0–7.5. Ligula yellow brown, with 5 teeth, with toothed margin concave, and with median tooth distinctly shorter than first lateral teeth. Paraglossa single, with weak inner branch; branch often appressed to main shaft. Paralabial plate convex, with 7 or 8 teeth; outer tooth smaller than adjacent tooth. Mandible with 1 inner tooth and 1 accessory tooth. Procercus about 4–5 times as long as wide.

**Remarks.**   Larvae of *Macropelopia* are similar to those of *Apsectrotanypus* (see remarks under *Apsectrotanypus*). In terms of number of species within the Tanypodinae this genus is second only to *Procladius* and *Ablabesmyia* (Fittkau 1962). Larvae are described by Roback (1955; as *Anatopynia* (*Psectrotanypus*)) and Roback (1978).

**Distribution.**   Ontario to New Brunswick.

**Habitat.**   Larvae of *Macropelopia* inhabit small flowing water and ponds, and according to Fittkau (1962) they are generally adapted to cool conditions.

## Genus *Procladius* Skuse

### Figs. 41–44

*Procladius* Skuse, 1889:283; Roback 1957:48; Fittkau 1962:88.
*Psilotanypus* Kieffer, 1906a:318; Fittkau 1962:88.

Color yellow to brown; head capsule yellow brown. A.R. 4.5–8.0. Ligula yellow brown with 5 black teeth; toothed margin concave, with teeth variable in length but median tooth usually shorter than first lateral teeth and these shorter than outer lateral teeth. Paraglossa broad, with long apical branch, with 1–4 inner spines or branches and 4–8 outer spines or branches. Mandible with 1 inner tooth and 1 accessory tooth; inner tooth usually larger than accessory tooth. Paralabial plate straight to weakly convex, with 7 or 8 teeth; outer tooth smaller than adjacent tooth. Procercus about 2–4 times as long as wide.

**Remarks.** The ligula with five black teeth and paraglossa with long apical branch and shorter inner and outer branches will distinguish the larvae of *Procladius* from those of the other genera of Macropelopiini. Diagnostic characters for the subgenera *Procladius* and *Psilotanypus* have not been established; Fittkau (1962) brings them out together in his larval key. However, in reared material available to us, larvae associated with adults of *P. (Psilotanypus)* have a pecten hypopharyngis with 4–8 teeth per side and those associated with adults of *P. (Procladius)* have 11 to about 20 teeth. This distinction needs confirmation.

Larvae of most North American species are unknown. Roback (1957) gives a key for three species, *P. adrumbatus* (as *P. bellus*), *P. culiciformis*, and *P. riparius*, based primarily on the shape of the paraglossa. He states that this is the only character that shows promise of eventually separating the larvae.

**Distribution.** Widespread, including Arctic Archipelago.

**Habitat.** The larvae of *Procladius* occurs in all types of freshwater, permanent or temporary, oxygen-rich to oxygen-poor, warm to cool, still to flowing, deep to shallow, etc. They are generally found in mud bottoms of ponds and lakes, and in slower sections or side areas of rivers and streams (Paine and Gaufin 1956; Paloumpis and Starrett 1960; Wurtz and Roback 1955; Miller 1941; Sublette 1957; Buckley and Sublette 1964; Roback 1971).

## Genus *Psectrotanypus* Kieffer

Figs. 45–47

*Psectrotanypus* Kieffer, 1909:42; Fittkau 1962:129.
*Psectrotanypus* (*Psectrotanypus*); Roback 1971:98.

Color greenish, yellow brown or reddish; head capsule yellowish. A.R. 6–7.5. Ligula yellow brown, with 4 teeth; toothed margin straight. Paraglossa weakly sclerotized, multibranched; apical branches longest, with outer branches progressively shorter; branches usually single; apical branches sometimes bifid. Paralabial plate straight to convex, with 6–8 teeth; outer tooth smaller than adjacent tooth. Mandible with row of 6–8 inner teeth and 1 accessory tooth; dorsal surface without or with 1 or 2 small spines at level of inner teeth. Procercus about 5–8 times as long as wide.

**Remarks.** The larvae of *Psectrotanypus* and *Derotanypus* are similar (see remarks under *Derotanypus*). Larvae are described by Roback (1955, 1957; as *Anatopynia* (*Psectrotanypus*), 1978).

**Distribution.** British Columbia and Yukon Territory to Newfoundland.

**Habitat.** Larvae of *Psectrotanypus* prefer warm, oxygen-poor and organically rich small bodies of still water (Fittkau 1962; Roback 1971). Some species are tolerant of polluted conditions and others inhabit bog pools.

## Tribe Pentaneurini

Figs. 1, 5, 48–78

Color usually yellowish, brownish, or greenish, rarely red. Head capsule usually smooth, rarely granular; head ratio 1.6–2.5. Antenna at least one-half as long as head, except about one-third in *Natarsia*. Ligula with 5 teeth; anterior one-quarter to one-third, including teeth, dark brown to black; remainder yellow brown. Paraglossa bifid, with inner branch shorter than apical branch. Pecten hypopharyngis usually with 7–20 teeth per side, rarely with more than 20. Paralabial plate and teeth absent. Mandible with 1 inner tooth and 1 accessory tooth; base not enlarged. Maxillary palpus usually with 1 basal segment, occasionally with 2–6 segments. Abdominal segments without lateral setal fringe; body setae usually simple, sometimes plumose. Procercus about 2.5–6.0 times as long as wide, bearing 7 or 8 anal setae. Two pairs of anal tubules present; tubules usually shorter than posterior parapods, rarely longer. Seta on basal half of posterior parapod usually simple, sometimes thickened or spinulose.

**Remarks.** The narrow elongate head capsule, absence of a lateral abdominal setal fringe and paralabial plates or teeth, and a five-toothed ligula distinguishes the larvae of Pentaneurini.

Fittkau (1962) recognized 18 genera in the tribe Pentaneurini. Prior to Fittkau's revision most North American species were placed in *Pentaneura* or *Ablabesmyia* (e.g., Roback 1957). Beck and Beck (1966) followed the classification of Fittkau (1962) and described larvae belonging to *Nilotanypus* and *Paramerina* for the first time. Roback (1971) essentially followed Fittkau. He placed *Rheopelopia* as a subgenus of *Thienemannimyia*, erected subgenera in *Conchapelopia*, *Arctopelopia*, and *Ablabesmyia*, and described two new genera, *Telopelopia* and *Cantopelopia*. Only the genus *Telmatopelopia*, recognized by Fittkau, is not found in North America. The larvae of *Telopelopia* and *Cantopelopia* are unknown. All genera except *Telopelopia*, *Xenopelopia*, and *Krenopelopia* occur in Canada. *Krenopelopia* occurs in Alaska and is expected to be found in adjacent areas of Canada.

## Key to the genera of Pentaneurini

1. Maxillary palpus with 2–6 basal segments (Figs. 50–51, 70) ............ 2
   Maxillary palpus with 1 basal segment (Fig. 67) ...................... 4
2. Posterior parapods with 1–4 claws dark brown, remainder yellowish. Toothed margin of ligula concave (Fig. 48) ................................
   ........................ ***Ablabesmyia* Johannsen** (in part) (p. 41)

Posterior parapods with all claws yellowish. Toothed margin of ligula straight (Fig. 68) .................................................................... 3
3. Posterior parapods with all claws simple. Abdominal segments with simple and plumose setae ........ ***Ablabesmyia* Johannsen** (in part) (p. 41)
Posterior parapods with several claws pectinate. Abdominal segments with simple or forked setae ................ ***Paramerina* Fittkau** (p. 46)
4. Body with longitudinal sinuous wrinkles (Fig. 54). Anal tubules horn-shaped, about one-seventh to one-fifth as long as posterior parapods. Head capsule granular ..................... ***Guttipelopia* Fittkau** (p. 41)
Body smooth. Anal tubules at least one-quarter as long as posterior parapods. Head capsule smooth, rarely granular .......................... 5
5. Antenna about one-third as long as head. Head ratio 1.6–1.7 ........... ............................................. ***Natarsia* Fittkau** (p. 44)
Antenna at least one-half as long as head. Head ratio usually greater than 1.7 ..................................................................... 6
6. Second antennal segment distinctly darker than first segment (see Fig. 55) ..................................................................... 7
Second antennal segment not darker than first segment .............. 8
7. Middle tooth of ligula longer than first lateral teeth (Fig. 56). Seta arising from basal part of posterior parapod spinulose (Fig. 58) ............ ............................................. ***Labrundinia* Fittkau** (p. 43)
Middle tooth of ligula shorter than first lateral teeth (Fig. 61). Seta arising from basal part of posterior parapod smooth ....................... ............................................. ***Monopelopia* Fittkau** (p. 44)
8. Anal tubules longer than posterior parapods ....................... 9
Anal tubules shorter than posterior parapods ...................... 10
9. Posterior parapod with all claws yellowish; 1 claw pectinate. Procercus about 3 times as long as wide, partly infuscate .. ***Nilotanypus* Kieffer** (p. 45)
Posterior parapod with 1 brown claw, rest yellowish; all claws simple. Procercus about 6 times as long as wide, completely infuscate, strongly contrasting with body color .......... ***Pentaneura* Philippi** (p. 46)
10. First lateral tooth of ligula outcurved (Figs. 73, 75) ................ 11
First lateral tooth of ligula straight (Figs. 59, 77) .................. 12
11. Basal segment of maxillary palpus about 4 times as long as wide. A.R. usually greater than 4.0 ......... ***Thienemannimyia* Fittkau** complex (p. 47)
Basal segment of maxillary palpus about 7 times as long as wide A.R. less than 4.0 ............................... ***Trissopelopia* Kieffer** (p. 47)
12. Toothed margin of ligula concave. Body white ...................... ............................................. ***Krenopelopia* Fittkau** (p. 42)
Toothed margin of ligula straight (Figs. 59, 77). Body pale yellow, yellow brown, or reddish .......................................... 13
13. Mandible length greater than 110 μ. Posterior parapod with at least 1 claw with basal spine ..................... ***Zavrelimyia* Fittkau** (p. 48)
Mandible length less than 90 μ. Posterior parapod with all claws simple .. ............................................. ***Larsia* Fittkau** (p. 43)

## Clé des genres de Pentaneurini

1. Palpe maxillaire formé de 2 à 6 articles basaux (fig. 50, 51, 70) ....... 2
Palpe maxillaire formé de 1 article basal (fig. 67) .................... 4
2. Parapodes postérieurs avec de 1 à 4 griffes brun foncé, le reste étant jaunâtre. Bord denté de la ligule concave (fig. 48) ........................ .................... ***Ablabesmyia* Johannsen** (en partie) (p. 41)

|      | Parapodes postérieurs avec toutes les griffes jaunâtres. Bord denté de la ligule droit (fig. 68) .......................................... 3 |
| ---- | --- |
| 3.   | Parapodes postérieurs avec toutes les griffes simples. Segments abdominaux avec soies simples et plumeuses ................................... ........................ ***Ablabesmyia* Johannsen** (en partie) (p. 41) |
|      | Parapodes postérieurs avec plusieurs griffes pectinées. Segments abdominaux avec soies simples ou fourchues .. ***Paramerina* Fittkau** (p. 46) |
| 4.   | Corps avec rides longitudinales sinueuses (fig. 54). Tubules anaux en forme de corne, d'une longueur variant approximativement entre le septième et le cinquième de celle des parapodes postérieurs. Capsule céphalique granulaire .......................... ***Guttipelopia* Fittkau** (p. 41) |
|      | Corps lisse. Tubules anaux d'une longueur au moins égale à un quart de celle des parapodes postérieurs. Capsule céphalique lisse, rarement granulaire ................................................................. 5 |
| 5.   | Antenne d'une longueur égale à environ un tiers de celle de la tête. Rapport céphalique de 1,6 à 1,7 ................. ***Natarsia* Fittkau** (p. 44) |
|      | Antenne d'une longueur égale à au moins la moitié de celle de la tête. Rapport céphalique habituellement supérieur à 1,7 ...................... 6 |
| 6.   | Deuxième article antennaire nettement plus foncé que le premier (*voir* fig. 55) ............................................................ 7 |
|      | Deuxième article antennaire pas plus foncé que le premier .......... 8 |
| 7.   | Dent du milieu de la ligule plus longue que les premières dents latérales (fig. 56). Soie spinuleuse se dressant sur la partie proximale du parapode postérieur (fig. 58) ................... ***Labrundinia* Fittkau** (p. 43) |
|      | Dent du milieu de la ligule plus courte que les premières dents latérales (fig. 61). Soie lisse se dressant de la partie proximale du parapode postérieur ............................ ***Monopelopia* Fittkau** (p. 44) |
| 8.   | Tubules anaux plus longs que les parapodes postérieurs ............. 9 |
|      | Tubules anaux plus courts que les parapodes postérieurs .......... 10 |
| 9.   | Parapode postérieur avec toutes les griffes jaunâtres; 1 griffe pectinée. Procerque environ 3 fois plus long que large, partiellement enfumé .... ............................... ***Nilotanypus* Kieffer** (p. 45) |
|      | Parapode postérieur avec 1 griffe brune, le reste étant jaunâtre; toutes les griffes simples. Procerque environ 6 fois plus long que large, complètement enfumé, contrastant fortement avec la couleur du corps ...... ................................ ***Pentaneura* Philippi** (p. 46) |
| 10.  | Première dent latérale de la ligule courbée vers l'extérieur (fig. 73, 75) .. ............................................................ 11 |
|      | Première dent latérale de la ligule droite (fig. 59, 77) ............. 12 |
| 11.  | Article proximal du palpe maxillaire environ 4 fois plus long que large. R.A. habituellement supérieur à 4,0 ................................... ....................... ***Thienemannimyia* Fittkau** complexe (p. 47) |
|      | Article proximal du palpe maxillaire environ 7 fois plus long que large. R.A. inférieur à 4,0 ..................... ***Trissopelopia* Kieffer** (p. 47) |
| 12.  | Bord denté de la ligule concave. Corps blanc ...................... ................................. ***Krenopelopia* Fittkau** (p. 42) |
|      | Bord denté de la ligule droite (fig. 59, 77). Corps jaune pâle, brun jaune ou rougeâtre ..................................................... 13 |
| 13.  | Longueur de la mandibule supérieure à 110 μ. Parapode postérieur avec au moins 1 griffe munie d'une épine basale ........................... .................................. ***Zavrelimyia* Fittkau** (p. 48) |
|      | Longueur de la mandibule inférieure à 90 μ. Parapode postérieur avec toutes les griffes simples ....................... ***Larsia* Fittkau** (p. 43) |

# Genus *Ablabesmyia* Johannsen

Figs. 48–51

*Ablabesmyia* Johannsen, 1905:135; Fittkau 1962:416; Roback 1971:354.

Color yellow brown often mottled with dark brown; head capsule brown. Head ratio 1.7–2.0. A.R. 3.8–12.0. Ligula usually with concave toothed margin; margin rarely straight; first lateral tooth straight or outcurved. Mandible with well developed inner and accessory teeth. Maxillary palpus with 2–6 basal segments. Abdominal segments with both simple and plumose setae. Procercus about 3.0–6.5 times as long as wide. Anal tubules slender, at most one-half as long as posterior parapods. Posterior parapod with 1–4 brown claws that strongly contrast in color with remaining yellow claws; most claws simple, but several pectinate; rarely all claws unicolorous and simple.

**Remarks.** Generally the combination of brown claws on the posterior parapods and a maxillary palpus with two or more basal segments will distinguish the larvae of *Ablabesmyia* from those of all other Pentaneurini. However, several species which have unicolorous simple claws, a two-segmented maxillary palpus, and a ligula with straight toothed margin, are difficult to distinguish from *Paramerina*. The abdominal setae in *Paramerina* are simple or forked, never plumose, and several of the smaller claws are bifid or pectinate.

Roback (1971) erected two subgenera *Ablabesmyia* (*Ablabesmyia*) and *Ablabesmyia* (*Karelia*), based on adult characteristics. The larvae of too few species are known to define the characteristics of the two subgenera. Larvae are described by Malloch (1915), Johannsen (1937a), Roback (1957), Sublette (1964), and Beck and Beck (1966).

**Distribution.** Widespread south of treeline.

**Habitat.** The larvae of *Ablabesmyia* prefer warm shallow still water or pools in flowing water, and are usually associated with soft substrate with or without aquatic vegetation. In flowing water they are more common in pools than in riffle areas. Some species can tolerate low oxygen-concentrations (Fittkau 1962).

# Genus *Guttipelopia* Fittkau

Figs. 52–54

*Guttipelopia* Fittkau, 1962:251; Roback 1971:257.

Body greenish with white flecks; head capsule brownish. Head capsule granular; head ratio about 2.0. A.R. 6.0–7.3. Ligula with concave toothed margin; first lateral tooth straight. Mandible with inner tooth larger than accessory tooth. Maxillary palpus with 1 basal segment about 3–4 times as long as wide. Body with sinuate longitudinal wrinkles. Procercus about 3 times as long as wide. Anal tubules short, about one-seventh to one-fifth as long as posterior parapods. Posterior parapod with 2 or 3 brown claws that contrast in color with remaining yellow claws; several claws pectinate; rest simple.

**Remarks.** The larvae of *Guttipelopia* are unique among the Tanypodinae in having longitudinal wrinkles on the body surface. These usually cannot be seen on larval exuvia, but the granular head capsule and short anal tubules are also diagnostic. Larvae are described by Chernovskii (1949) and Beck and Beck (1966).

**Distribution.** District of Mackenzie; Manitoba; and southern Ontario.

**Habitat.** Larvae of *Guttipelopia* are generally warm-adapted, living in small stagnant water bodies (Fittkau 1962; Beck and Beck 1966). The one Canadian species is apparently able to withstand cooler conditions and inhabits quieter reaches of flowing water.

## Genus *Krenopelopia* Fittkau

*Krenopelopia* Fittkau, 1962:262; Roback 1971:274.

Body white; head capsule yellow. Head ratio about 1.8. A.R. about 3.0. Ligula with concave toothed margin, with median and first lateral teeth subequal in size and shorter than outer lateral teeth, and with first lateral tooth straight. Mandible with large blunt inner tooth and smaller pointed accessory tooth. Maxillary palpus with 1 basal segment about 3 times as long as wide. Procercus about 3 times as long as wide. Anal tubules less than half length of posterior parapods.

**Remarks.** The larvae of *Krenopelopia* are similar to those of *Zavrelimyia* and *Trissopelopia*. The ligula with the median and first lateral teeth smaller than the outer lateral teeth and straight first lateral teeth, the mandible with the inner tooth distinctly larger than the accessory tooth, and the white body should distinguish them from those of the latter two genera.

The foregoing description is based on those by Fittkau (1962) and Zavřel and Thienemann (1919; as *Pelopia, minima* group) as no larvae of *Krenopelopia* have been collected in North America. The North American records are based on adults (Roback 1971) and a pupa (Fittkau 1962).

**Distribution.** Not known from Canada but occurs in Alaska and northwestern United States.

**Habitat.** Larvae of *Krenopelopia* are cool-adapted, living in flowing water and shallow areas of lakes (Fittkau 1962).

## Genus *Labrundinia* Fittkau

Figs. 55–58

*Labrundinia* Fittkau, 1962:372; Roback 1971:275.

Body green; head capsule light brown, sometimes marked with brown or black. Head capsule smooth or granular; head ratio about 2.1. Second antennal segment brown, at least basally; A.R. 2.0–2.5. Ligula with convex toothed margin; middle tooth usually much longer than first lateral teeth and these usually shorter than outer lateral teeth. Mandible with well developed inner and accessory teeth. Maxillary palpus with 1 basal segment about 3 times as long as wide. Procercus 2–6 times as long as wide. Anal tubules about three-quarters length of posterior parapods. Posterior parapod with yellowish bifid, pectinate, or simple claws; seta on basal half of parapod spinulose.

**Remarks.** Larvae of *Labrundinia*, *Pentaneura*, and *Nilotanypus* all have anal tubules three-quarters or more the length of the posterior parapods; about twice as long in *Pentaneura* and *Nilotanypus* and about three-quarters in *Labrundinia*. In all three genera the seta arising from the basal part of the posterior parapod is thickened but only bears short spines in *Labrundinia*. The presence of a brown or partly brown second antennal segment should also distinguish the larvae of *Labrundinia*, except from *Monopelopia* (see key, couplet 6). Larvae of all North American species apparently have this character (Beck and Beck 1966). However, in European species it is concolorous with the rest of the antennal segments (Fittkau 1962). Larvae are described by Beck and Beck (1966).

**Distribution.** British Columbia and District of Mackenzie to Quebec.

**Habitat.** Larvae of *Labrundinia* inhabit shallow bodies of still water and slower reaches of flowing water (Fittkau 1962; Beck and Beck 1966).

## Genus *Larsia* Fittkau

Figs. 59–60

*Larsia* Fittkau, 1962:339; Roback 1971:259.

Body pale to translucent yellow, with faint brownish marbling; head capsule darker. A.R. 3.7–4.6. Ligula with straight toothed margin; first lateral teeth straight. Mandible with well developed inner and accessory teeth; length less than 90 μ. Maxillary palpus with 1 basal segment about 4 times as long as wide. Anal tubules less than one-half length of posterior parapods. Posterior parapod with simple yellowish claws.

**Remarks.** Except for their smaller size (expressed here as length of mandible) the larvae of *Larsia* are difficult to separate from those of *Zavrelimyia* (see key, couplet 13). According to Beck and Beck (1966) the second antennal segment of *Larsia berneri* is brown, but in all Canadian specimens seen by us this segment is concolorous with the rest of the antenna. Larvae are described by Beck and Beck (1966).

**Distribution.** Widespread south of treeline.

**Habitat.** Larvae of *Larsia* live in a wide variety of habitats including shallow areas of cold lakes, mountain streams, and shallow warm ponds or slow flowing water (Fittkau 1962; Beck and Beck 1966).

## Genus *Monopelopia* Fittkau

Figs. 61–62

*Monopelopia* Fittkau, 1962:394; Roback 1971:280.

Body pale green to yellow; head capsule yellow. Head ratio about 2.0. Second antennal segment dark brown; A.R. 3.0–3.8. Ligula with concave toothed margin, and with first lateral tooth straight. Mandible with triangular inner and accessory teeth. Maxillary palpus with 1 basal segment about 3–4 times as long as wide. Procercus about 3–4 times as long as wide. Anal tubules less than one-half length of posterior parapods. Posterior parapod claws all yellow brown or 1 claw brown and rest yellow brown; toothed or pectinate claws present.

**Remarks.** As in *Labrundinia* the second antennal segment is darker in color than the other segments. The larvae of *Labrundinia* and *Monopelopia* may be separated by the characters given in the key. Larvae are described by Beck and Beck (1966).

**Distribution.** British Columbia and Alberta; southeastern Ontario to New Brunswick.

**Habitat.** Larvae of *Monopelopia* typically inhabit warm, shallow, organically rich still water (Fittkau 1962; Beck and Beck 1966), but Canadian species are found in slower reaches of small bodies of flowing water.

## Genus *Natarsia* Fittkau

Figs. 63–64

*Natarsia* Fittkau, 1962:151; Roback 1971:107.

Body red; head capsule yellow brown. Head ratio 1.6–1.7. Antenna about one-third length of head capsule; A.R. about 3.0. Ligula with weakly concave toothed margin; first lateral tooth straight. Mandible with inner tooth larger than accessory tooth. Maxillary palpus with 1 basal segment 3–4 times as long as wide. Procercus about 5 times as long as wide. Posterior parapod with yellowish claws; several claws pectinate, most simple.

**Remarks.** In the Pentaneurini only the larvae of *Natarsia* and *Zavrelimyia* are red but the antenna of *Zavrelimyia* are one-half or more the head length. Larvae of *Natarsia* resemble those of Macropelopiini but lack paralabial plates and lateral abdominal setal fringes. Larvae are described by Johannsen (1937a) and Roback (1957)—both as *Anatopynia* (*Macropelopia*), and Roback (1978).

**Distribution.** British Columbia; Yukon Territory; District of Mackenzie; southeastern Ontario and southern Quebec.

**Habitat.** Larvae of *Natarsia* inhabit quieter reaches of small bodies of flowing water (Fittkau 1962).

## Genus *Nilotanypus* Kieffer

Figs. 65–67

*Nilotanypus* Kieffer, 1923:191; Fittkau 1962:405; Roback 1971:281.

Body pale; head capsule yellow brown. Head ratio about 2.3. A.R. about 3.0. Ligula with almost straight toothed margin; middle tooth usually slightly longer than first lateral teeth; first lateral tooth straight. Mandible with inner tooth slightly larger than accessory tooth. Maxillary palpus with 1 basal segment about 2.5–3.0 times as long as wide. Procercus darker than body, about 3 times as long as wide. Anal tubules about twice as long as posterior parapods. Posterior parapod with yellowish claws; 1 claw pectinate, rest simple.

**Remarks.** The larvae of *Nilotanypus* are similar to those of *Labrundinia* and *Pentaneura* (see remarks under these genera). Larvae are described by Beck and Beck (1966).

**Distribution.** Alberta; District of Mackenzie; Ontario to New Brunswick.

**Habitat.** Larvae of *Nilotanypus* inhabit moderately cool to warm flowing water (Fittkau 1962; Beck and Beck 1966). They prefer riffle areas.

## Genus *Paramerina* Fittkau

Figs. 68–70

*Paramerina* Fittkau, 1962:317; Roback 1971:271.

Body pale yellow, with reddish brown marbling; head capsule pale yellow brown. Head ratio about 2.0. A.R. 2.5–3.0. Ligula with straight to slightly concave toothed margin; first lateral tooth straight. Mandible with inner tooth slightly larger than accessory tooth. Maxillary palpus with 2 basal segments. Procercus about 3 times as long as wide. Anal tubules about one-half length of posterior parapods. Posterior parapod claws pale yellow brown; several claws finely spined on both margins.

**Remarks.** In the Pentaneurini, *Paramerina* and *Ablabesmyia* are the only genera that have a maxillary palpus with more than one basal segment (see remarks under *Ablabesmyia*). Larvae are described by Beck and Beck (1966) and Roback (1972).

**Distribution.** British Columbia to District of Mackenzie and Saskatchewan; southeastern Ontario to New Brunswick.

**Habitat.** Larvae of *Paramerina* inhabit shallow regions of still water and quieter reaches of flowing water (Fittkau 1962).

## Genus *Pentaneura* Philippi

Figs. 71–72

*Pentaneura* Philippi, 1865:629; Fittkau 1962:364; Roback 1971:268.

Color unknown. Head ratio about 2.1. A.R. 3.7–4.5. Ligula with slightly concave to straight toothed margin; first lateral teeth outcurved. Mandible with inner tooth larger than accessory tooth. Maxillary palpus with 1 basal segment about 7 times as long as wide. Procercus dark, about 6 times as long as wide. Anal tubules about twice as long as posterior parapods. Posterior parapod with 1 brown claw, remainder yellowish; all claws simple; seta on base thickened, brownish, not spinulose.

**Remarks.** The foregoing description is based on Beck and Beck's (1966) description of *Pentaneura inculta* Beck & Beck and one reared specimen from southern Ontario. Larvae of *Pentaneura* are similar to those of *Labrundinia* and *Nilotanypus*. In addition to the characters mentioned in remarks under *Labrundinia*, the presence of long dark procerci will distinguish the larvae of *Pentaneura*.

**Distribution.** Southeastern Ontario.

**Habitat.** Larvae of *Pentaneura* inhabit small streams, quieter reaches of larger bodies of flowing water, and shallow regions of still water (Fittkau 1962; Beck and Beck 1966).

## *Thienemannimyia* complex

Figs. 73–74

Color variable, usually pale yellow brown or green brown; head capsule yellow brown, yellow, or pale brown. Head ratio 1.60–1.80. A.R. 3.5–6.0. Ligula with concave toothed margin; first lateral tooth outcurved. Mandible with indistinct inner and accessory teeth, sometimes with only inner tooth. Maxillary palpus with 1 basal segment about 2.5–4.5 times as long as wide. Procercus about 2–3 times as long as wide, frequently darker in color than body. Anal tubules about one-half length of posterior prolegs. Posterior parapod claws all yellow brown or with several claws darker; usually several claws pectinate, rest simple.

**Remarks.** Larvae of the *Thienemannimyia* complex are distinguished by a high A.R., usually greater than 4.0, outcurved first lateral teeth of the ligula, and minute inner and accessory teeth on the mandible. They are similar to the larvae of *Trissopelopia* but can be separated by the length of the basal segment of the maxillary palpus (see key, couplet 11). Also the inner and accessory teeth of the mandible of *Trissopelopia* are larger than in the *Thienemannimyia* complex.

Fittkau (1962) included *Thienemannimyia*, *Arctopelopia*, *Conchapelopia*, and *Rheopelopia* in this complex. Roback (1971) added *Xenopelopia*. Although larvae of one or two species in each genus, except *Xenopelopia*, have been described (Johannsen 1905; Roback 1957 (both as *Pentaneura*); Beck and Beck 1966) current knowledge is insufficient to permit generic separation.

**Distribution.** Widespread throughout Canada; *Arctopelopia* occurs in the Arctic Archipelago.

**Habitat.** Larvae of the *Thienemannimyia* complex occur in both still and flowing water of all types (Fittkau 1962). Generally the complex is cool-adapted, living in well-oxygenated habitats and appearing to prefer flowing water to still water.

## Genus *Trissopelopia* Kieffer

Figs. 75–76

*Trissopelopia* Kieffer, 1923:178, 179; Fittkau 1962:353; Roback 1971:267.

Body pale yellow green or yellow brown, with brown marbling; head capsule yellow brown. Head ratio 1.8–2.0. A.R. 3.5–4.0. Ligula with concave toothed margin; first lateral tooth outcurved. Mandible with small subequal inner and accessory teeth. Maxillary palpus with 1 basal segment 7.0–7.5 times as long as wide. Procercus about 3 times as long as wide. Anal tubules about one-half length of posterior parapods. Posterior parapod claws yellow brown, with most simple but with several longer claws spinulose.

**Remarks.** Larvae of *Trissopelopia* resemble those of the *Thienemannimyia* complex (see remarks under *Thienemannimyia* complex).

**Distribution.** Alberta; District of Mackenzie; southeastern Ontario to Newfoundland.

**Habitat.** Larvae of *Trissopelopia* generally are cool-adapted, living in well-oxygenated flowing water and shallow areas of still water (Fittkau 1962).

## Genus *Zavrelimyia* Fittkau

Figs. 77–78

*Zavrelimyia* Fittkau, 1962:285; Roback 1971:264.

Body yellow brown to reddish, partly mottled with brown; head capsule pale brown. Head ratio 1.9–2.0. A.R. 2.4–2.9. Ligula with straight toothed margin; first lateral tooth straight. Mandible with well developed, triangular inner and accessory teeth; length greater than 110 μ. Maxillary palpus with 1 basal segment 3.5–4.5 times as long as wide. Procercus 2.5–4.0 times as long as wide. Anal tubules about one-half length of posterior parapods. Posterior parapod claws yellowish; 1 claw bifid, most of rest pectinate.

**Remarks.** The larvae of *Zavrelimyia* are similar to those of *Paramerina* but the maxillary palpus of the latter has two basal segments. Some species have reddish colored larvae, which may be confused with *Natarsia* (see remarks under *Natarsia*). They are also similar to the larvae of *Larsia*, especially in the structure of the ligula; however, these two genera may be separated by the characters given in the key. Larvae are described by Beck and Beck (1966).

**Distribution.** Southern British Columbia and Alberta; Manitoba to Newfoundland.

**Habitat.** Larvae of *Zavrelimyia* generally occur in warm, shallow bodies of water, both still and flowing. Some can withstand moderate reduction in oxygen-concentration (Fittkau 1962).

# Subfamily Podonominae

Figs. 79–87

Body brownish, bluish, or greyish blue; head capsule brown. Antenna 5-segmented; first segment with ring organ; blade on apex of first segment, usually shorter than combined length of terminal segments; second segment with apical style (= Lauterborn's organ?); style sometimes segmented; third (and sometimes fourth) segment annulate. Hypostoma with 1 median and 7–15 pairs of lateral teeth. Paralabial plate absent or indistinct. Mandible with 5–11 teeth, usually 5–7; seta subdentalis present but often indistinct and sometimes segmented; seta interna consisting of a row of 15–30 setae; pecten mandibularis absent. Premandible absent. SI and SII enlarged, sickle-shaped, borne on well developed socles. Labral lamellae present. Pecten epipharyngis consisting of 3 or 5 scales, often difficult to discriminate from chaetulae laterales of ungula. Abdominal segments with simple setae; eighth segment with or without pair of spiracular rings. Procercus elongate, greater than 5 times as long as wide, bearing 5–15 anal setae. Two pairs of anal tubules present. Posterior parapods well developed.

**Remarks.** The foregoing description is based on the larvae of northern hemisphere species of Podonominae. It does not encompass the structural diversity exhibited by the southern species (Brundin 1966), such as a four-segmented antenna, non-annulated third antennal segment, fleshy procercus without apical anal setae, and procercus less than five times as long as wide. However, the enlarged sickle-shaped SI and SII articulated to well developed socles will distinguish all larvae of the Podonominae from those of the other Chironomidae (Brundin 1966).

The subfamily has been revised by Brundin (1966), and Wirth and Sublett (1970) have reviewed the adults of the Podonominae of North America. The classification used here follows that of Brundin (1966). Two tribes are recognized, Podonomini and Boreochlini.

**Distribution.** The subfamily Podonominae reaches its greatest development in the southern part of the southern hemisphere. About 150 species in 8 genera occur in the southern hemisphere, whereas there are about 18 species in 5 genera in the northern hemisphere. Except for one species of *Parochlus* (tribe Podonomini) all the northern Podonominae belong to the tribe Boreochlini. There are no species in common between the two hemispheres, and all occur in the genus *Parochlus* (Brundin 1966; Wirth and Sublette 1970).

All the northern hemisphere genera, except *Paraboreochlus*, occur in North America, and in Canada. Within Canada they generally occur in mountain and northern streams and bogs.

**Biology.** Larvae of Podonominae are generally cold-adapted, living in well-oxygenated small bodies of flowing water (Brundin 1966).

Occurrence in warm flowing water or small bodies of still water, such as bog-pools, is rare. Many species are able to live within a wide range of temperatures, but a general characteristic of members of this subfamily is the ability to complete development at low temperatures. The larvae are free-living, feeding on algae (mainly diatoms) and detritus.

## Key to the tribes of Podonominae

1. Procercus unicolored. Antenna shorter than mandibular length .........
................ Tribe **Podonomini**; *Parochlus* **Enderlein** (p. 50)
Procercus with posterior part distinctly darker than anterior part. Antenna longer than mandibular length ......... Tribe **Boreochlini** (p. 51)

## Clé des tribus de Podonominæ

1. Procerque unicolore. Antenne plus courte que la mandibule ............
.............. Tribu des **Podonomini**; *Parochlus* **Enderlein** (p. 50)
Procerque avec partie postérieure nettement plus foncée que la partie antérieure. Antenne plus longue que la mandibule ..................
............................... Tribu des **Boreochlini** (p. 51)

## Tribe Podonomini

### Figs. 79–80

Only one species, *Parochlus kiefferi* (Garrett), is found in the northern hemisphere (Brundin 1966). The following description, based on *P. kiefferi*, defines the characters of the tribe Podonomini. It does not include the structural variation as given by Brundin (1966) for the genus.

## Genus *Parochlus* Enderlein

### Figs. 79–80

*Parochlus* Enderlein, 1912:109; Brundin 1966:109.
*Paratanytarsus* Garrett, 1925:8.

Antenna shorter than mandibular length; ring organ on basal half of first segment; third segment annulate. Hypostoma with 1 median and 7 pairs of lateral teeth; median tooth longer and broader than first lateral teeth. Mandible with 7 teeth, with outer subapical tooth appressed to longer apical tooth, and with 5 inner teeth. Pecten epipharyngis consisting of 3 scales. Abdominal setae similar in size; eighth segment without spiracular rings. Procercus unicolored bearing 7 or 8 anal setae.

**Remarks.** The characters given in the key will distinguish the larvae of the one Canadian species of *Parochlus*, and of Podonomini, from those of the Boreochlini. *Parochlus* belongs to the *araucana* group (Brundin 1966). Larvae are described by Brundin (1966).

**Distribution.** Southern British Columbia; southern Ontario to New Brunswick.

**Habitat.** Larvae of *Parochlus* inhabit small, cool flowing water, usually associated with moss (Brundin 1966).

## Tribe Boreochlini

### Figs. 81–87

Antenna longer than mandible; third segment annulate, also fourth in *Boreochlus*. Hypostoma with 1 median and 7–15 pairs of lateral teeth. Mandible with 5–11 teeth. Pecten epipharyngis consisting of 3 or 5 scales. Abdominal segments with setae similar in size, or with each segment with 1 pair larger than others; eighth abdominal segment with or without spiracular rings. Procercus black to blackish brown posteriorly, paler anteriorly, bearing 5–15 anal setae.

**Remarks.** Larvae of this tribe may be distinguished by the characters given in the key. Four genera, *Boreochlus*, *Lasiodiamesa*, *Paraboreochlus*, and *Trichotanypus* occur in the northern hemisphere. *Paraboreochlus* has not been recorded from North America.

## Key to the genera of Boreochlini

1. Abdominal segments each with a pair of long, dark ventral setae; lateral teeth of hypostoma deeply recessed (Fig. 85) .................................................... ***Trichotanypus* Kieffer** (p. 53)
   Abdominal segments with setae similar in size; eighth segment with a pair of spiracular rings. Median and first lateral teeth of hypostoma not recessed (Fig. 81) .................................................................................................. 2
2. Hypostoma with 7 pairs of lateral teeth. Procercus with 5 anal setae. Third and fourth antennal segments annulate .................................................... ***Boreochlus* Edwards** (p. 52)
   Hypostoma with 12–15 pairs of lateral teeth. Procercus with 11–15 anal setae. Only third antennal segment annulate (Fig. 84) .................................................... ***Lasiodiamesa* Kieffer** (p. 52)

## Clé des genres de Boreochlini

1. Segments abdominaux avec chacun une paire de longues soies ventrales foncées. Dents latérales de l'hypostome profondément renfoncées (fig. 85) .......................... ***Trichotanypus* Kieffer** (p. 53)

51

Segments abdominaux avec soies de taille semblable; huitième segment avec une paire d'anneaux stigmatiques. Dent médiane et premières dents latérales de l'hypostome non renfoncées (fig. 81) .................. 2

2. Hypostome avec 7 paires de dents latérales. Procerque avec 5 soies anales. Troisième et quatrième articles antennaires annelés ................
.................................... ***Boreochlus* Edwards** (p. 52)
Hypostome avec de 12 à 15 paires de dents latérales. Procerque avec de 11 à 15 soies anales. Troisième article antennaire annelé seulement (fig. 84)
.................................... ***Lasiodiamesa* Kieffer** (p. 52)

## Genus *Boreochlus* Edwards

*Boreochlus* Edwards *in* Edwards and Thienemann, 1938:152; Brundin 1966:299.

Head capsule parallel-sided. Antenna with third and fourth segments annulate, elongate, individually longer than second segment. Hypostoma with 7 pairs of lateral teeth. Pecten epipharyngis consisting of 3 scales. Abdominal segments without differentiated setae; eighth segment with a pair of spiracular rings. Procercus bearing 5 short anal setae.

**Remarks.** The foregoing description is based on Brundin (1966) and Thienemann (1938; as *Podonomus* (? *Paratanytarsus* sp. B)). No North American specimens were available for study but the elongate annulated third and fourth antennal segments distinguish the larvae of this genus. Brundin (1966) divided *Boreochlus* into two groups; *thienemanni* group and *burmanicus* group, both of which occur in Canada.

**Distribution.** British Columbia; Alberta; southern Ontario to New Brunswick.

**Habitat.** Larvae of *Boreochlus* inhabit cool springs and mountain streams (Brundin 1966).

## Genus *Lasiodiamesa* Kieffer

Figs. 81–84

*Syndiamesa* (*Lasiodiamesa*) Kieffer, 1924*a*:46.
*Lasiodiamesa*; Edwards 1937:102; Brundin 1966:315.
*Linaceus* Garrett, 1925:9.

Head slightly widened posteriorly. Antenna with third segment annulate, longer than second segment and with fourth segment very short, non-annulate; ring organ on middle third of first segment. Hypostoma with 1 median and 12–15 pairs of lateral teeth. Mandible with 11 teeth composed of an outer subapical tooth appressed to apical tooth and 9 inner teeth; third inner tooth enlarged. Pecten epipharyngis consisting of

5 scales covered by chaetulae laterales. Abdominal segments without differentiated setae; eighth segment with a pair of spiracular rings. Procercus about 12 times as long as wide, bearing 11–15 anal setae. Anal tubules longer than posterior parapods.

**Remarks.** The large number of teeth on the hypostoma will distinguish the larvae of *Lasiodiamesa*. Sæther (1969) described the larvae of two species found in Canada, both with 15 pairs of lateral teeth on the hypostoma. Thienemann (1937a) and Brundin (1966) both recorded 12 pairs of lateral teeth but Sæther (1969) noted that the last three pairs of the species described by him are small and could be easily overlooked.

**Distribution.** British Columbia and Yukon Territory to Saskatchewan and District of Keewatin; southern Ontario to New Brunswick.

**Habitat.** Larvae of *Lasiodiamesa* mainly occur in *Sphagnum* bogpools, but also occur in pools in other types of bogs and in rock pools in granite crevices (Brundin 1966; Sæther 1969). Unlike other podonomid larvae they may inhabit waters with temperatures up to 24°C.

## Genus *Trichotanypus* Kieffer

Figs. 85–87

*Trichotanypus* Kieffer, 1906a:319; Brundin 1966:310.

Head capsule widened posteriorly. Antenna with third segment annulate, with fourth segment non-annulate, with segments decreasing consecutively in size, and with ring organ near base of first segment. Hypostoma with 1 median and 12 pairs of lateral teeth; median tooth simple or apically emarginated, and first lateral teeth deeply recessed. Mandible with 5 or 6 teeth evenly spaced around apex. Pecten epipharyngis consisting of 5 scales covered by chaetulae laterales. Abdominal segments each with a pair of long dark ventral setae; eighth abdominal segment lacking spiracular rings. Procercus about 6 times as long as wide, bearing 6 anal setae. Anal tubules shorter than posterior parapods.

**Remarks.** The hypostoma with deeply recessed median and first lateral teeth will distinguish the larvae of *Trichotanypus*.

**Distribution.** Northwest Territories, widespread in Arctic Archipelago.

**Habitat.** The cold-adapted larvae of *Trichotanypus* live among moss and algae in cool pools, springs, streams, and shallow areas of lakes (Thienemann 1939; Brundin 1966).

## Subfamily Chironominae

Figs. 2, 6–7, 88–252

Body usually red; head capsule brownish. Eye spots, usually 2, rarely 3, usually lying one above another or with dorsal eye spot anterior to ventral eye spot. Antenna non-retractile, arising directly from head capsule or mounted on antennal tubercle, 4- to 8-segmented; blade present; ring organ on first segment; Lauterborn's organs usually opposite on apex of second segment, rarely alternate on apices of second and third segments or arising from shaft of second segment; third segment nonannulate. Hypostoma usually convex, always toothed. Paralabial plate large, nearly always striated, without setae. Mandible with 1 apical and 3–5 inner teeth; seta subdentalis present; seta interna rarely absent, apicodorsal tooth and pecten mandibularis present or absent. Premandible with 1–7 teeth. Maxillary palpus usually shorter than wide. SI and SII simple, plumose, or pectinate. Pectinate labral lamellae present or absent. Pecten epipharyngis consisting of 3 or more scales or a toothed bar. Anterior parapods separated, with apical crown of claws. Abdominal setae simple or plumose. Anal end bearing pair of procerci with anal setae, 2 pairs of anal tubules, and separate posterior parapods with apical crown of claws; procerci and anal tubules rarely absent; ventral tubules sometimes present.

**Remarks.** The presence of two eye spots on each side of the head capsule will usually distinguish larvae of the Chironominae. Larvae of the Tanypodinae and most Orthocladiinae have a single eye spot. In the Orthocladiinae with two eye spots or one partly divided eye spot, the larger eye spot or part lies posterior to the smaller eye spot or part. In the Chironominae the eye spots are usually subequal in size and lie one above the other. Most larvae of Chironominae are some shade of red, and as this color rarely occurs in other subfamilies it may be used for initial separation. Also the presence of large striated paralabial plates is diagnostic. This character is absent only in the genus *Stenochironomus*, which has a distinctive hypostoma (Fig. 204).

Three tribes, Chironomini, Pseudochironomini, and Tanytarsini are recognized (Sæther 1977*b*). The generic classification follows that of Hamilton et al. (1969) with modifications in the *Harnischia* complex (Sæther 1977*a*).

*Graceus* is the only genus occurring in Canada with unknown larvae. Several genera, for example *Goeldichironomus* and *Nilodorum*, recorded from southern United States are not expected to occur in Canada. The genera *Gillotia*, *Microchironomus*, and *Pedionomus*, which occur farther north in the United States, may occur in Canada and therefore are included in the key.

**Distribution.** The subfamily Chironominae is worldwide in distribution occurring in all major geographical regions except Antarctica. It occurs throughout Canada although there is a decrease in number of species from north to south and from warm to cold habitats. Most genera of the Tanytarsini occur in the arctic region, whereas only a few genera of Chironomini and no Pseudochironomini occur there.

**Biology.** Larvae of Chironominae are generally warm adapted, living in all types of freshwater. They are common in shallow warm still and flowing waters, but also occur to a lesser extent in cool waters. Some species inhabit brackish or terrestrial saline waters. There are no terrestrial species. Some species can withstand severe oxygen depletion and strongly polluted conditions. Cases of substrate cemented together by salivary gland secretion are built by the majority of species; only a few are free-living or tunnel in living or dead plant material.

Most of the larvae feed on small plants and animals and detritus, although a few are predaceous attacking larger invertebrates. Feeding mechanisms involving the use of salivary gland secretion have been described by Walshe (1951). Nets of various types are spun to entrap food particles carried by water currents or the sticky secretion is spread on surrounding substrate. In both cases the secretion is eaten with entrapped food particles.

## Key to the tribes of Chironominae

1. Antenna usually long, 5-segmented, arising from an antennal tubercle at least as long as wide (Fig. 7); first antennal segment curved; Lauterborn's organs prominent, opposite each other on apex of second segment, and sometimes mounted on long petioles (Fig. 7). Abdominal segments usually with plumose setae .............. Tribe **Tanytarsini** (p. 92)
Antenna usually shorter, 4- to 8-segmented; antennal tubercle rarely present, if present, then shorter than wide (Fig. 6); first antennal segment straight (except in *Pagastiella*); Lauterborn's organs usually smaller, either opposite each other on apex of second segment or on apices of second and third segments, never mounted on long petioles. Abdominal segments usually with only simple setae ................................. 2
2. Posterior parapod with 2 rows of claws arranged in a horseshoe shape. Median apices of straplike paralabial plates contiguous or very narrowly separated (Fig. 216) .................................................
..... Tribe **Pseudochironomini**; *Pseudochironomus* **Malloch** (p. 91)
Posterior parapod with claws irregularly arranged. Median apices of paralabial plates usually separated by more than width of median or median pair of teeth of hypostoma; if narrowly separated or contiguous then plates not straplike ................... Tribe **Chironomini** (p. 56)

## Clé des tribus de Chironominæ

1. Antenne habituellement longue, formée de 5 articles, dressée sur un tubercule antennaire au moins aussi long que large (fig. 7); premier article antennaire courbé; organes de Lauterborn proéminents, mutuellement opposés sur l'apex du deuxième article et parfois montés sur de long pétioles (fig. 7). Segments abdominaux habituellement avec des soies plumeuses .......................... Tribu des **Tanytarsini** (p. 92)
   Antenne habituellement plus courte, formée de 4 à 8 articles; tubercule antennaire rarement présent mais, le cas échéant, plus court que large (fig. 6); premier article antennaire droit (sauf chez *Pagastiella*); organes de Lauterborn habituellement plus petits, soit mutuellement opposés sur l'apex du deuxième article ou sur l'apex des deuxième et troisième articles, jamais montés sur de longs pétioles. Segments abdominaux avec habituellement des soies simples seulement ...................... 2
2. Parapode postérieur avec 2 rangés de griffes disposées en forme de fer à cheval. Apex médians des plaques paralabiales rubanés contigus ou très peu séparés (fig. 216) ................................................................
   Tribu des **Pseudochironomini;** *Pseudochironomus* **Malloch** (p. 91)
   Parapode postérieur avec griffes disposées irrégulièrement. Apex médians des plaques paralabiales habituellement séparés par plus de la largeur d'une dent ou d'une paire de dents médianes de l'hypostome; s'ils sont très peu séparés ou contigus, alors plaques non rubanées ...........
   ................................................. Tribu des **Chironomini** (p. 56)

## Tribe Chironomini

Figs. 2, 6, 88–215

Head with two, rarely three, eye spots. Antenna rarely arising from an antennal tubercle but, if tubercle present, then tubercle shorter than wide; 4- to 8-segmented; first segment nearly always straight; blade usually on apex of first segment; Lauterborn's organs usually on apex of second segment. Paralabial plate nearly always striated; median apices usually separated by more than width of median tooth or median pair of teeth. Mandible with or without apicodorsal tooth and pecten mandibularis; seta interna rarely absent. SI and SII simple, pectinate, or plumose. Pectinate labral lamellae present or absent. Abdominal segments nearly always with only simple setae (plumose only in *Microtendipes*). Eighth abdominal segment usually without ventral tubules. Posterior parapod with claws irregularly arranged.

**Remarks.** Larvae of Chironomini are quite diverse compared to those of the other two tribes. Except for a few genera that share some key characters with the other tribes, most Chironomini may be easily distinguished by the combination of a straight first antennal segment, abdomen with only simple setae, and irregularly arranged claws on the posterior parapods. Larvae of *Pagastiella* with a curved first antennal segment can be mistaken for a species of Tanytarsini. The median apices of the paralabial plates of *Lauterborniella* and *Xenochironomus* are contiguous or narrowly separated as in the Pseudochironomini. Some species of *Microtendipes* have plumose setae as well as simple setae on the abdominal segments as in the Tanytarsini. However, these can be separated by the other key characters or by the structure of the hypostoma.

## Key to the genera of Chironomini

1. One or 2 pairs of ventral tubules present (Fig. 2) .................... 2
   Ventral tubules absent ............................................. 5
2. Two pairs of ventral tubules present ..............................
   ........................... **Chironomus Meigen** (in part) (p. 64)
   One pair of ventral tubules present .................................. 3
3. Premandible with 5 or 6 teeth ......... **Kiefferulus Goetghebuer** (p. 80)
   Premandible bifid .................................................. 4
4. Mandible with serrated inner margin. Third antennal segment distinctly shorter than fourth ............ **Einfeldia Kieffer** (in part) (p. 67)
   Mandible with smooth inner margin. Third antennal segment usually as long as or longer than fourth .... **Glyptotendipes Kieffer** (in part) (p. 69)
5. Median apices of paralabial plates fused or, if separate, then by less than width of median or median pair of teeth (Fig. 168) .............. 6
   Median apices of paralabial plates separated by more than width of median or median pair of teeth of hypostoma (Fig. 6) ....................... 7
6. Antenna 6-segmented. Hypostoma with an even number of teeth .......
   ................................... **Lauterborniella Bause** (p. 81)
   Antenna 5-segmented. Hypostoma with an odd number of teeth ........
   ................................. **Xenochironomus Kieffer** (p. 89)
7. Hypostoma pale, with apically truncated teeth (Fig. 175). First antennal segment slightly curved. Toothed projection dorsal to antennal base present (Fig. 182) .................... **Pagastiella Brundin** (p. 83)
   Hypostoma darker, teeth apically rounded or pointed. First antennal segment straight. Toothed projection dorsal to antennal base absent ... 8
8. Labral lamellae absent or vestigial. SI simple .........................
   ............. **Harnischia Kieffer** complex (p. 70) ............. 9
   Labral lamellae clearly present (as in Fig. 6). SI plumose or pectinate ..24
9. Seven anterior body segments subdivided, giving appearance of a total of 20 segments present .................... **Chernovskiia Sæther** (p. 73)
   Anterior body segments not subdivided ........................... 10
10. Lateral teeth of hypostoma arched anterolaterally from light-colored and dome-shaped median area (Figs. 124, 135) .................... 11
    Lateral teeth of hypostoma arched posterolaterally or extending straight laterally, or if slightly arched anterolaterally, then median area not dome-shaped (Fig. 140); median area usually concolorous with lateral teeth, rarely lighter in color .................................. 13
11. Antenna 7- or 8-segmented; blade arising from third segment (Fig. 139) .
    ................................. **Demicryptochironomus Lenz** (p. 76)
    Antenna 5- or 6-segmented; blade arising from second segment ..... 12
12. Hypostoma with 7 pairs of lateral teeth. SI bristlelike and reduced .....
    ............................................. ***Gillotia* Kieffer**
    Hypostoma with 5 pairs of lateral teeth (Fig. 124). SI bladelike and about one-half length of SII .......... **Cryptochironomus Kieffer** (p. 74)

57

13. Antenna 7- or 8-segmented ........................................ 14
    Antenna 5- or 6-segmented ........................................ 15
14. Hypostoma with broad, often trifid, median tooth and 4 pairs of lateral teeth. Blade arising from proximal part of third antennal segment ........
    .................................................... ***Beckidia* Sæther** (p. 72)
    Hypostoma with 2 median and 5 or 6 pairs of lateral teeth (Fig. 154). Blade arising from distal part of second antennal segment ...............
    .................................................... ***Robackia* Sæther** (p. 79)
15. Hypostoma with median tooth lighter in color than lateral teeth ..... 16
    Hypostoma unicolorous ............................................ 20
16. Median tooth of hypostoma broadly triangular (Fig. 116) ...............
    ................................................ ***Acalcarella* Shilova** (p. 71)
    Median tooth of hypostoma not triangular, or 2 median teeth present (Figs. 132, 140, 148, 149, 159) ........................................ 17
17. Second and third antennal segments subequal in length; antenna 5-segmented. Basal segment of maxillary palpus about four times as long as wide .............................. ***Harnischia* Kieffer** (p. 77)
    Second antennal segment much longer than third segment; antenna 5- or 6-segmented. Basal segment of maxillary palpus at most three times as long as wide ................................................... 18
18. Antenna 6-segmented. Premandible with 3 subequal teeth (Fig. 161) ....
    .................................................... ***Saetheria* Jackson** (p. 79)
    Antenna 5-segmented. Premandible with 4–6 teeth decreasing in size apically to laterally or first two subequal and larger than remaining teeth (Figs. 133, 152) ........................................................ 19
19. Basal two-thirds of second antennal segment unsclerotized (Fig. 134) ....
    .................................................... ***Cyphomella* Sæther** (p. 75)
    Second antennal segment fully sclerotized ...........................
    ............................ ***Paracladopelma* Harnisch** (in part) (p. 78)
20. Hypostoma with double- or usually single-pointed median tooth and 6 or 7 pairs of lateral teeth; outer 1 to 3 lateral teeth not enlarged (Fig. 142). Anterior margin of paralabial plate usually strongly crenulated (Fig. 143). Pecten epipharyngis consisting of convex plate with pointed teeth
    .................................................... ***Parachironomus* Lenz** (p. 77)
    Hypostoma with trifid, medially notched, or broadly rounded median tooth; outer 1 to 3 lateral teeth distinctly enlarged (Figs. 120, 130). Anterior margin of paralabial plate at most weakly crenulate. Pecten epipharyngis consisting of 2 or 3 scales partly or completely fused to each other ..
    ................................................................ 21
21. Antennal blade longer than combined length of terminal segments. Median tooth of hypostoma trifid ............... ***Microchironomus* Kieffer**
    Antennal blade shorter than combined length of terminal segments. Median tooth of hypostoma trifid, medially notched or broadly rounded ... 22
22. Fifth and sixth lateral teeth of hypostoma equal in size ..................
    ............................ ***Paracladopelma* Harnisch** (in part) (p. 78)
    Sixth lateral tooth of hypostoma much larger than fifth (Figs. 120, 130)
    ................................................................ 23
23. Median tooth of hypostoma trifid or broadly rounded (Fig. 130) ........
    .................................................... ***Cryptotendipes* Lenz** (p. 75)
    Median tooth of hypostoma medially notched or double (Fig. 121) ......
    .................................................... ***Cladopelma* Kieffer** (p. 73)
24. Antenna 6-segmented; 1 Lauterborn's organ on each of second and third segments ........................................................ 25

Antenna 5-segmented or, rarely, 4-segmented; Lauterborn's organs opposite each other on apex of second segment, sometimes indistinct ..... 29
25. Hypostoma with light-colored median tooth or teeth ............... 26
Hypostoma uniformly dark colored ............................... 28
26. Hypostoma with median tooth broad, dome-shaped, sometimes serrated anteriorly, longer than lateral teeth (Fig. 184) ..................... ................................ *Paralauterborniella* Lenz (p. 84)
Hypostoma with median and at least first lateral teeth recessed (Figs. 172, 187, 188) ..................................................... 27
27. Hypostoma with only median teeth light colored; first lateral tooth shorter than median or second lateral teeth (Fig. 172)....................... ................................... *Microtendipes* Kieffer (p. 81)
Hypostoma with median and first lateral teeth light colored; median, first and second lateral teeth shorter than third lateral tooth (Figs. 187–188) ................................ *Paratendipes* Kieffer (p. 85)
28. First lateral tooth of hypostoma longer than median and second lateral teeth (Fig. 207). Antenna about as long as mandible, or shorter .......... ................................. *Stictochironomus* Kieffer (p. 89)
Second lateral tooth of hypostoma longer than first and third lateral teeth (Fig. 179). Antenna longer than mandible .. *Omisus* Townes (p. 83)
29. Hypostoma concave, with 9 or 10 large dark teeth (Fig. 204) ............ ................................ *Stenochironomus* Kieffer (p. 88)
Hypostoma convex or straight, with more than 10 teeth ............. 30
30. Mandible with 4 inner teeth on a common lobelike base (Fig. 177). Hypostoma with 2–4 minute apical projections on median tooth (Fig. 176) ... ................................... *Nilothauma* Kieffer (p. 82)
Mandible with inner teeth in usual regular row. Hypostoma variable, but never with minute apical projections on median tooth or teeth ... 31
31. Anterior margin of paralabial plate crenulate ...................... 32
Anterior margin of paralabial plate smooth ....................... 33
32. Paralabial plate one to one and one-half times as wide as long. Pecten epipharyngis with less than 7 teeth ... *Dicrotendipes* Kieffer (p. 66)
Paralabial plate more than twice as wide as long. Pecten epipharyngis with more than 7 teeth .......... *Glyptotendipes* Kieffer (in part) (p. 69)
33. Pecten epipharyngis divided into 3 parts (Figs. 193, 201). Hypostoma almost always with even number of teeth ............................ 34
Pecten epipharyngis not divided, with more or less continuous row of equal or unequal teeth (Fig. 101). Hypostoma with odd number of teeth .. ..................................................... 39
34. Hypostoma with odd number of teeth ............................ ........................ *Phaenopsectra* Kieffer (in part) (p. 85)
Hypostoma with even number of teeth ........................... 35
35. First lateral tooth of hypostoma much shorter than median or second lateral teeth (Fig. 197) .............. *Polypedilum* Kieffer (in part) (p. 86)
First lateral tooth of hypostoma longer than or about same length as median and second lateral teeth (Figs. 104–105, 190–191, 196) .......... 36
36. First lateral tooth of hypostoma about same length as median and second lateral teeth (Fig. 196) ........................................ 37
First lateral tooth of hypostoma longer than second lateral tooth and usually longer than median teeth (Figs. 104–105, 190–191) ............. 38
37. Pecten epipharyngis consisting of 3 undivided scales ................. ................................... *Pedionomus* Sublette

Pecten epipharyngis consisting of 3 scales, each with 3 or more teeth (Fig. 201) ........................ ***Polypedilum*** **Kieffer** (in part) (p. 86)
38. Paralabial plate with striations in central band stronger and more widely spaced than those in band along anterior margin (Fig. 106). SII plumose with 1 strong central branch bearing smaller branches (Fig. 109) ....
........................ ***Endochironomus*** **Kieffer** (p. 68)
Paralabial plate with striations in central band weaker and more closely spaced than those in band along anterior margin (Fig. 192). SII plumose with all branches thin and more or less equal in size (Fig. 194) ......
........................ ***Phaenopsectra*** **Kieffer** (in part) (p. 85)
39. Median tooth of hypostoma dome-shaped, sometimes weakly notched laterally (Fig. 111) .............. ***Glyptotendipes*** **Kieffer** (in part) (p. 69)
Median tooth of hypostoma tripartite (Fig. 99) ...................... 40
40. Pecten epipharyngis with 3-7 teeth .. ***Einfeldia*** **Kieffer** (in part) (p. 67)
Pecten epipharyngis with more than 7 teeth ...........................
........................ ***Chironomus*** **Meigen** (in part) (p. 64)

## Clé des genres de Chironomini

1. Une ou 2 paires de tubules ventraux présentes (fig. 2) ................ 2
Tubules ventraux absents ............................................. 5
2. Deux paires de tubules ventraux présentes ..............................
........................ ***Chironomus*** **Meigen** (en partie) (p. 64)
Une paire de tubules ventraux présente ............................. 3
3. Prémandibule avec 5 ou 6 dents ....... ***Kiefferulus*** **Gœtghebuer** (p. 80)
Prémandibule bifide .................................................. 4
4. Mandibule à bord interne serratiforme. Troisième segment antennaire nettement plus court que le quatrième ............................................
........................ ***Einfeldia*** **Kieffer** (en partie) (p. 67)
Mandibule à bord interne lisse. Troisième segment antennaire habituellement aussi long ou plus long que le quatrième ............................................
........................ ***Glyptotendipes*** **Kieffer** (en partie) (p. 69)
5. Apex médians des plaques paralabiales soudés ou, si séparés, alors par moins de la largeur de la dent ou de la paire de dents médianes (fig. 168)..
.................................................................... 6
Apex médians des plaques paralabiales séparés par plus de la largeur de la dent ou de la paire de dents médianes de l'hypostome (fig. 6) ..... 7
6. Antenne à 6 articles. Hypostome à nombre pair de dents ...............
........................ ***Lauterborniella*** **Bause** (p. 81)
Antenne à 5 articles. Hypostome à nombre impair de dents ...........
........................ ***Xenochironomus*** **Kieffer** (p. 89)
7. Hypostome pâle, à dents tronquées apicalement (fig. 175). Premier article antennaire légèrement courbé. Appendice denté derrière la base de l'antenne présent (fig. 182) ........... ***Pagastiella*** **Brundin** (p. 83)
Hypostome plus foncé, dents arrondies ou pointues à l'extrémité. Premier article antennaire droit. Appendice denté derrière la base de l'antenne absent ............................................................. 8
8. Lamelles labrales absentes ou vestigiales. SI simple ....................
............. ***Harnischia*** **Kieffer** complexe (p. 70) ............ 9
Lamelles labrales nettement présentes (comme dans la figure 6). SI plumeuse ou pectinée ...................................................... 24

9.  Sept segments antérieurs du corps subdivisés, donnant l'apparence de 20 segments au total ................ **Chernovskiia Sæther** (p. 73)
    Segments antérieurs du corps non subdivisés ...................... 10
10. Dents latérales de l'hypostome arquées antérolatéralement à partir de la partie médiane pâle et en forme de dôme (fig. 124, 135) ........ 11
    Dents latérales de l'hypostome arquées postérolatéralement ou s'étendant droites latéralement, ou si légèrement arquées antérolatéralement, alors partie médiane non en forme de dôme (fig. 140); partie médiane habituellement de la même couleur que les dents latérales, rarement plus pâle ....................................................... 13
11. Antenne à 7 ou 8 articles; lame faisant saillie du troisième article (fig. 139) ............................. **Demicryptochironomus Lenz** (p. 76)
    Antenne à 5 ou 6 articles; lame faisant saillie du deuxième article .... 12
12. Hypostome avec 7 paires de dents latérales. SI sétiforme et réduite ..... ................................................. **Gillotia Kieffer**
    Hypostome avec 5 paires de dents latérales (fig. 124). SI lamiforme et environ moitié moins longue que SII ..... **Cryptochironomus Kieffer** (p. 74)
13. Antenne à 7 ou 8 articles ....................................... 14
    Antenne à 5 ou 6 articles ....................................... 15
14. Hypostome avec une dent médiane large, souvent trifide, et 4 paires de dents latérales. Lame faisant saillie de la partie proximale du troisième article antennaire ............................. **Beckidia Sæther** (p. 72)
    Hypostome avec 2 dents médianes et 5 ou 6 paires de dents latérales (fig. 154). Lame faisant saillie de la partie distale du deuxième article antennaire ............................................ **Robackia Sæther** (p. 79)
15. Hypostome à dent médiane plus pâle que les dents latérales ......... 16
    Hypostome unicolore ........................................... 20
16. Dent médiane de l'hypostome largement triangulaire (fig. 116) ......... ................................................. **Acalcarella Shilova** (p. 71)
    Dent médiane de l'hypostome non triangulaire, ou 2 dents médianes présentes (fig. 132, 140, 148, 149, 159) ............................ 17
17. Deuxième et troisième articles antennaires à peu près de la même longueur; antenne à 5 articles. Article basal du palpe maxillaire environ 4 fois plus long que large ....................... **Harnischia Kieffer** (p. 77)
    Deuxième article antennaire beaucoup plus long que le troisième; antenne à 5 ou 6 articles. Article basal du palpe maxillaire au moins trois fois plus long que large ............................................. 18
18. Antenne à 6 articles. Prémandibule avec 3 dents à peu près égales (fig. 161) ............................. **Sætheria Jackson** (p. 79)
    Antenne à 5 articles. Prémandibule avec de 4 à 6 dents allant en diminuant de l'apex aux côtés ou deux premières dents à peu près égales et plus grosses que les autres (fig. 133, 152) ............................ 19
19. Deux tiers proximaux du deuxième article antennaire non sclérotisés (fig. 134) .......................... **Cyphomella Sæther** (p. 75)
    Deuxième article antennaire entièrement sclérotisé ..................... ................... **Paracladopelma Harnisch** (en partie) (p. 78)
20. Hypostome avec une dent médiane à pointe double ou habituellement simple et 6 ou 7 paires de dents latérales; de 1 à 3 dents latérales externes non élargies (fig. 142). Bord antérieur de la plaque paralabiale habituellement fortement crénelé (fig. 143). Peigne épipharyngéal consistant en une plaque convexe avec dents pointues ........................ ................................. **Parachironomus Lenz** (p. 77)

Hypostome avec une dent médiane trifide, entaillée au centre ou largement arrondie; de 1 à 3 dents latérales externes nettement élargies (fig. 120, 130). Bord antérieur de la plaque paralabiale au plus légèrement crénelé. Peigne épipharyngéal consistant en 2 ou 3 écailles partiellement ou complètement soudées les unes aux autres .................. 21

21. Lame antennaire plus longue que la longueur combinée des articles terminaux. Dent médiane de l'hypostome trifide ....................... ................................................ ***Microchironomus* Kieffer**
Lame antennaire plus courte que la longueur combinée des articles terminaux. Dent médiane de l'hypostome trifide, entaillée au centre ou largement arrondie ......................................... 22

22. Cinquième et sixième dents latérales de l'hypostome de même grosseur .. ...................... ***Paracladopelma* Harnisch** (en partie) (p. 78)
Sixième dent latérale de l'hypostome plus grosse que la cinquième (fig. 120, 130) ...................................................... 23

23. Dent médiane de l'hypostome trifide ou largement arrondie (fig. 130) ... .................................. ***Cryptotendipes* Lenz** (p. 75)
Dent médiane de l'hypostome entaillée au centre ou double (fig. 121) ... .................................. ***Cladopelma* Kieffer** (p. 73)

24. Antenne à 6 articles; 1 organe de Lauterborn sur chacun des deuxième et troisième articles ........................................... 25
Antenne à 5 articles, rarement, à 4; organes de Lauterborn opposés mutuellement à l'apex du deuxième article, parfois indistincts ........... 29

25. Hypostome avec un ou des dents médianes pâles .................. 26
Hypostome uniformément foncé ............................... 28

26. Hypostome à dent médiane large, en forme de dôme, parfois serratiforme vers l'avant, plus longue que les dents latérales (fig. 184) ......... ................................. ***Paralauterborniella* Lenz** (p. 84)
Hypostome avec les dents médianes et au moins les premières dents latérales renfoncées (fig. 172, 187, 188) ............................. 27

27. Hypostome à dents médianes, seulement, légèrement colorées; première dent latérale plus courte que les dents médianes ou que les deuxièmes dents latérales (fig. 172) ............ ***Microtendipes* Kieffer** (p. 81)
Hypostome à dents médianes et à premières dents latérales légèrement colorées; dents médianes et premières et deuxièmes dents latérales plus courtes que la troisième dent latérale (fig. 187, 188) ................ ................................. ***Paratendipes* Kieffer** (p. 85)

28. Première dent latérale de l'hypostome plus longue que les dents médianes et les deuxièmes dents latérales (fig. 207). Antenne à peu près aussi longue que la mandibule ou plus courte .. ***Stictochironomus* Kieffer** (p. 89)
Deuxième dent latérale de l'hypostome plus longue que les première et troisième dents latérales (fig. 179). Antenne plus longue que la mandibule ................................. ***Omisus* Townes** (p. 83)

29. Hypostome concave avec 9 ou 10 grosses dents foncées (fig. 204) ....... .................................. ***Stenochironomus* Kieffer** (p. 88)
Hypostome convexe ou droit, avec plus de 10 dents ................ 30

30. Mandibule avec 4 dents internes sur une base commune en forme de lobe (fig. 177). Hypostome avec de 2 à 4 minuscules appendices apicaux sur la dent médiane (fig. 176) .............. ***Nilothauma* Kieffer** (p. 82)
Mandibule avec les dents internes disposées en rangée régulière habituelle. Hypostome variable, mais toujours sans minuscules appendices apicaux sur la ou les dents médianes ................................ 31

31. Bord antérieur de la plaque paralabiale crénelé .................... 32
    Bord antérieur de la plaque paralabiale lisse ....................... 33
32. Plaque paralabiale d'aussi large à une fois et demi plus large que longue. Peigne épipharyngéal avec moins de 7 dents ......................
    .................................... ***Dicrotendipes*** **Kieffer** (p. 66)
    Plaque paralabiale plus de deux fois plus large que longue. Peigne épipharyngéal avec plus de 7 dents ........................................
    ........................ ***Glyptotendipes*** **Kieffer** (en partie) (p. 69)
33. Peigne épipharyngéal divisé en 3 parties (fig. 193, 201). Hypostome avec presque toujours un nombre pair de dents ..................... 34
    Peigne épipharyngéal non divisé, avec une rangée plus ou moins continue de dents égales ou inégales (fig. 101). Hypostome avec un nombre impair de dents ......................................................... 39
34. Hypostome avec un nombre impair de dents .........................
    ........................ ***Phænopsectra*** **Kieffer** (en partie) (p. 85)
    Hypostome avec un nombre pair de dents ........................ 35
35. Première dent latérale de l'hypostome beaucoup plus courte que les dents médianes ou les deuxièmes dents latérales (fig. 197) ................
    ........................... ***Polypedilum*** **Kieffer** (en partie) (p. 86)
    Première dent latérale de l'hypostome plus longue ou environ de la même longueur que les dents médianes et les deuxièmes dents latérales (fig. 104, 105, 190, 191, 196) ................................... 36
36. Première dent latérale de l'hypostome environ de la même longueur que les dents médianes et les deuxièmes dents latérales (fig. 196) ........ 37
    Première dent latérale de l'hypostome plus longue que la deuxième dent latérale et habituellement plus longue que les dents médianes (fig. 104, 105, 190, 191) ............................................... 38
37. Peigne épipharyngéal consistant en 3 écailles non divisées .............
    ........................................... ***Pedionomus*** **Sublette**
    Peigne épipharyngéal consistant en 3 écailles, chacune portant 3 dents ou plus (fig. 201) .............. ***Polypedilum*** **Kieffer** (en partie) (p. 86)
38. Plaque paralabiale avec striations de la bande centrale plus fortes et plus largement espacées que celles de la bande longeant le bord antérieur (fig. 106). SII plumeuse avec 1 forte ramification centrale portant des ramifications plus petites (fig. 109) ..............................
    .................................. ***Endochironomus*** **Kieffer** (p. 68)
    Plaque paralabiale avec striations de la bande centrale plus faibles et plus rapprochées que celles de la bande longeant le bord antérieur (fig. 192). SII plumeuse avec toutes les ramifications minces et plus ou moins égales (fig. 194) ................ ***Phænopsectra*** **Kieffer** (en partie) (p. 85)
39. Dent médiane de l'hypostome en forme de dôme, parfois faiblement entaillée latéralement (fig. 111) .... ***Glyptotendipes*** **Kieffer** (en partie) (p. 69)
    Dent médiane de l'hypostome tripartite (fig. 99) ................... 40
40. Peigne épipharyngéal avec de 3 à 7 dents ............................
    ........................... ***Einfeldia*** **Kieffer** (en partie) (p. 67)
    Peigne épipharyngéal avec plus de 7 dents .........................
    ........................ ***Chironomus*** **Meigen** (en partie) (p. 64)

# Genus *Chironomus* Meigen

Figs. 2, 6, 88–93

*Chironomus* Meigen, 1803:260.
*Tendipes* Meigen, 1800:17.
*Tendipes (Tendipes)*; Townes 1945:101.
*Camptochironomus* Kieffer, 1918a:45.

Antenna 5-segmented, with third segment usually shorter than, rarely as long as fourth; ring organ position variable; blade shorter than combined length of terminal segments; Lauterborn's organs on apex of second segment. Hypostoma with 1 broad tripartite median and 6 pairs of lateral teeth. Paralabial plate with usually smooth, rarely crenulate, anterior margin; median apex curved posteriorly. Mandible with 2 or 3 inner teeth; apicodorsal tooth and pecten mandibularis present; inner margin with spines. Premandible bifid. SI and SII usually pectinate; SI sometimes plumose. Labral lamellae present. Pecten epipharyngis consisting of a bar with 10–20 teeth. Procercus present, bearing anal setae. Seventh abdominal segment with or without posterolateral projection. Eighth abdominal segment with two pairs of ventral tubules or none. Two pairs of anal tubules present.

**Remarks.** The tripartite median tooth of the hypostoma and the antenna with the third segment shorter than the fourth segment will distinguish most larvae of *Chironomus* from those of the rest of the Chironomini. Larvae of *Kiefferulus* and some species of *Einfeldia* have a weakly developed tripartite median tooth, but generally the median tooth is only notched laterally as in *Dicrotendipes* and some species of *Glyptotendipes*. In these genera the third antennal segment usually is as long as or longer than the fourth segment. Other than in *Goeldichironomus* (which does not occur in Canada), *Chironomus* is the only genus with larvae having two pairs of ventral tubules on the eighth abdominal segment, although these are absent in some species groups.

Three subgenera, *Chironomus (Chironomus)*, *Chironomus (Chaetolabis)*, and *Chironomus (Wirthiella)*, appear in North American literature. *Chironomus (Wirthiella)* (Sublette 1960) does not occur in Canada. The separation between *Chironomus (Chironomus)* and *Chironomus (Chaetolabis)* is not well defined. The characters given in the following key (couplet 6) will separate most specimens. However, the number of teeth on the pecten epipharyngis is variable, sometimes less than 18 are present in *Chaetolabis*. If the teeth are not worn then the shape of the teeth as described in couplet 6 will afford separation of the two genera. A fourth subgenus, *Chironomus (Camptochironomus)* has sometimes been used. Larvae of this subgenus are similar to those of the *plumosus* group and are distinguished by having the frontoclypeal apotome darker than adjacent parts of the head capsule. Formerly the genus included many subgenera that are now treated as genera, e.g., *Dicrotendipes, Cryptochironomus, Einfeldia, Kiefferulus, Endochironomus, Nilodorum,* and *Xenochironomus* (see Sublette and Sublette 1965).

European taxonomists have divided the genus into a number of species groups that have not been used much in North America. Lenz (1954–62) gives a description of the groups and a larval key. The groups are primarily defined upon the presence or absence and the relative length of the ventral tubules. Although the length of the tubules may vary according to the physiological state of the larvae or with environmental conditions, the species groups are useful for rough sorting larvae of *Chironomus*. A key to subgenera and species groups, adapted from Lenz (1954–62), is given below.

Larvae are described by Johannsen (1937b), Roback (1957), Curry (1961), Sublette and Sublette (1974a, 1974b), and Wulker et al. (1971).

## Key to the subgenera and species groups of *Chironomus*

1. Eighth abdominal segment with two pairs of ventral tubules (Fig. 2) ... 2
   Eighth abdominal segment without ventral tubules ..............
   .............. ***Chironomus* (*Chironomus*)** (in part) .............. 9
2. Ventral tubules as long as or longer than length of eighth abdominal segment
   .............................................................. 3
   Ventral tubules shorter than width of eighth abdominal segment ........
   .............. ***Chironomus* (*Chironomus*)** (in part) .............. 8
3. Seventh abdominal segment with posterolateral projection ..............
   .............. ***Chironomus* (*Chironomus*)** (in part) .............. 4
   Seventh abdominal segment without posterolateral projection ......... 6
4. Anterior margin of paralabial plate crenulate ............ ***staegeri*** group
   Anterior margin of paralabial plate smooth (Fig. 89) ................ 5
5. Larva more than 20 mm long ........................ ***plumosus*** group
   Larva less than 18 mm long ..................... ***annularius*** group
6. Pecten epipharyngis with 18 or more teeth; teeth varying irregularly in length and width (Fig. 92) .............. ***Chironomus* (*Chaetolabis*)**
   Pecten epipharyngis with less than 18 teeth; teeth equal in size or becoming progressively smaller laterally (Fig. 91) ............................
   .............. ***Chironomus* (*Chironomus*)** (in part) .............. 7
7. Ring organ of basal half, usually basal third, of first antennal segment. Larva about 12 mm long ................................ ***thummi*** group
   Ring organ on about middle of first antennal segment. Larva longer than 15 mm .......................................... ***anthracinus*** group
8. Seventh abdominal segment with posterolateral projection ..............
   ............................................ ***semireductus*** group
   Seventh abdominal segment without posterolateral projection ..........
   .............................................. ***halophilus*** group
9. Seventh abdominal segment with posterolateral projection ..............
   ................................................ ***reductus*** group
   Seventh abdominal segment without posterolateral projection ..........
   ............................................... ***salinarius*** group

## Clé des sous-genres et des groupes d'espèces de *Chironomus*

1. Huitième segment abdominal avec deux paires de tubules ventraux (fig. 2) .................................................................. 2
   Huitième segment abdominal sans tubules ventraux .................... 
   ............ ***Chironomus (Chironomus)*** (en partie) ............ 9
2. Tubules ventraux aussi longs ou plus longs que la longueur du huitième segment abdominal ........................................... 3
   Tubules ventraux plus courts que la largeur du huitième segment abdominal ............ ***Chironomus (Chironomus)*** (en partie) ............ 8
3. Septième segment abdominal avec un appendice postérolatéral ........
   ............ ***Chironomus (Chironomus*** (en partie ) ............ 4
   Septième segment abdominal sans appendice postérolatéral ........... 6
4. Bord antérieur de la plaque paralabiale crénelé ...... groupe des ***stægeri***
   Bord antérieur de la plaque paralabiale lisse (fig. 89) ................. 5
5. Larve de plus de 20 mm de longueur ............ groupe des ***plumosus***
   Larve de moins de 18 mm de longueur .......... groupe des ***annularius***
6. Peigne épipharyngéal avec 18 dents ou plus; dents de longueur et de largeur variant irrégulièrement (fig. 92) ............ ***Chironomus (Chætolabis)***
   Peigne épipharyngéal avec moins de 18 dents; dents égales de taille ou allant en diminuant progressivement latéralement (fig. 91) ................
   ............ ***Chironomus (Chironomus)*** (en partie) ............ 7
7. Organe annulaire situé sur la moitié proximale, habituellement le tiers proximal, du premier article antennaire. Larve d'environ 12 mm de longueur .................................................. groupe des ***thummi***
   Organe annulaire situé à peu près au milieu du premier article antennaire. Larve dépassant 15 mm de longueur ....... groupe des ***anthracinus***
8. Septième segment abdominal avec un appendice postérolatéral ........
   .................................................. groupe des ***semireductus***
   Septième segment abdominal sans appendice postérolatéral ............
   .................................................. groupe des ***halophilus***
9. Septième segment abdominal avec appendice postérolatéral ............
   .................................................. groupe des ***reductus***
   Septième segment abdominal sans appendice postérolatéral ............
   .................................................. groupe des ***salinarius***

**Distribution.** Widespread, occurring north of treeline.

**Habitat.** Larvae of *Chironomus* occur in all types of freshwater, although they are commoner in still water and slower reaches of flowing water. Some species can tolerate low oxygen and strongly polluted conditions (Thienemann 1954; Lenz 1954–62; Curry 1961).

## Genus *Dicrotendipes* Kieffer

Figs. 94–98

*Dicrotendipes* Kieffer, 1913*d*:23.
*Limnochironomus* Kieffer, 1920*b*:166.
*Tendipes* (*Limnochironomus*); Townes 1945:102.

Antenna 5-segmented; with third segment slightly shorter than or equal in length to fourth segment; ring organ on basal third of first segment; blade shorter than combined length of terminal segments; Lauterborn's organs on apex of second segment. Hypostoma with 1 median and 6 pairs of lateral teeth; median tooth entire or notched laterally; first and second lateral teeth frequently appressed. Paralabial plate short, one to one and one-half times as wide as long, with anterior margin crenulate, and with median apex curved posteriorly. Mandible with 3 inner teeth; apicodorsal tooth and pecten mandibularis present; inner margin smooth. Premandible bifid. SI plumose; SII simple. Labral lamellae present. Pecten epipharyngis consisting of a bar with 3–7 teeth. Procercus present, bearing anal setae. Ventral tubules absent. Two pairs of anal tubules present.

**Remarks.** The paralabial plate, with a crenulated anterior margin and less than one and one-half times as wide as long, will distinguish most *Dicrotendipes*. The paralabial plates of *D. californicus* are smooth (Webb 1972), but this species does not occur in Canada. Some species of *Glyptotendipes* that have a similar hypostoma also have crenulated paralabial plates, but the plates are more than one and one-half times as wide as long. Larvae are described by Johannsen (1937*b*; as *Chironomus* (*Chironomus*) group *Limnochironomus*), Lenz (1954–62; as *Limnochironomus*), Curry (1961), and Webb (1972).

**Distribution.** Widespread south of treeline.

**Habitat.** Larvae of *Dicrotendipes* inhabit shallow regions of still water of all sizes and quieter reaches of flowing water (Lenz 1954–62; as *Limnochironomus*).

## Genus *Einfeldia* Kieffer

Figs. 99–103

*Einfeldia* Kieffer, 1924*b*:393.
*Tendipes* (*Einfeldia*); Townes 1945:111.

Antenna 5-segmented, with third segment slightly shorter than or as long as fourth segment; ring organ on basal half, usually on basal third, of first segment; blade shorter than combined length of terminal segments; Lauterborn's organs on apex of second segment. Hypostoma with 1 median and 6 pairs of lateral teeth; median tooth tripartite or entire and dome-shaped with weak lateral notches. Paralabial plate about twice as long as wide, with anterior margin smooth, and with median apex curved posteriorly. Mandible with 2 or 3 inner teeth; apicodorsal tooth and pecten mandibularis present; inner margin with spines. Premandible bifid. SI plumose; SII simple. Labral lamellae present. Pecten epipharyngis consisting of a bar bearing 4–8 teeth or 16–20 teeth; teeth may be irregular or in multiple rows. Procercus present, bearing anal setae. One pair of ventral tubules present or absent. Two pairs of anal tubules present.

**Remarks.** Larvae of *Einfeldia* with one pair of ventral tubules are easily distinguished from those of *Chironomus* with two pairs of tubules, but are difficult to separate from some *Glyptotendipes*. Usually the presence of spines on the inner margin of the mandible of *Einfeldia* will distinguish it from *Glyptotendipes*, but these are sometimes weak or worn. The relative lengths of the third and fourth antennal segments (see key, couplet 4) will usually afford separation, but sometimes the two segments are subequal in length. Therefore it is not always possible to distinguish the larvae of *Einfeldia* from those of *Glyptotendipes* with one pair of ventral tubules. The larvae without ventral tubules can be distinguished by the characters given in the key. Larvae are described by Lenz (1954–62), Curry (1961), Beck and Beck (1970), and Oliver (1971a).

**Distribution.** Widespread south of treeline.

**Habitat.** Larvae of *Einfeldia* generally occur in small bodies and shallow regions of still water and in quieter reaches of flowing water (Lenz 1954–62; Danks 1971b).

## Genus *Endochironomus* Kieffer

Figs. 104–110

*Endochironomus* Kieffer, 1918b:69.
*Tanytarsus (Endochironomus)*; Townes 1945:64.

Antenna 5-segmented, with third segment equal to or longer than fourth; ring organ on basal third of first segment; Lauterborn's organs on apex of second segment. Hypostoma with 1 or 2 median and 7 pairs of lateral teeth; first lateral tooth longer than second and usually longer than median teeth. Paralabial plate with 2 bands of striations; striations in central band usually longer and more widely spaced than striations in band along anterior margin; plate about one and one-half times as wide as long, with smooth anterior margin, and with median apex curved anteriorly. Mandible with 3 inner teeth; pecten mandibularis and apicodorsal tooth present. Premandible bifid. SI pectinate; SII plumose, composed of a main branch with smaller lateral branches. Labral lamellae present. Pecten epipharyngis weakly divided into 3 scales, each part bearing teeth. Ventral tubules absent. Procercus present, bearing anal setae. Two pairs of anal tubules present.

**Remarks.** Larvae of *Endochironomus* and *Phaenopsectra* cannot always be separated with certainty. The characters given in the key (couplet 38) will afford separation in most cases. However, in some *Endochironomus* the central band of striations is stronger but the striations are placed closer together than those along the anterior margin. Also, in some *Phaenopsectra* the two bands are almost equal in development.

Kalugina (1961) recognized three larval groups: *nymphoides*, *signaticornis*, and *dispar* groups. The hypostoma of the *nymphoides* and *dispar* groups has two median teeth. In the *nymphoides* group the base of the median teeth is on the same level as the base of the lateral teeth, whereas in the *dispar* group it is anterior. In the *signaticornis* group, which has not been found in Canada, the hypostoma has a single median tooth. However, as Kalugina (1961) states there is no consensus on the division of the genus into groups. Larvae are described by Johannsen (1937b; as *Chironomus* (*Endochironomus*)), Berg (1950), Lenz (1954–62), Kalugina (1961).

**Distribution.**   Widespread south of treeline.

**Habitat.**   Larvae of *Endochironomus* live in shallow still water and quieter reaches of flowing water. They are usually associated with aquatic vegetation; living among plant residues on the bottom, on dead or living plants, or in dead parts (Lenz 1955; Kalugina 1961). Some species bore into soft parts of living plants (Berg 1950; Kalugina 1961).

## Genus *Glyptotendipes* Kieffer

Figs. 111–115

*Glyptotendipes* Kieffer, 1913b:225; Townes 1945:136.
*Demeijerea* Kruseman, 1933:154.
*Phytotendipes* Goetghebuer, 1937–54:14.

Two, sometimes 3 eye spots present. Antenna 5-segmented, with third segment as long as or longer than fourth segment; ring organ on basal half, usually on basal third, of first segment; Lauterborn's organs on apex of second segment. Hypostoma with 1 median and 6 pairs of lateral teeth; median tooth usually broad, dome-shaped, with lateral incisions. Paralabial plate usually more than two and one-half times as long as wide, occasionally about one to one and one-half times, with anterior margin smooth or crenulate, and with medial apex curved posteriorly. Mandible with 3 inner teeth; apicodorsal tooth and pecten mandibularis present; inner margin smooth. Premandible bifid. SI plumose; SII simple. Labral lamellae present. Pecten epipharyngis consisting of a bar with variable number of teeth, usually more than 18; teeth often irregular in size, sometimes in multiple rows. Procercus present, bearing anal setae. One pair of ventral tubules present or absent. Two pairs of anal tubules present.

**Remarks.**   Most larvae of *Glyptotendipes* may be distinguished by the characters given in the key, but some are difficult to separate from some *Dicrotendipes* or *Einfeldia* (see remarks under these genera).

Three subgenera are recognized: *Glyptotendipes* (*Glyptotendipes*), *Glyptotendipes* (*Demeijerea*), and *Glyptotendipes* (*Phytotendipes*). The following key gives the distinguishing characters of each subgenus. Larvae are described by Johannsen (1937b; as *Chironomus* (*Glyptotendipes*)), Lenz (1954–62), and Berg (1950).

## Key to the subgenera of *Glyptotendipes*

1. Ventral tubules absent ................. **Glyptotendipes (Glyptotendipes)**
   Ventral tubules present ........................................... 2
2. Ventral tubules longer than posterior parapods and apically tapering. Three eye spots present .................... **Glyptotendipes (Demeijerea)**
   Ventral tubules shorter, rarely as long as posterior parapods and more or less of equal thickness throughout. Two eye spots present ..............
   ............................................ **Glyptotendipes (Phytotendipes)**

## Clé des sous-genres de *Glyptotendipes*

1. Tubules ventraux absents .............. **Glyptotendipes (Glyptotendipes)**
   Tubules ventraux présents ......................................... 2
2. Tubules ventraux plus longs que les parapodes postérieurs et allant en s'amincissant vers l'extrémité. Trois stemmates présents ............
   ............................................ **Glyptotendipes (Demeijerea)**
   Tubules ventraux plus courts, rarement aussi longs que les parapodes postérieurs et plus ou moins de la même épaisseur sur toute leur longueur. Deux stemmates présents ............ **Glyptotendipes (Phytotendipes)**

**Distribution.** Widespread south of treeline.

**Habitat.** Larvae of *Glyptotendipes* inhabit all types of still water, slower reaches of flowing water, and, rarely, brackish water (Lenz 1954–62). Many species mine in dead and living plant parts (Lenz 1954–62; Berg 1950), and a few in bryozoans (Neff and Benfield 1970) or freshwater sponges (Wundsch 1943). The free-living tube building species generally occur in areas with rooted aquatic plants or among green algae filaments.

## *Harnischia* complex

Figs. 116–162

The following genera of the *Harnischia* complex occur in Canada:

| | | |
|---|---|---|
| *Acalcarella* | *Cryptotendipes* | *Parachironomus* |
| *Beckidia* | *Cyphomella* | *Paracladopelma* |
| *Chernovskiia* | *Demicryptochironomus* | *Robackia* |
| *Cladopelma* | *Harnischia* | *Saetheria* |
| *Cryptochironomus* | | |

Separation of many of the genera in the *Harnischia* complex is difficult and often depends on small hard-to-see characters. Therefore the complex is treated here as a unit. A modified version of Sæther's (1977a) key is given in the key to genera. Sæther's (1977a) paper should be used in conjunction with this one, especially for illustrations, because the amount of material suitable for microphotographs was limited.

According to Sæther (1977a) the only character that will distinguish all members of the *Harnischia* complex from other Chironomini is the composition of the pecten epipharyngis. Except in *Parachironomus*, it consists of a single plate or two or three tightly appressed scales with or without points. The pecten epipharyngis in *Parachironomus* consists of a convex plate with three or more pointed and transparent teeth. However, this character is very difficult to interpret and there are several other characters that will distinguish the complex from the majority of the Chironomini. The SI is simple; in other Chironomini, except in *Pagastiella*, it is plumose or pectinate. Also the labral lamellae are absent or vestigial. They consist of one or two transverse pectinate or toothed structures in other Chironomini.

Townes (1945) placed all species of the complex occurring in North America in either *Cryptochironomus* or *Harnischia*. Until the publications of Beck and Beck (1969) and Sæther (1971a, 1977a) this system was followed, except by a few authors, e.g., Sublette and Sublette (1965), who placed the two genera as subgenera of *Chironomus*. Sæther's classification generally follows that of Lenz (1954–62).

The genera of the complex fall into three groups (Sæther 1977a). One includes only *Parachironomus*, a second includes *Cryptotendipes*, *Microchironomus*, and *Cladopelma*, and a third includes the remaining genera. Of these groups *Parachironomus* is distinctive with the pecten epipharyngis consisting of a toothed convex plate and the hypostoma with 13, 15, or 16 teeth of more or less equal size or consecutively smaller from medial to lateral. *Parachironomus* also has three eye spots. The second group (*Cryptotendipes*, *Microchironomus*, and *Cladopelma*) is characterized by having the three outer teeth of the hypostoma enlarged. There is difficulty in distinguishing the larvae of the three genera in this group from one another. The third and largest group is diverse but the hypostoma of some genera has a central light-colored dome-shaped area. Within this group the larvae of *Cryptochironomus*, *Gillotia*, and *Demicryptochironomus* are distinctive in that the lateral teeth arch anterolaterally on either side of a central dome-shaped area.

## Genus *Acalcarella* Shilova

Figs. 116–118

*Acalcarella* Shilova, 1955:319; Sæther 1977a:87.

Antenna 5-segmented; ring organ on basal third of first segment; blade longer than combined length of terminal segments. Hypostoma with 1 broad triangular pointed median and 7 pairs of lateral teeth. Paralabial plate two to three times as long as wide, with anterior margin crenulate. Mandible with 3 inner teeth; apicodorsal tooth and pecten mandibularis absent. Premandible with 1 or 2 long apical teeth and 2–4 shorter teeth. SI and SII simple. Labral lamellae vestigial. Pecten epipharyngis consisting of a single plate bearing several points. Maxillary palpus more than two and one-half times as long as wide; basal segment about as long as first segment of antenna. Procercus present, bearing anal setae. Ventral tubules absent. Two pairs of short triangular anal tubules present.

**Remarks.** Sæther (1977a) in his key to the genera of the *Harnischia* complex states that the anal tubules of *Acalcarella* are vestigial. Larvae from the Northwest Territories, on which the foregoing description is based, have short but well developed anal tubules. In all other characters, the larvae agree with the description given by Shilova (1955) and Lenz (1954–62) except that the premandible has two long apical teeth, and will key to *Acalcarella* in Sæther's (1977a) key. *Acalcarella* may be separated from the other genera in the *Harnischia* complex by the shape of the hypostoma; the median tooth is broad, triangular, and pointed.

**Distribution.** District of Mackenzie.

**Habitat.** Larvae of *Acalcarella* inhabit lakes in the Mackenzie River Delta. In Asia they inhabit delta areas, lagoons, and irrigation canals and ditches (Lenz 1954–62).

## Genus *Beckidia* Sæther

*Beckidia* Sæther, 1979:315.
*Beckiella* Sæther, 1977a:119, not Grandjean 1964:695.

Antenna 7-segmented; ring organ position unknown; blade arising from proximal part of third segment, shorter than combined length of terminal segments. Hypostoma with 1 broad weakly trifid median and 4 pairs of lateral teeth. Paralabial plate about as long as wide, with striations strong, and with anterior margin crenulate. Mandible with 3 inner teeth; apicodorsal tooth and pecten mandibularis absent. Premandible bifid or trifid. SI and SII simple. Labral lamellae absent. Pecten epipharyngis consisting of 1 single plate. Procercus absent; pair of anal setae present. Ventral tubules absent. Dorsal pair of anal tubules long, tapering, about same size as posterior parapods; ventral pair strongly reduced, hidden between bases of posterior parapods.

**Remarks.** The larvae of *Beckidia* and *Robackia* are similar but may be distinguished by the shape of the hypostoma and position of the antennal blade (see key, couplet 14). In addition the larva of *Beckidia* lacks procerci and has only one pair of anal setae, whereas in *Robackia* the low procerci each bear about six anal setae. The foregoing description is taken from Sæther (1977a).

**Distribution.**   Saskatchewan.

**Habitat.**   Larvae of *Beckidia* inhabit sandy substrates of large rivers (Sæther 1977a).

## Genus *Chernovskiia* Sæther

Fig. 119

*Chernovskiia* Sæther, 1977a:107.

Antenna 8-segmented; ring organ on basal half of first segment; blade arising from third or fourth segment, and shorter than combined length of terminal segments. Hypostoma concave, with 4 or 5 pairs of weak lateral teeth; median area without teeth. Paralabial plates about as long as wide, with striations distinct, and with anterior margin crenulate. Mandible without apicodorsal tooth or pecten mandibularis. Premandible with 2 apical teeth and 2 shorter inner teeth. SI and SII simple. Labral lamellae absent. Basal segment of maxillary palpus longer than combined length of first three antennal segments. Body with 20 more or less distinct segments (seven anterior abdominal segments subdivided). Procercus low, with short anal setae. Ventral tubules absent. Two pairs of long digitiform anal tubules present.

**Remarks.**   Larvae of *Chernovskiia* are distinctive, with the body having 20 apparent segments due to subdivision of the first seven abdominal segments (Sæther 1977a). Sæther in his key to genera of the *Harnischia* complex includes two larval types, "*Cryptochironomus*" *macropodus* Ljachov and "*Cryptochironomus*" sp. Pagast, with the same body characteristic. These have not been reported from North America. Larvae of Chernovskiia are described by Sæther (1977a).

**Distribution.**   Saskatchewan.

**Habitat.**   Larvae of *Chernovskiia* inhabit sandy substrates of large rivers and lakes (Sæther 1977a).

## Genus *Cladopelma* Kieffer

Figs. 120–123

*Cladopelma* Kieffer, 1921a:63.
*Cryptocladopelma* Lenz, 1941a:34.

Antenna 5-segmented; ring organ on basal half of first segment; blade arising from apex of first segment and shorter than combined length of terminal segments. Hypostoma with 1 notched or weakly doubled median and 6 pairs of lateral teeth; sixth lateral tooth much larger than fifth lateral tooth. Paralabial plate about twice as long as wide, with median apex pointed, and with anterior margin smooth to weakly crenulate. Mandible with 3 flat inner teeth; apicodorsal tooth and pecten mandibularis absent. Premandible bifid. SI and SII simple and bristlelike. Labral lamellae absent. Pecten epipharyngis consisting of a single plate. Basal segment of maxillary palpus less than two and one-half times as long as wide and less than one-half length of first antennal segment. Procercus present, bearing anal setae. Ventral tubules absent. Two pairs of short anal tubules present.

**Remarks.** The hypostoma with a notched or partly divided median tooth will separate most larvae of *Cladopelma* from those of *Cryptotendipes* and *Microchironomus*, which also have an enlarged sixth lateral tooth. However, Sæther (1977a) states that the median tooth can be entire. Larvae are described by Beck and Beck (1969) as *Harnischia*.

**Distribution.** British Columbia and Yukon Territory to Quebec.

**Habitat.** Larvae of *Cladopelma* inhabit small bodies of flowing water (Lenz 1954–62; Beck and Beck 1969) and occur also in shallow regions of still water.

## Genus *Cryptochironomus* Kieffer

Figs. 124–129

*Cryptochironomus* Kieffer, 1918a:46; Townes 1945:96.

Antenna 5-segmented; ring organ on distal third of first segment; blade arising from distal part of second segment, shorter than combined length of terminal segments. Hypostoma with 1 light-colored dome-shaped median and 5 pairs of lateral teeth; median tooth usually smooth, sometimes with 2 small median projections present; lateral teeth arched anterolaterally on either side of median tooth. Paralabial plate narrow, about three to four times as long as wide; median apex blunt, somewhat arched posteriorly; anterior margin smooth. Mandible with 2 inner teeth; apicodorsal tooth and pecten mandibularis absent. Premandible with 4–6 teeth. SI and SII simple. Labral lamellae absent. Pecten epipharyngis triangular, consisting of a long median and a pair of shorter toothed lateral scales. Basal segment of maxillary palpus at least twice as long as wide. Ventral tubules absent. Procercus present, bearing anal setae. Two pairs of anal tubules present.

**Remarks.** Larvae of *Cryptochironomus*, *Demicryptochironomus*, and *Gillotia* all have a hypostoma with lateral teeth arching anterolaterally on either side of a light-colored dome-shaped median tooth. *Cryptochironomus* has five pairs of lateral teeth, whereas the other two genera each have seven pairs. Larvae are described by Johannsen (1937*b*; as *Chironomus* (*Chironomus*) *Cryptochironomus* group) and Curry (1958).

**Distribution.**  Widespread south of treeline.

**Habitat.**  Larvae of *Cryptochironomus* inhabit shallower regions of flowing and standing water, often in sandy substrates (Lenz 1954–62; Curry 1958).

## Genus *Cryptotendipes* Lenz

### Figs. 130–131

*Cryptotendipes* Lenz, 1941*a*:34; Sæther 1977*a*:95.

Antenna 5-segmented; ring organ on basal half of first segment; blade arising from apex of first segment, shorter than combined length of terminal segments. Hypostoma with 1 weak tripartite or 1 single, broadly rounded median tooth and 6 or 7 pairs of lateral teeth; sixth lateral tooth much larger than fifth lateral tooth. Paralabial plate about two and one-half times as wide as long, with median apex pointed, and with anterior margin weakly crenulate. Mandible with 2 or 3 inner teeth, lacking apicodorsal tooth and pecten mandibularis. Premandible bifid. SI and SII simple. Labral lamellae absent. Pecten epipharyngis consisting of 2 or 3 fused scales. Basal segment of maxillary palpus less than two and one-half times as long as wide and less than one-half as long as first antennal segment. Procercus present, bearing anal setae. Ventral tubules absent. Two pairs of anal tubules present.

**Remarks.**  Larvae of *Cryptotendipes* and *Cladopelma* are difficult to separate from each other. The hypostomal characters given in the key (couplet 23) will distinguish most larvae of the two genera from each other, but Sæther (1977*a*) states that an undescribed species of *Cladopelma* has an undivided and unnotched median tooth. Larvae are described by Beck and Beck (1969).

**Distribution.**  British Columbia; Yukon Territory; District of Mackenzie; Manitoba to Quebec.

**Habitat.**  Larvae of *Cryptotendipes* inhabit shallow areas of standing and flowing water, often in sandy substrates (Lenz 1954–62).

## Genus *Cyphomella* Sæther

### Figs. 132–134

*Cyphomella* Sæther, 1977*a*:101.

Antenna 5-segmented with second segment longer than third; ring organ on basal half of first segment; blade on apex of first segment, shorter to longer than combined length of terminal segments; basal two-thirds of second segment unsclerotized. Hypostoma with 1 wide, convex, pale median tooth with lateral notches and with 6 pairs of darker lateral teeth. Paralabial plate about two and one-half times as long as wide, with striations distinct, with median apex blunt, and with anterior margin smooth. Mandible with 3 pointed inner teeth; apicodorsal tooth and pecten mandibularis absent. Premandible with 2 long apical and 2 short inner teeth. SI and SII simple. Labral lamellae absent. Pecten epipharyngis consisting of 1 single-rounded scale. Basal segment of maxillary palpus about two and one-third times as long as wide and about three and one-half times as long as first antennal segment. Procercus present, bearing anal setae. Ventral tubules absent. Two pairs of anal tubules present.

**Remarks.** The larvae of *Cyphomella* are difficult to separate from *Paracladopelma*, but having the basal two-thirds of the second antennal segment unsclerotized is diagnostic. The larvae are described by Sæther (1977a).

**Distribution.** Yukon Territory; District of Mackenzie; Saskatchewan; Manitoba.

**Habitat.** Larval habitat of *Cyphomella* unknown, but pupae were found in large bodies of flowing water (Sæther 1977a).

## Genus *Demicryptochironomus* Lenz

Figs. 135–139

*Demicryptochironomus* Lenz, 1941a:34.

Antenna 7- to 8-segmented, with third segment generally longer than second; ring organ on distal half of first segment; blade arising from side of third segment and shorter to longer than combined length of terminal segments. Hypostoma with 1 light-colored dome-shaped median and 7 pairs of lateral teeth; median tooth smooth; lateral teeth arched anterolaterally on either side of median tooth. Paralabial plate narrow, about three to four times as long as wide, with lateral apex curved dorsally, and with anterior margin smooth. Mandible with 2 inner teeth; apicodorsal tooth and pecten mandibularis absent. Premandible with 4 or 5 apical teeth. SI and SII simple. Labral lamellae absent. Pecten epipharyngis consisting of 1 long median and 1 pair of smaller lateral scales. Basal segment of maxillary palpus at least three times as long as wide. Procercus present, bearing anal setae. Ventral tubules absent. Two pairs of anal tubules present.

**Remarks.** Larvae of *Demicryptochironomus* are similar to those of *Cryptochironomus* (see remarks under *Cryptochironomus*) and *Gillotia*. They may be distinguished from *Gillotia* by having seven to eight antennal segments. Larvae are described by Lenz (1954–62).

**Distribution.** Yukon Territory; District of Mackenzie; Saskatchewan; Ontario to New Brunswick.

**Habitat.** Larvae of *Demicryptochironomus* inhabit large bodies of standing water (Lenz 1954–62) and quieter reaches of large bodies of flowing water.

## Genus *Harnischia* Kieffer

### Figs. 140–141

*Harnischia* Kieffer, 1921a:69; Townes 1945:147 (in part); Sæther 1971a:347; 1977a:84.

Antenna 5-segmented, with second and third antennal segments subequal in length; ring organ on distal half of basal segment; blade arising from apex of first segment, longer than combined length of terminal segments. Hypostoma with 1 broad, low median and 7 pairs of lateral teeth; anterior margin more or less straight. Paralabial plate about two and one-half times as long as wide, with median apex bluntly pointed, and with anterior margin smooth. Mandible with 2 inner teeth; apicodorsal tooth and pecten mandibularis absent. Premandible with 5 teeth. SI and SII simple. Labral lamellae absent. Pecten epipharyngis consisting of 1 single plate with several points. Basal segment of maxillary palpus about four times as long as wide. Procercus present, bearing anal setae. Ventral tubules absent. Two pairs of anal tubules present.

**Remarks.** The hypostoma of *Harnischia* and *Paracladopelma* are similar but that of *Paracladopelma* is more convex and the median tooth is usually notched. Also the second antennal segment is much longer than the third in *Paracladopelma*. This character will also distinguish the larvae of *Cyphomella* from those of *Harnischia*, which have the second and third antennal segments subequal in length. Larvae are described by Sæther (1971a).

**Distribution.** Widespread south of treeline, but not recorded from Maritime Provinces or Newfoundland.

**Habitat.** Larvae of *Harnischia* inhabit large bodies of still water (Sæther 1971a).

## Genus *Parachironomus* Lenz

### Figs. 142–147

*Parachironomus* Lenz, 1921:160.

Three eye spots present. Antenna 5-segmented; ring organ on basal half of first segment; blade arising from apex of first segment, shorter than combined length of terminal segments. Hypostoma with 13 or 15, rarely 16, pointed teeth; all teeth about same size to consecutively smaller laterally. Paralabial plate about twice as long as wide, with median apex bluntly pointed, and with anterior margin crenulate. Mandible with 2 inner teeth; apicodorsal tooth and pecten mandibularis absent. Premandible bifid. SI and SII simple. Labral lamellae absent. Pecten epipharyngis consisting of a convex plate with three or more pointed transparent teeth. Basal segment of maxillary palpus more than twice as long as wide. Procercus present, bearing anal setae. Ventral tubules absent. Two pairs of anal tubules present.

**Remarks.** The evenly concave hypostoma with teeth of more or less the same size will distinguish the larvae of *Parachironomus* from all other members of the *Harnischia* complex. The pecten epipharyngis with three or more transparent teeth is also distinctive. Larvae are described by Lenz (1954–62) and Beck and Beck (1969).

**Distribution.** British Columbia and Yukon Territory to Ontario.

**Habitat.** Larvae of *Parachironomus* inhabit ponds, lakes, and slowly flowing water (Lenz 1954–62).

## Genus *Paracladopelma* Harnisch

Figs. 148–152

*Paracladopelma* Harnisch, 1924:304; Sæther 1977a:84; Jackson 1977:1324.

Antenna 5-segmented, with second segment much longer than third; ring organ on basal third of first segment; blade arising from apex of first segment, shorter than combined length of terminal segments. Hypostoma with 1 broad, low, usually light-colored, sometimes notched median and 7 pairs of lateral teeth, rarely unicolorous; anterior margin weakly convex. Paralabial plate about two and one-half times as long as wide, with median apex pointed, and with anterior margin smooth. Mandible with 2 or 3 inner teeth; apicodorsal tooth absent; weak pecten mandibularis present. Premandible with 5 or 6 apical teeth. SI and SII simple. Labral lamellae absent. Pecten epipharyngis consisting of 1 small, triangular scale. Basal segment of maxillary palpus three to four times as long as wide. Procercus present, bearing anal setae. Ventral tubules absent. Two pairs of anal tubules present.

**Remarks.** Larvae of *Paracladopelma* are similar to those of *Harnischia* and *Saetheria* (see remarks under *Harnischia* and *Saetheria*). Larvae are described by Lenz (1954–62), Beck and Beck (1969), and Jackson (1977).

**Distribution.** Widespread south of treeline, but not recorded from Maritime Provinces.

**Habitat.** Larvae of *Paracladopelma* inhabit large bodies of still water (Lenz 1954–62; Jackson 1977).

## Genus *Robackia* Sæther

Figs. 153–157

*Robackia* Sæther, 1977*a*:123.

Antenna 7-segmented; ring organ on middle third of first segment; blade arising from distal part of second segment, shorter than combined length of terminal segments. Hypostoma with 1 pair of median teeth and 5 or 6 pairs of lateral teeth; anterior margin straight to concave. Paralabial plate about as wide as long; median apex blunt; striations distinct; anterior margin irregular. Mandible with 2 or 3 inner teeth; posterior tooth notched; apicodorsal tooth and pecten mandibularis absent. Premandible with 4 apical teeth. SI and SII simple. Labral lamellae absent. Pecten epipharyngis consisting of 1 single plate. Basal segment of maxillary palpus more than three times as long as wide. Procercus present, bearing anal setae. Ventral tubules absent. Two pairs of anal tubules present.

**Remarks.** The narrow, hypostoma with small teeth flanked by paralabial plates with irregular anterior margins will distinguish the larvae of *Robackia*. Sæther (1977*a*) states that the blade arises from the third antennal segment, but in our specimens it arises from the second. Larvae are described by Sæther (1977*a*).

**Distribution.** Alberta and Yukon Territory to Ontario.

**Habitat.** Larvae of *Robackia* inhabit sandy substrates of large bodies of flowing water (Sæther 1977*a*).

## Genus *Saetheria* Jackson

Figs. 159–162

*Saetheria* Jackson, 1977:1325

Antenna 6-segmented, with second segment shorter than third segment; ring organ on basal third of first segment; blade arising from apex of second segment, equal to or shorter than combined length of terminal segments. Hypostoma with 1 broad, undivided median and 6 or 7 pairs of lateral teeth; median tooth lighter or unicolorous with lateral teeth. Paralabial plate one to two times as long as wide, with median apex pointed, and with anterior margin smooth or weakly crenulated. Mandible with 2 inner teeth; apicodorsal tooth absent; weak pecten mandibularis present. Premandible with 3 teeth. SI and SII simple. Labral lamellae absent. Pecten epipharyngis consisting of 1 small, triangular

79

plate. Basal segment of maxillary palpus two to three times as long as wide. Procercus present, bearing anal setae. Ventral tubules absent. Two pairs of anal tubules present.

**Remarks.** Larvae of *Saetheria* and *Paracladopelma* are separable from each other only by antennal and premandibular characters (see key, couplet 18). In addition to the characters given in the key the second antennal segment is shorter than the third in *Saetheria*. The reverse occurs in *Paracladopelma*. Also the two segments are weakly separated in *Saetheria* but clearly differentiated in *Paracladopelma*. Larvae are described by Jackson (1977).

**Distribution.** Yukon Territory; District of Mackenzie; southern Ontario.

**Habitat.** Larvae of *Saetheria* inhabit flowing water (Jackson 1977).

## Genus *Kiefferulus* Goetghebuer

Figs. 163–167

*Kiefferulus* Goetghebuer, 1922:40.
*Tendipes* (*Kiefferulus*); Townes 1945:110.

Antenna 5-segmented, with third segment as long as or longer than fourth segment; ring organ on middle third of first segment; blade shorter than combined length of terminal segments; Lauterborn's organs on apex of second segment. Hypostoma with 1 median and 6 pairs of lateral teeth; median tooth notched laterally, sometimes appearing weakly tripartite; second lateral tooth small, appressed to first lateral tooth. Paralabial plate with crenulate anterior margin, and with median apex curved posteriorly. Mandible with 3 inner teeth; apicodorsal tooth and pecten mandibularis present. Premandible broad, with five or six teeth. SI plumose; SII simple. Labral lamellae present. Pecten epipharyngis consisting of a bar with row of 18 or more irregular teeth. One pair of ventral tubules present. Procercus present, bearing anal setae. Two pairs of anal tubules present.

**Remarks.** Larvae of *Kiefferulus* are similar to those of *Chironomus*, *Einfeldia*, *Dicrotendipes*, and *Glyptotendipes* but may be distinguished by having one pair of ventral tubules and a broad premandible with five or six teeth. Larvae are described by Lenz (1954–62) and Curry (1961).

**Distribution.** British Columbia; Manitoba to Ontario.

**Habitat.** Larvae of *Kiefferulus* appear to prefer shallow warm bodies of still and flowing water (Lenz 1954–62; Lehmann 1969).

## Genus *Lauterborniella* Bause

Figs. 168–171

*Lauterborniella* Bause, 1914:120; Townes 1945:19.
*Zavreliella* Kieffer, 1920a:334.

Antenna 6-segmented, arising from a low antennal tubercle, with second segment shorter than third; first segment with ring organ on basal third and with prominent seta on distal half; blade shorter than combined length of terminal segments. Lauterborn's organs large, alternate on apices of second and third segments. Hypostoma with 1 pair of median and 6 pairs of lateral teeth. Seta posterior to hypostoma simple or plumose. Paralabial plate trapezoid, with bluntly pointed median apices that almost meet at midline; anterior margin smooth. Mandible with 2 inner teeth; apicodorsal tooth and pecten mandibularis present. Premandible with 3 apical teeth. SI and SII pectinate. Labral lamellae present. Pecten epipharyngis consisting of 3 toothed scales. Procercus present, bearing anal setae. Eighth abdominal segment with dorsomedial conical projection. Ventral tubules absent. Two pairs of anal tubules present.

**Remarks.** Larvae of *Lauterborniella* have trapezoid paralabial plates almost meeting at the midline, long antennae, and a conical projection on eighth abdominal segment. Superficially the long antennae and the large Lauterborn's organs resemble those of the Tanytarsini, but the paralabial plates in the latter are generally straplike.

North American authors (e.g., Townes 1945; Roback 1957) consider *Zavreliella* as a synonym of *Lauterborniella*, but European authors treat them as separate genera (e.g., Fittkau et al. 1967). The pair of setae posterior to the hypostoma is plumose in the *Lauterborniella* group, whereas it is simple in the *Zavreliella* group. In all other Chironomini this pair of setae is simple. Larvae are described by Bause (1914) and Roback (1957).

**Distribution.** District of Mackenzie; eastern Ontario; southern Quebec.

**Habitat.** Larvae of *Lauterborniella* inhabit small bodies of slow flowing water and small bodies of still water (Brundin 1949; Roback 1957).

## Genus *Microtendipes* Kieffer

Figs. 172–175

*Microtendipes* Kieffer, 1915:70; Townes 1945:22.

Antenna 6-segmented, with second segment usually shorter than third; ring organ on basal third of first segment; blade longer than combined length of terminal segments; Lauterborn's organs alternate on apex or distal part of second segment and apex of third segment. Hypostoma with 1 pair of light-colored median and 6 pairs of lateral teeth; median and first lateral teeth recessed; small tooth sometimes present between median teeth; first lateral tooth smaller than either median or second lateral teeth. Median apices of paralabial plate curved anteriorly, with anterior margin smooth. Mandible with 3 inner teeth; apicodorsal tooth and pecten mandibularis present. Premandible bifid. SI pectinate; SII simple. Labral lamellae present. Pecten epipharyngis consisting of 3 scales, each usually toothed or subdivided. Abdominal segments with 1 pair of plumose setae plus regular simple setae. Procercus present, bearing anal setae. Ventral tubules absent. Two pairs of anal tubules present.

**Remarks.** The combination of a six-segmented antenna and a hypostoma with two light-colored median teeth is distinctive. Larvae are described by Johannsen (1937b; as *Chironomus* (*Microtendipes*)), Lenz (1954–62), and Roback (1953, 1957).

**Distribution.** Widespread south of treeline.

**Habitat.** Larvae of *Microtendipes* prefer warm habitats, generally the shallow areas of still water of all sizes or the inshore and slower reaches of flowing water (Lenz 1954–62; Townes 1945; Roback 1953, 1957). Larvae also occur in riffle areas of streams (Coffman 1967).

## Genus *Nilothauma* Kieffer

Figs. 176–178

*Nilothauma* Kieffer, 1921b:270.
*Kribioxenus*, authors, nec Kieffer 1921d:29.

Antenna 5-segmented, with third segment shorter than second or fourth segments; ring organ on distal third of first segment; blade shorter than combined length of terminal segments; Lauterborn's organs on apex of second segment. Hypostoma with 1 median and 5 pairs of lateral teeth; median tooth broad, with 2–4 minute apical projections. Paralabial plate with rounded median apex ending about level of second lateral tooth of hypostoma; anterior margin smooth. Mandible with 4 inner teeth on a common lobelike base; apicodorsal tooth and pecten mandibularis present. Premandible with 4 apical teeth. SI plumose; SII simple. Labral lamellae present. Pecten epipharyngis consisting of 3 smooth scales. Procercus present, bearing anal setae. Ventral tubules absent. Two pairs of anal tubules present.

**Remarks.** The mandible of *Nilothauma*, with the inner teeth forming a compound tooth, is distinctive. Larvae are described by Brundin (1949) and Lenz (1954–62)—both as *Kribioxenus*.

**Distribution.** British Columbia; District of Mackenzie; southeastern Ontario to Newfoundland.

**Habitat.** Larvae of *Nilothauma* inhabit still and flowing water (Brundin 1949).

## Genus *Omisus* Townes

Figs. 179–180

*Omisus* Townes, 1945:27; Beck and Beck 1970:30.

Antenna 6-segmented, with second segment slightly shorter than third; ring organ on basal half of first segment; blade shorter than combined length of terminal segments; Lauterborn's organs large, alternate on apices of second and third segments. Hypostoma with 1 pair of median and 6 pairs of lateral teeth; second lateral tooth longer than either median or first lateral tooth; teeth uniformly dark. Paralabial plate with pointed median apex, slightly curved posteriorly, with anterior margin smooth. Mandible with 2 inner teeth; apicodorsal tooth and pecten mandibularis present. Premandible bifid. SI and SII weakly plumose. Labral lamellae present. Pecten epipharyngis consisting of 3 toothed scales. Procercus present, bearing anal setae. Ventral tubules absent. Two pairs of anal tubules present.

**Remarks.** The hypostoma of *Omisus* is similar to that of *Microtendipes* except that the median teeth are not lighter in color than the lateral teeth. In several aspects the larvae of *Omisus* resemble that of *Lauterborniella*, e.g., long six-segmented antennae with alternate and large Lauterborn's organs, general shape and coloration of the hypostoma, and narrowed paralabial plates. However, the larvae of the latter have a number of distinctive characters that afford easy separation (see remarks under *Lauterborniella*). The larvae of the only North American species are described by Beck and Beck (1970). They give an antennal ratio for a five-segmented antenna; however, their Fig. 16 illustrates a six-segmented antenna. A reared specimen in the Canadian National Collection has a six-segmented antenna.

**Distribution.** Manitoba to southern Quebec.

**Habitat.** Larvae of *Omisus* inhabit shallow regions of still water.

## Genus *Pagastiella* Brundin

Figs. 181–182

*Pagastiella* Brundin, 1949:840.

Antenna 5-segmented, with first segment slightly curved; third segment shorter than second or fourth segments; ring organ on basal third of first segment; blade longer than combined length of terminal segments; Lauterborn's organs on apex of second segment; curved comb present dorsal to antennal base. Hypostoma with 1 pair of median and 14 pairs of lateral teeth; first lateral teeth much smaller than median or second lateral teeth. Paralabial plate with median apex bluntly pointed; anterior margin smooth. Mandible with 4 or 5 inner teeth; 2 apicodorsal teeth and pecten mandibularis present. Premandible trifid. SI and SII long and simple. Labral lamellae vestigial. Procercus present, bearing anal setae. Ventral tubules absent. Two pairs of anal tubules present.

**Remarks.** Among the Chironomini the larvae of *Pagastiella* are distinctive in that the first antennal segment is slightly curved and a toothed projection is present dorsal to the base of the antenna. In these characters they resemble some Tanytarsini, but in this tribe, the toothed projection, when present, is borne on a distinct antennal tubercle. Also the abdominal setae are plumose in most Tanytarsini. Larvae are described by Brundin (1949) and Webb (1969).

**Distribution.** British Columbia; District of Mackenzie; Manitoba to Quebec.

**Habitat.** Larvae of *Pagastiella* inhabit standing water (Brundin 1949; Webb 1969).

## Genus *Paralauterborniella* Lenz

Figs. 184–186

*Paralauterborniella* Lenz, 1941*b*:48.
*Apedilum* Townes, 1945:32.

Antenna 6-segmented; ring organ on middle third of first segment; blade on apex of first segment and shorter than combined length of terminal segments; Lauterborn's organs alternate on apices of second and third segments. Hypostoma with 1 broad, light-colored, dome-shaped median tooth and 6 pairs of lateral teeth. Paralabial plate with a pointed, anteriorly curved median apex; anterior margin smooth. Mandible with 3 inner teeth; apicodorsal tooth and pecten mandibularis absent. Premandible bifid. SI plumose; SII simple. Labral lamellae present. Pecten epipharyngis consisting of 2 pointed scales. Procercus present, bearing anal setae. Ventral tubules absent. Two pairs of anal tubules present.

**Remarks.** The foregoing description is based on reared specimens of *Paralauterborniella nigrohalterale* (Malloch) and on descriptions by Lenz (1954–62). The dome-shaped median tooth will distinguish it from other larvae of Chironomini with six-segmented antennae. Beck and Beck (1970) described the larvae of *P. nigrohalterale* and of *P. elachista* (Townes). The larvae of *P. elachista* are quite different from the description given here. The hypostoma has two pale median teeth and six pairs of lateral teeth. The first lateral tooth is smaller than and appressed to the second lateral tooth. It is similar to that of *Microtendipes*, but may be distinguished by having the paralabial plates almost meeting at the midline of the head capsule and by having a trifid premandible. *P. elachista* has not been recorded from Canada, but should it occur its larvae would key to *Lauterborniella*. The larvae of this genus have a number of distinctive characteristics that will afford easy separation (see remarks under *Lauterborniella*).

**Distribution.** Alberta and Yukon Territory to New Brunswick.

**Habitat.** Larvae of *Paralauterborniella* inhabit shallower regions of still water (Lenz 1954–62).

## Genus *Paratendipes* Kieffer

Figs. 187–189

*Paratendipes* Kieffer, 1911a:41; Townes 1945:27.

Antenna 6-segmented; ring organ on middle third of first segment; blade shorter than terminal segments; Lauterborn's organs alternate on apices of second and third segments. Hypostoma with 15 or 16 teeth; median 5 or 6 teeth recessed, and median 2–4 teeth lighter in color than remainder of teeth. Paralabial plate with pointed, anteriorly curved median apex; anterior margin smooth. Mandible with 3 inner teeth; apicodorsal tooth and pecten mandibularis present. Premandible bifid. SI and SII plumose. Labral lamellae present. Pecten epipharyngis consisting of short, irregularly toothed bar. Procercus present, bearing anal setae. Ventral tubules absent. Two pairs of anal tubules present.

**Remarks.** The hypostoma with five or six median teeth shorter than the adjacent teeth will distinguish the larvae of *Paratendipes* from those of other Chironomini with six-segmented antennae. Larvae are described by Lenz (1954–62) and Johannsen (1937b; as *Chironomus (Paratendipes)*).

**Distribution.** Widespread south of treeline.

**Habitat.** Larvae of *Paratendipes* live in both still and flowing water (Lenz 1954–62).

## Genus *Phaenopsectra* Kieffer

Figs. 190–195

*Phaenopsectra* Kieffer, 1921b:274.
*Tanytarsus (Tanytarsus)*; Townes 1945:71.
*Tanytarsus (Tribelos)*; Townes 1945:66.

Antenna 5-segmented; ring organ on basal half of first segment; blade length variable; Lauterborn's organs opposite on apex of second segment. Hypostoma with 2 median and 7 pairs of lateral teeth; first lateral tooth longer than median or second lateral tooth; second lateral tooth may be much smaller than either first or third lateral tooth. Paralabial plate with 2 bands of striations; striations of band along anterior margin usually stronger and more widely spaced than those of band on central area; median apex pointed, strongly curved anteriorly; anterior margin smooth. Mandible with 3 or 4 inner teeth; apicodorsal and pecten mandibularis present. Premandible bifid. SI and SII plumose; all branches of SII about equal in size. Labral lamellae present. Pecten epipharyngis consisting of 3 scales, each bearing uneven teeth. Procercus present, bearing anal setae. Ventral tubules absent. Two pairs of anal tubules present.

**Remarks.** In some species of *Phaenopsectra* the two bands of striations on the paralabial plates are almost equal in development. These are difficult to separate from *Endochironomus*, but the shape of the SII will distinguish the larvae of the two genera (see key, couplet 38).

In North American literature the generic name *Sergentia* is frequently treated as a synonym of *Phaenopsectra*, but the names are used separately in European literature. Further work is required to determine the status of the two groups. *Phaenopsectra* has been divided into two subgenera, *Phaenopsectra* (*Phaenopsectra*) and *Phaenopsectra* (*Tribelos*). This is the most common usage, although Sæther (1977b) gives *Tribelos* full generic status. Larvae assigned to *Phaenopsectra* or *Tribelos* are similar and their separation is difficult. Some larvae of *Tribelos* have a hypostoma with the second lateral tooth much smaller than either the first or third lateral teeth.

Larvae are described by Andersen (1937; as *Sergentia*) and by Johannsen (1937b; as *Pentapedilum* (*Phaenopsectra*)).

**Distribution.** Widespread; occurs north of treeline.

**Habitat.** Larvae of *Phaenopsectra* inhabit both still and flowing water, including the deep region of cold lakes (Andersen 1937; Lenz 1954–62).

## Genus *Polypedilum* Kieffer

Figs. 196–203

*Polypedilum* Kieffer, 1913c:15; Townes 1945:36.
*Pentapedilum* Kieffer, 1913c:25.

Antenna 5-segmented; relative lengths of second, third, and fourth segments variable; ring organ position and blade length variable; Lauterborn's organs on apex of second segment. Hypostoma usually convex, with 2 median and 6 or 7 pairs of lateral teeth; first lateral tooth usually shorter than median and second lateral teeth; all teeth sometimes of even length; hypostoma sometimes concave (median area recessed), with 10 or less teeth. Paralabial plate with pointed or rounded apex, usually curved anteriorly, with or without posterior extension; anterior margin smooth. Mandible with 2 inner teeth; apicodorsal tooth and pecten mandibularis present. Premandible usually bifid, rarely simple. SI and SII plumose. Labral lamellae present. Pecten epipharyngis consisting of 3 toothed scales. Procercus present, bearing anal setae. Ventral tubules absent. Two pairs of anal tubules present.

**Remarks.** Larvae of *Polypedilum* are variable, and most of those with a convex hypostoma are similar to those of *Endochironomus* and *Phaenopsectra*. They may be distinguished by the relative length of the median, first, and second lateral teeth (see key, couplets 35 and 36). The larvae of *Pedionomus* (not recorded from Canada) are similar to that of *Polypedilum* (*Pentapedilum*), but apparently may be distinguished by the composition of the pecten epipharyngis (see key, couplet 37). Larvae are described by Johannsen (1937b; as *Chironomus* (*Polypedilum*)), Berg (1950), Hauber (1947), Lenz (1954–62), Bryce (1960), and Maschwitz (1975).

Larvae of the three subgenera *Polypedilum* (*Polypedilum*), *Polypedilum* (*Pentapedilum*), and *Polypedilum* (*Tripodura*) may be separated by the characters given in the following key. The antennal characters will separate most larvae of the subgenera *Polypedilum* (*Polypedilum*) and *Polypedilum* (*Tripodura*). However, several *Polypedilum* (*Polypedilum*) have the third antennal segment one-third or less as long as the second segment. These may be distinguished from *Polypedilum* (*Tripodura*) by having a paralabial plate without a posterior extension.

## Key to the subgenera of *Polypedilum*

1. Hypostoma with recessed median area. Premandible simple ............
   .............................. ***Polypedilum* (*Polypedilum*)** (in part)
   Hypostoma convex. Premandible bifid ............................. 2
2. First lateral tooth of hypostoma about equal in length to median and second lateral teeth (Fig. 196) ................ ***Polypedilum* (*Pentapedilum*)**
   First lateral tooth of hypostoma shorter than median and second lateral teeth (Fig. 197) ................................................... 3
3. Third antennal segment much shorter than either second or fourth segment (Fig. 203) ............................. ***Polypedilum* (*Tripodura*)**
   Antennal segments 2–4 consecutively smaller (Fig. 200) or, if third segment equal to or shorter than fourth, then third more than one-half length of second ....................... ***Polypedilum* (*Polypedilum*)** (in part)

## Clé des sous-genres de *Polypedilum*

1. Hypostome avec région médiane renfoncée. Prémandibule simple .......
   ............................ ***Polypedilum* (*Polypedilum*)** (en partie)
   Hypostome convexe. Prémandibule bifide ........................... 2

2. Première dent latérale de l'hypostome à peu près de la même longueur que les dents médianes et les deuxièmes dents latérales (fig. 196) ........
............................................ ***Polypedilum (Pentapedilum)***
Première dent latérale de l'hypostome plus courte que les dents médianes et les deuxièmes dents latérales (fig. 197) ......................... 3
3. Troisième article antennaire beaucoup plus court que le deuxième ou le quatrième article (fig. 203) ............... ***Polypedilum (Tripodura)***
Articles antennaires de 2 à 4 successivement plus courts (fig. 200) ou, si le troisième article est égal ou plus court que le quatrième, alors sa longueur est plus de la moitié de la longueur du deuxième .................
............................ ***Polypedilum (Polypedilum)*** (en partie)

**Distribution.** Widespread south of treeline.

**Habitat.** Larvae of *Polypedilum* inhabit all types of freshwater and occur in brackish water (Lenz 1954–62; Maschwitz 1975). Most live in bottom substrate, often associated with plant debris, but some tunnel in living plant tissue (Berg 1950).

## Genus *Stenochironomus* Kieffer

Figs. 204–206

*Stenochironomus* Kieffer, 1919:44; Townes 1945:84.

Body with flat enlarged thorax and long slender abdomen. Antenna 5-segmented; ring organ on basal third of first segment; blade short, about equal to length of second segment; Lauterborn's organs small, on apex of second segment. Hypostoma concave, with 2 median and 4 pairs of lateral teeth; teeth strongly sclerotized and more or less equal in size. Paralabial plate without striations, with median apex blunt, and with anterior margin smooth. Mandible subtriangular, with 2 inner teeth; apicodorsal tooth present; pecten mandibularis and seta interna absent. Premandible with 2 or 3 apical teeth. SI coarsely plumose; SII simple. Labral lamellae present. Pecten epipharyngis weak, consisting of 2 toothed scales. Procercus present, bearing anal setae. Ventral tubules absent. Two pairs of anal tubules present.

**Remarks.** The flattened enlarged thorax and long slender abdomen will distinguish the larvae of *Stenochironomus* from all other Chironominae. The concave hypostoma with 10 teeth and unstriated paralabial plates are also distinctive. Larvae are described by Lenz (1954–62) and Beck and Beck (1970).

**Distribution.** Alberta and District of Mackenzie to New Brunswick.

**Habitat.** Larvae of *Stenochironomus* mine in living and dead plant parts such as wood, stalks of *Phragmites* and *Scirpus*, and water lily leaves (Johannsen 1937*b*; as *Chironomus* (*Stenochironomus*)); Lenz (1954–62); Kalugina (1959).

# Genus *Stictochironomus* Kieffer

Figs. 207–210

*Stictochironomus* Kieffer, 1919:44.
*Tanytarsus (Stictochironomus)*; Townes 1945:77.

Antenna 6-segmented, ring organ on proximal third of first segment; blade on apex of first segment, with length variable; Lauterborn's organs alternate on apices of second and third segments. Hypostoma with 1 pair of median and 7 pairs of lateral teeth; first lateral teeth longer than either median or second lateral teeth; teeth uniformly dark. Apices of paralabial plates curved anteriorly; anterior margin smooth. Mandible with 2 inner teeth; apicodorsal tooth and pecten mandibularis present. Premandible bifid, with apical tooth attenuated to sharp point, and with inner tooth bluntly rounded. SI pectinate; SII simple. Labral lamellae present. Pecten epipharyngis consisting of 3 scales; each scale with 3–5 teeth. Procercus present, bearing anal setae. Ventral tubules absent. Two pairs of anal tubules present.

**Remarks.** Among the Chironomini with six-segmented antennae, only the larvae of *Stictochironomus*, *Omisus*, and *Lauterborniella* have a uniformly dark hypostoma. The larvae of the remaining genera have the median teeth lighter colored than the lateral teeth. Larvae of *Lauterborniella* are distinctive (see remarks under *Lauterborniella*), and that of *Omisus* may be distinguished by the characters given in the key (couplet 28). Larvae are described by Johannsen (1937b; as *Chironomus (Stictochironomus)*), Lenz (1954–62), and Roback (1957).

**Distribution.** Alberta and Yukon Territory to New Brunswick.

**Habitat.** Larvae of *Stictochironomus* inhabit lakes and to a lesser extent smaller bodies of still water and slower reaches of flowing water (Lenz 1954–62).

# Genus *Xenochironomus* Kieffer

Figs. 211–215

*Xenochironomus* Kieffer, 1921a:69; Townes 1945:91; Roback 1963:235.

Antenna 5-segmented; ring organ on basal third of first segment; blade shorter than combined length of terminal segments. Lauterborn's organs on apex of second segment. Hypostoma with 1 median and 7 pairs of lateral teeth; median tooth sometimes incised; first lateral tooth shorter than median or second lateral tooth; third lateral tooth sometimes shorter than second lateral tooth. Paralabial plates fused medially or median apices pointed, separated by less than width of median tooth of hypostoma. Mandible with 2 or 3 inner teeth; apicodorsal tooth and pecten mandibularis present. Premandible with 2 long apical teeth and 3–5 shorter teeth along inner margin. SI plumose; SII simple. Labral lamellae present. Pecten epipharyngis consisting of a bar bearing irregular teeth. Procercus present, bearing anal setae. Ventral tubules absent. Two pairs of anal tubules present.

**Remarks.** Larvae of *Xenochironomus* with medially fused or nearly fused paralabial plates and a hypostoma with an odd number of teeth are distinctive. In these characters they superficially resemble the Tanytarsini, but the paralabial plates of the latter are straplike. The premandible with 3–5 teeth along the inner margin is also distinctive.

Roback (1963, 1980) recognizes two subgenera, *Xenochironomus (Xenochironomus)* and *Xenochironomus (Axarus)*, which may be distinguished by the following key. Larvae of *Xenochironomus* are described by Lenz (1954–62) and Roback (1963).

## Key to the subgenera of *Xenochironomus*

1. Paralabial plates narrowly separated. First and third lateral teeth of hypostoma each shorter than second lateral tooth (Fig. 213) ............... 
   ............................ **Xenochironomus (Xenochironomus)**
   Paralabial plates fused medially (Fig. 214). Only first lateral tooth of hypostoma shorter than second lateral tooth (Fig. 212) ................... 
   ........................................ **Xenochironomus (Axarus)**

## Clé des sous-genres de *Xenochironomus*

1. Plaques paralabiales un peu séparés. Première et troisième dents latérales de l'hypostome plus courtes que la deuxième dent latérale (fig. 213) .... 
   ............................ **Xenochironomus (Xenochironomus)**
   Plaques paralabiales soudées par le milieu (fig. 214). Seule la première dent latérale de l'hypostome est plus courte que la deuxième dent latérale (fig. 212) ............................... **Xenochironomus (Axarus)**

**Distribution.** British Columbia and District of Mackenzie to New Brunswick.

**Habitat.** Larvae of *Xenochironomus (Xenochironomus)* live in sponges (Lenz 1954–62; Roback 1963). Larvae of *Xenochironomus (Axarus)* live in the substrate of lakes and quieter reaches of larger bodies of flowing water (Roback 1963).

# Tribe Pseudochironomini

Figs. 216–218

Only one genus, *Pseudochironomus*, of the tribe Pseudochironomini occurs in North America. Therefore the following description of *Pseudochironomus* will also serve for the tribal description.

## Genus *Pseudochironomus* Malloch

Figs. 216–218

*Pseudochironomus* Malloch, 1915:500; Townes 1945:14; Sæther 1977*a*:59.

Two eye spots present. Antenna 5-segmented, mounted on low tubercle, with small apicomesal point; ring organ on basal third of first segment; blade on apex of first segment, as long as or longer than combined length of terminal segments; Lauterborn's organs on apex of second segment. Hypostoma with 1 median and 4–6 pairs of lateral teeth; second lateral tooth often greatly reduced. Paralabial plate striated, straplike, with medial apices separated by less than width of median tooth of hypostoma, and with anterior margin smooth. Mandible with 3 inner teeth; apicodorsal tooth and pecten mandibularis present; seta subdentalis long. Premandible bifid. SI and SII plumose. Labral lamellae consisting of 2 pectinate lobes. Pecten epipharyngis consisting of 3 separate blunt scales. Procerus present, usually smooth, sometimes with spines, bearing 7–9 anal setae. Ventral tubules absent. Two pairs of short blunt anal tubules present. Posterior parapod claws arranged in two horseshoe-shaped rows.

**Remarks.** Larvae of *Pseudochironomus* (and of Pseudochironomini) are intermediate in structure between the Chironomini and Tanytarsini, but share more anatomical characteristics with the Tanytarsini. They have straplike paralabial plates, antennae borne on tubercles, and posterior parapod claws arranged in two horseshoe-shaped rows as in the Tanytarsini, but the first antennal segment is straight and the body setae are simple. The arrangement of the posterior parapod claws will distinguish the larvae of *Pseudochironomus* from those of the Chironomini. Larvae are described by Sæther (1977*a*).

**Distribution.** British Columbia and District of Mackenzie to New Brunswick.

**Habitat.** Larvae of *Pseudochironomus* inhabit shallower regions of still water and quieter reaches of large bodies of flowing water. They prefer sandy or gravelly substrate (Sæther 1977*a*).

# Tribe Tanytarsini

### Figs. 7, 219–252

Antenna 5-segmented, with first segment curved, ring organ and usually a seta present; blade on apex of first segment; Lauterborn's organs usually on apex of second segment, sometimes arising from different levels of segment, usually mounted on petioles; antennal tubercle present, with or without apicomesal projection. Hypostoma convex to straight, always with teeth, sometimes strongly curved dorsally. Paralabial plate striated, straplike, often separated medially from opposite plate by less than width of median tooth of hypostoma; anterior margin smooth. Mandible with 2 or 3 inner teeth; seta subdentalis extending to or beyond apex of apical tooth; apicodorsal tooth, pecten mandibularis, and seta interna present. Premandible with 2–5 apical teeth. SI usually pectinate, rarely plumose; SII usually pectinate, rarely simple or plumose. Labral lamellae consisting of transverse pectinate lobe or of two narrowly separated lobes. Pecten epipharyngis usually consisting of 3 toothed scales, sometimes of 3–5 simple teeth or transverse toothed bar. Bifid plumose setae nearly always present on caudolateral areas of abdominal segments. Procercus present, well developed, bearing 6–8 anal setae. Two pairs of anal tubules present. Ventral tubules absent. Posterior parapod well developed, bearing apical claws arranged in horseshoe-shaped pattern.

**Remarks.** Within the Chironominae the Tanytarsini form a relatively distinct group characterized by having long antennae with a curved first segment mounted on an antennal tubercle and with Lauterborn's organs usually mounted on petioles. Also the seta subdentalis extends to the level or beyond the apex of the apical tooth of the mandible, the abdominal segments bear bifid plumose setae, and the apical claws of the posterior parapods are arranged in two rows shaped like a horseshoe. A few Chironomini and the Pseudochironomini possess some of the characters of the Tanytarsini (see remarks under Chironomini and Pseudochironomini), but the combination of characters given in the key will generally afford separation of these from the Tanytarsini. The larvae of *Corynocera* lack plumose abdominal setae, but the dorsally curved hypostoma and the mandible without inner teeth are distinctive.

Two groups of Tanytarsini may be distinguished by the structure of the paralabial plates. The plates of *Constempellina*, *Stempellina*, *Stempellinella*, and *Zavrelia* have pointed median apices that are widely separated, reaching about the level of the second or third lateral tooth of the hypostoma. The remaining genera have plates with blunt median apices almost meeting at the midline of the head and they are more straplike than those of the above group.

The larvae of the Tanytarsini are poorly known and there has been no good recent revision involving the larvae. A number of genera, based primarily on adults and sometimes also pupae, are recognized in Europe (see Fittkau et al. 1967; Reiss 1968, 1969a, 1969b). The key to genera and the following descriptions have been primarily based on reared Canadian material, and will certainly be modified as more material becomes available.

## Key to the genera of Tanytarsini

1. Pointed median apices of paralabial plates separated by more than width of median tooth of hypostoma (Fig. 244) .......................... 2
   Blunt median apices of paralabial plates almost meeting at midline (Figs. 249–250) ...................................................... 6
2. Mandible without inner teeth (Fig. 228) .. ***Corynocera* Zetterstedt** (p. 96)
   Mandible with inner teeth .......................................... 3
3. Antennal tubercle with palmate apicomesal projection bearing 3 or more points (Fig. 242) ..................... ***Stempellina* Bause** (p. 98)
   Antennal tubercle with a simple pointed apicomesal projection (Figs. 225, 246) ................................................................ 4
4. Lauterborn's organs situated on apex of second antennal segment (as in Fig. 235) ............................ ***Constempellina* Brundin** (p. 95)
   Lauterborn's organs situated at different levels on second antennal segment (Fig. 245) ........................................................ 5
5. Distal Lauterborn's organ situated on shaft of second antennal segment; second segment about equal to combined length of third to fifth segments ........................... ***Zavrelia* Kieffer** (p. 100)
   Distal Lauterborn's organ situated on apex of second antennal segment (Fig. 245); second segment clearly longer than combined length of third to fifth segments ..................... ***Stempellinella* Brundin** (p. 99)
6. Antennal tubercle with apicomesal projection (Fig. 232) ..................
   ........................................ ***Micropsectra* Kieffer** (p. 96)
   ........................................ ***Lauterbornia* Kieffer**
   Antennal tubercle without apicomesal projection (Fig. 251) ........... 7
7. Pecten epipharyngis consisting of a row of 10–16 simple scales (Fig. 240)
   ........................................ ***Rheotanytarsus* Bause** (p. 97)
   Pecten epipharyngis consisting of 3 or 5 simple (Fig. 237) or serrated scales
   ................................................................ 8
8. Petiole of Lauterborn's organ longer than combined length of third to fifth antennal segments ..................... ***Tanytarsus* Wulp** (p. 99)
   Petiole of Lauterborn's organ absent or much shorter than combined length of third to fifth segments ...................................... 9
9. Pecten epipharyngis consisting of 3 serrated scales ......................
   ........................................ ***Cladotanytarsus* Kieffer** (p. 94)
   Pecten epipharyngis consisting of 3 or 5 simple scales ..................
   ........................................ ***Paratanytarsus* Bause** (p. 97)

## Clé des genres de Tanytarsini

1. Apex médians pointus des plaques paralabiales séparés par un espace supérieur à la largeur de la dent médiane de l'hypostome (fig. 244) ... 2
   Apex médians obtus des plaques paralabiales se joignant presque à la ligne médiane (fig. 249, 250) .......................................... 6

2. Mandibule sans dents internes (fig. 228) .. ***Corynocera* Zetterstedt** (p. 96)
   Mandibule avec dents internes ................................... 3
3. Tubercule antennaire avec appendice apicomésal palmé portant 3 pointes ou plus (fig. 242) ........................ ***Stempellina* Bause** (p. 98)
   Tubercule antennaire avec appendice apicomésal pointu simple (fig. 225, 246) ................................................................... 4
4. Organes de Lauterborn situés à l'apex du deuxième article antennaire (comme à la figure 235) ............... ***Constempellina* Brundin** (p. 95)
   Organes de Lauterborn situés à différents niveaux sur le deuxième article antennaire (fig. 245) .................................................. 5
5. Organe de Lauterborn distal situé sur la gaine du deuxième article antennaire; deuxième article à peu près égal à la longueur combiné du troisième au cinquième articles ........... ***Zavrelia* Kieffer** (p. 100)
   Organe de Lauterborn distal situé à l'apex du deuxième article antennaire (fig. 245); deuxième article nettement plus long que la longueur combinée du troisième au cinquième articles ...........................
   ................................ ***Stempellinella* Brundin** (p. 99)
6. Tubercule antennaire avec appendice apicomésal (fig. 232) .............
   ................................ ***Micropsectra* Kieffer** (p. 96)
   ................................ ***Lauterbornia* Kieffer**
   Tubercule antennaire sans appendice apicomésal (fig. 251) ........... 7
7. Peigne épipharyngéal consistant en un rangée de 10 à 16 écailles simples (fig. 240) ........................ ***Rheotanytarsus* Bause** (p. 97)
   Peigne épipharyngéal consistant en 3 ou 5 écailles simples (fig. 237) ou serratiformes .................................................. 8
8. Pétiole de l'organe de Lauterborn plus long que la longueur combinée des troisième au cinquième articles antennaires ........................
   ................................ ***Tanytarsus* Wulp** (p. 99)
   Pétiole de l'organe de Lauterborn absent ou beaucoup plus court que la longueur combinée des troisième au cinquième articles ........... 9
9. Peigne épipharyngéal consistant en 3 écailles serratiformes ........
   ................................ ***Cladotanytarsus* Kieffer** (p. 94)
   Peigne épipharyngéal consistant en 3 ou 5 écailles simples ...........
   ................................ ***Paratanytarsus* Bause** (p. 97)

## Genus *Cladotanytarsus* Kieffer

Figs. 219–223

*Cladotanytarsus* Kieffer, 1922a:100.

Antenna with third segment longer than fourth; blade shorter than combined length of terminal segments; Lauterborn's organs large, with petiole distinct but shorter than length of a Lauterborn's organ; antennal tubercle without apicomesal projection. Hypostoma with 1 simple or trifid median and 4–6 pairs of lateral teeth. Paralabial plate with bluntly rounded median apices that almost meet at head midline. Mandible with 3 inner teeth. Premandible with 3 or more apical teeth. SI pectinate; SII distally pectinate. Pecten epipharyngis consisting of 3 serrated scales. Abdomen with short plumose setae. Procercus present, bearing anal setae.

**Remarks.** The larvae of *Cladotanytarsus* and *Paratanytarsus* are similar. Apparently they can be distinguished by the composition of the pecten epipharyngis (see key, couplet 9), but this distinction is based on few specimens. The Lauterborn's organs of *Cladotanytarsus* are much larger than those of *Paratanytarsus*, a fact easily seen when specimens of the two genera are compared side by side. Larvae are described by Thienemann (1929), Krüger (1938), and Roback (1957; as *Calopsectra*, *Atanytarsus* group).

**Distribution.** Widespread south of treeline.

**Habitat.** Larvae of *Cladotanytarsus* inhabit still and flowing water (Roback 1957).

## Genus *Constempellina* Brundin

Figs. 224–226

*Constempellina* Brundin, 1947:82; Brundin 1948:19.

Antenna with second segment less than one-half as long as combined length of third to fifth segments; blade as long as or longer than combined length of terminal segments; Lauterborn's organs on apex of second segment, mounted on petioles longer than third segment but shorter than combined length of third to fifth segments; antennal tubercle with simple apicomesal projection. Frontoclypeal apotome with pair of coarsely plumose setae. Hypostoma with 1 trifid median and 6 pairs of lateral teeth, and with median tooth slightly shorter than lateral teeth. Paralabial plates with pointed median apices that are separated medially by more than width of median tooth of hypostoma. Mandible with 2 inner teeth. Premandible bifid. SI pectinate; SII distally pectinate. Pecten epipharyngis consisting of 3 toothed scales. Abdomen with short plumose setae. Procercus present, smooth, bearing 6 anal setae; anal setae simple or apically divided into 2 or 3 branches.

**Remarks.** The larvae of *Constempellina* and those of *Stempellina*, *Stempellinella*, and *Zavrelia* are distinguished from the other Tanytarsini by having paralabial plates with pointed median apices that are widely separated and end about the level of the second or third lateral tooth of the hypostoma. All build sturdy cylindrical cases similar to those of *Heterotanytarsus* and *Abiskomyia*. The larvae of the four genera are mainly distinguished by antennal characters (see key, couplets 2–5). Larvae are described by Brundin (1948).

**Distribution.** Yukon Territory; District of Mackenzie; northern Manitoba.

**Habitat.** Larvae of *Constempellina* inhabit still water (Brundin 1948).

## Genus *Corynocera* Zetterstedt

Figs. 227–228

*Corynocera* Zetterstedt, 1838:856.

Antenna with segments consecutively smaller; blade shorter than combined length of terminal segments; Lauterborn's organs borne on petioles longer than combined length of terminal segments; antennal tubercle without apicomesal projection. Hypostoma curved dorsally, with 1 median and 2 pairs of lateral teeth; only median and first lateral teeth visible in ventral view. Paralabial plates with blunt median apices separated by less than width of median tooth of hypostoma. Mandible apex shallowly bilobed, without inner teeth. Premandible with 4 or 5 apical teeth. SI pectinate; SII simple. Pecten epipharyngis consisting of 3 toothed scales. Abdominal segments without visible setae. Procercus bearing 7 or 8 anal setae.

**Remarks.** Larvae of *Corynocera* are unusual in that the second lateral teeth of the hypostoma are curved dorsally and are usually obscured by the first lateral teeth in a ventral view. The mandible lacks inner teeth and unlike in other Tanytarsini the abdomen apparently lacks both simple and plumose setae. The larvae are described by Hirvenoja (1961).

**Distribution.** Districts of Mackenzie and Keewatin; southern Alberta.

**Habitat.** Larvae of *Corynocera* inhabit small shallow bodies of still water (Hirvenoja 1961).

## Genus *Micropsectra* Kieffer

Figs. 229–232

*Micropsectra* Kieffer, 1909:50.

Antenna with segments consecutively smaller; blade shorter than combined length of terminal segments; Lauterborn's organs mounted on petioles longer than combined length of third to fifth segments; antennal tubercle with straight or curved apicomesal projection. Hypostoma with 1 single or tripartite median and 5 pairs of lateral teeth. Paralabial plates with blunt median apices that almost meet at midline. Mandible with 2 or 3 inner teeth. Premandible with 2–4 apical teeth. SI pectinate; SII apically pectinate. Pecten epipharyngis consisting of 3 toothed scales. Abdominal segments with plumose setae. Procercus bearing 7 or 8 anal setae.

**Remarks.** Among the larvae of Tanytarsini with paralabial plate apices almost meeting at the midline, only *Micropsectra* has an antennal tubercle bearing an apicomesal projection. This is the only character that will distinguish it from the larvae of *Tanytarsus*. In the key, *Lauterbornia* is brought out with *Micropsectra*. Larvae of these two genera are indistinguishable, except that in *Lauterbornia* the SII appears to be plumose not pectinate as in *Micropsectra*: Larvae are described by Johannsen (1937b; as *Tanytarsus* (*Micropsectra*)) and Roback (1957).

**Distribution.** Widespread, occurring north of treeline.

**Habitat.** Larvae of *Micropsectra* inhabit both still and flowing water (Thienemann 1954).

## Genus *Paratanytarsus* Bause

Figs. 233–237

*Paratanytarsus* Bause, 1914:120.

Antenna with segments consecutively smaller; blade as long as or longer than combined length of terminal segments; Lauterborn's organs small, with petiole present or absent, but if present, then shorter than length of a Lauterborn's organ; antennal tubercle without apicomesal projection. Hypostoma with 1 simple or trifid median and 5 pairs of lateral teeth. Paralabial plates with bluntly rounded median apices that almost meet at midline. Mandible with 2 inner teeth. Premandible bifid. SI pectinate; SII pectinate. Pecten epipharyngis consisting of 3 or 5 blunt smooth teeth. Abdominal segments with plumose setae. Procercus bearing 8 anal setae.

**Remarks.** For comments on distinguishing the larvae of *Paratanytarsus* and *Cladotanytarsus* see remarks under *Cladotanytarsus*. Larvae are described by Thienemann (1951; as *Ditanytarsus*), and Roback (1957).

**Distribution.** Widespread, occurring north of treeline.

**Habitat.** Larvae of *Paratanytarsus* inhabit large bodies of both still and flowing water (Thienemann 1954).

## Genus *Rheotanytarsus* Bause

Figs. 238–240

*Syntanytarsus* (*Rheotanytarsus*) Bause, 1914:120.
*Rheotanytarsus* Kieffer, 1921d:34.

Antenna with segments consecutively smaller; blade slightly shorter to slightly longer than combined length of terminal segments; Lauterborn's organs mounted on petioles as long as or longer than third segment; antennal tubercle without apicomesal projection. Hypostoma with 1 simple to tripartite median and 5 pairs of lateral teeth. Paralabial plates with blunt median apices that almost meet at midline. Mandible with 2 inner teeth. Premandible bifid. SI plumose; SII plumose. Pecten epipharyngis consisting of transverse row of 10–16 simple teeth. Abdominal segments with plumose setae. Procercus bearing 8 anal setae.

**Remarks.** Among the Tanytarsini with paralabial plates almost meeting at the midline, the length of the petioles of the Lauterborn's organs are intermediate in *Rheotanytarsus*. They are about equal to the combined length of the third to fifth antennal segments, whereas in *Micropsectra* and *Tanytarsus* they are much longer, and in *Cladotanytarsus* and *Paratanytarsus* much shorter, if present. The shape of the pecten epipharyngis appears to be distinctive but this character is based on very few specimens. Members of the "*exiguus* group" build a case with two or more filaments around the opening (Scott 1967). Larvae are described by Thienemann (1929) and Roback (1957).

**Distribution.**   Yukon Territory and Alberta to New Brunswick.

**Habitat.**   Larvae of *Rheotanytarsus* inhabit flowing water (Scott 1967; Roback 1957).

## Genus *Stempellina* Bause

Figs. 241–243

*Stempellina* Bause, 1914:120; Kieffer 1921*b*:276.

Antenna with second segment less than one-half as long as combined length of third to fifth segments; blade as long as or longer than combined length of terminal segments; Lauterborn's organs mounted on petioles longer than third segment but shorter than combined length of third to fifth segments; antennal tubercle with palmate or fan-shaped apicomesal projection. Hypostoma with 1 simple or tripartite median and 6 pairs of lateral teeth. Paralabial plates with pointed median apices that are separated by more than width of median tooth of hypostoma. Mandible with 2 inner teeth. Premandible with 2–4 apical teeth. SI pectinate; SII pectinate. Pecten epipharyngis consisting of 3 toothed scales. Abdominal segments with plumose setae. Procercus present, smooth or with spines, bearing 6 or 7 anal setae; anal setae simple, apically divided into 2 or 3 branches or coarsely thickened and bearing thornlike points.

**Remarks.**   Larvae of *Stempellina* with a palmate apicomesal projection on the antennal tubercle are distinctive. In some species the projection is not always strictly palmate in that the points are not arranged in one plane, but overlap one another. Larvae are described by Brundin (1948) and Johannsen (1937*b*; as *Tanytarsus* (*Stempellina*)).

**Distribution.**   Widespread south of treeline.

**Habitat.**   Larvae of *Stempellina* inhabit still water (Brundin 1948).

## Genus *Stempellinella* Brundin

Figs. 244–246

*Stempellinella* Brundin, 1947:87.

Antenna with segments consecutively smaller, with second segment longer than combined length of third to fifth segments; blade as long as or longer than combined length of terminal segments; Lauterborn's organs arising from different levels on second segment, mounted on petioles shorter than length of second segment; antennal tubercle with simple apicomesal projection. Hypostoma with 1 median and 6 pairs of lateral teeth. Paralabial plates with median apices pointed and separated by more than width of median tooth of hypostoma. Mandible with 3 inner teeth. Premandible bifid. SI pectinate; SII distally pectinate. Pecten epipharyngis consisting of 3 simple scales. Abdominal segments with plumose setae. Procercus bearing 4–6 simple anal setae.

**Remarks.** Larvae of *Stempellinella* and *Zavrelia* are only distinguished by the position of the distal Lauterborn's organ on the second antennal segment (see key, couplet 5). Larvae are described by Brundin (1948).

**Distribution.** Yukon Territory to northern Manitoba; Quebec; New Brunswick.

**Habitat.** Larvae of *Stempellinella* inhabit still water (Brundin 1948).

## Genus *Tanytarsus* Wulp

Figs. 247–252

*Tanytarsus* Wulp, 1874:134; Reiss and Fittkau 1971:75.

Antenna with segments consecutively smaller; blade shorter than combined length of terminal segments; Lauterborn's organs mounted on petioles longer than combined length of third to fifth segments; antennal tubercle without apicomesal projection. Hypostoma with 1 simple, bifid, or trifid median and 5 pairs of lateral teeth. Paralabial plates with blunt median apices that almost meet at midline. Mandible with 2 or 3 inner teeth. Premandible with 3–5 apical teeth. SI pectinate; SII apically pectinate. Pecten epipharyngis consisting of 3 toothed scales. Abdominal segments with plumose setae. Procercus smooth, with 6 or 7 anal setae.

**Remarks.** Larvae of *Tanytarsus* can only be separated from those of *Micropsectra* by the absence of an apicomesal projection on the antennal tubercle.

Townes (1945) regarded *Phaenopsectra*, *Endochironomus*, and *Stictochironomus* as subgenera of *Tanytarsus* in the tribe Chironomini. *Calopsectra* was used in place of *Tanytarsus* in the sense the genus is used here. The genus *Tanytarsus* has been employed in a broad sense encompassing some genera now treated separately (see Reiss and Fittkau (1971) for a review). Larvae are described by Zavřel (1926), Krüger (1945), Johannsen (1937*b*), and Roback (1957).

**Distribution.** Widespread, occurs north of treeline.

**Habitat.** Larvae of *Tanytarsus* mainly inhabit still water although a few species inhabit flowing water (Reiss and Fittkau 1971).

## Genus *Zavrelia* Kieffer

*Zavrelia* Kieffer *in* Bause, 1914:73.

Antenna with segments consecutively smaller, and with second segment about as long as combined length of third to fifth segments; blade longer than combined length of terminal segments; Lauterborn's organs arising from different levels on second segment, mounted on petioles shorter than length of second or third segment; antennal tubercle with simple apicomesal projection. Hypostoma with 1 simple median and 6 pairs of lateral teeth. Paralabial plate with blunt median apices separated by more than width of median tooth of hypostoma. Mandible with 3 inner teeth. Premandible bifid. SI pectinate; SII pectinate. Abdominal segments with plumose setae.

**Remarks.** Except for the antennal differences used in the key (couplet 5) the larvae of *Zavrelia* and *Stempellinella* are similar. No larvae were available for examination. The foregoing description is based on those of Bause (1914), Brundin (1948), and Hamilton (1965).

**Distribution.** Southern British Columbia; New Brunswick.

**Habitat.** Larvae of *Zavrelia* inhabit small bodies of still water and shallower regions of lakes (Brundin 1948).

## Subfamily Orthocladiinae

Figs. 3, 8, 253–424

Antenna nonretractile, rarely mounted on antennal tubercle, usually shorter in length than head capsule and generally shorter than mandible, 4- to 7-segmented, usually 5-segmented; ring organ on basal half of first segment, rarely located distal to midpoint of segment; Lauterborn's organs usually on apex of second segment, rarely alternate on apices of second and third segments or on apex of third segment, sometimes indistinct, vestigial, or only one present; third segment non-annulated; blade present. Hypostoma usually convex to triangular, nearly always toothed with teeth generally consecutively smaller from medial to lateral.

Paralabial plate variable in size, sometimes indistinct, without striations, and with or without setae. Mandible with 1 apical tooth and 3–5 inner teeth; seta interna present or absent; apicodorsal tooth and pecten mandibularis absent; seta subdentalis nearly always present, rarely absent; inner margin usually smooth, sometimes with spines; outer margin usually smooth, sometimes crenulate. Premandible with 1–4 teeth. Maxillary palpus usually shorter than wide. SI simple, bifid, plumose or palmate; SII usually simple, rarely bifid or plumose. Labral lamellae present or absent. Pecten epipharyngis nearly always consisting of 3 scales, sometimes appressed to form one scale or expressed as 5–7 scales. Prementohypopharyngeal complex entire, with scales or modified setae. Anterior parapods usually separate with apical crown of claws, sometimes fused. Abdominal setae usually simple, sometimes plumose or forked. Anal end usually bearing a pair of procerci with anal setae, 2 pairs of anal tubules, and a pair of posterior parapods each with apical crown of claws; one or more of these structures sometimes absent; anal end usually directed posteriorly, rarely directed ventrally.

**Remarks.** The larvae of the Orthocladiinae are the most structurally variable among chironomid subfamilies. Those with a reduced anal end are easily distinguished except from the Telmatogetoninae (see remarks under Telmatogetoninae). The general shape of those with normal anal ends resemble the Chironominae, thus differing from the elongate appearance of the Tanypodinae. Unmounted larvae can often be distinguished from the Chironominae by the arrangement of the eye spots (see remarks under Chironominae). Otherwise the absence of striated paralabial plates will afford separation between the two subfamilies. More difficulty is encountered in separating Orthocladiinae from some Diamesinae and from the Prodiamesinae. All larvae of Diamesinae, except *Protanypus*, have an annulated third antennal segment, which is absent in the Orthocladiinae. However, the larvae of *Protanypus* are very distinctive in that the head capsule bears many setae and has a pair of distinct posterolateral projections on the occipital margin. Also the labral lamellae of *Protanypus* consist of a transverse row of overlapping, apically serrated scales—no larvae of the Orthocladiinae have more than two labral lamellae. Some members of the Orthocladiinae with large paralabial plates bearing setae, e.g., *Diplocladius*, *Rheocricotopus*, *Nanocladius*, and *Psectrocladius*, might be confused with the Prodiamesinae but the latter have a 4-segmented antenna.

The classification used here follows that of Brundin (1956) and Hamilton et al. (1969) as modified by Sæther (1977b). Tribes have been used in the Orthocladiinae (see Brundin 1956), but the subfamily is not well known and Sæther's (1977b) recommendation "...it is better at present to keep the subfamily undivided." is followed here.

**Distribution.** The Orthocladiinae is worldwide in distribution and widespread within Canada. It is the dominant subfamily north of treeline, and although numbers of species decrease proportionally southward it is common in southern Canada.

**Biology.** Larvae of Orthocladiinae are generally cool-adapted, living in a greater variety of habitats than any other subfamily. They inhabit all types of freshwater, brackish and saline waters, and semiaquatic to terrestrial habitats. A number live in warm habitats but the subfamily is commoner in cool well-oxygenated waters. Others mine in living or dead plant tissue. The extent of case building is not known for most genera but it is probable that, except for a few free-living forms, some sort of case is built. A few build strong cylindrical cases of substrate particles cemented with silk, but generally the cases tend to be more loosely built than those of the Chironomini.

Most species feed on small plants and animals, and on detritus, although a few are predaceous attacking larger invertebrates; others eat living aquatic plant tissue. It is not known if any species use silk in feeding.

## Key to the genera of Orthocladiinae

1. Anal end with procerci bearing anal setae (Figs. 299, 306). Anterior parapods separated, rarely fused ................................... 2
   Anal end without procerci; anal seta, if present, arising directly from body wall. Anterior parapods fused (Fig. 3) ......................... 44
2. Antenna longer than one-half length of head capsule, usually longer than head capsule ................................................. 3
   Antenna shorter than one-half length of head capsule, generally not longer than length of mandible ........................................ 5
3. Anterior parapods with apical and preapical groups of claws (as in Fig. 256). Hypostoma with median and first lateral teeth deeply recessed (Fig. 313) ........................................ *Heterotanytarsus* **Spärck** (p. 126)
   Anterior parapods with only apical group of claws. Hypostoma with both median and first lateral teeth not recessed (Figs. 270, 415–416) ... 4
4. Antenna 4-segmented, longer than head capsule (Fig. 273) ............ ........................................ *Corynoneura* **Winnertz** (p. 116)
   Antenna 5-segmented, shorter than head capsule (Fig. 419) ........... ........................................ *Thienemanniella* **Kieffer** (p. 146)
5. Antenna tubercle present with apicomesal projection (Fig. 255). Anterior parapods with apical and preapical groups of claws (Fig. 256) ...... ........................................ *Abiskomyia* **Edwards** (p. 110)
   Antenna tubercle absent. Anterior parapods with only apical group of claws ..................................................................... 6
6. SI simple (Figs. 268, 291) ............................................ 7
   SI bifid (Figs. 8, 345, 361), plumose (Figs. 296, 322), or palmate (Fig. 392) ..................................................................... 14
7. Paralabial plate with setae ........................................... 8
   Paralabial plate without setae[5] ..................................... 9

---

[5]One species of *Nanocladius* has a small plate lateral to the hypostoma and dorsal to the paralabial plate bearing a tuft of setae. This should not be confused with setae arising from the ventral surface of a paralabial plate.

8. Mandible with long spine on inner margin (Fig. 410); seta interna absent. Abdominal segments with both simple and plumose setae (Fig. 411) . . . . . . . . . . . . . . . . . . . . . . . . . . . . *Synorthocladius* **Thienemann** (p. 145)
   Mandible with smooth inner margin; seta interna present. Abdominal segments with only simple setae . . . . . . . . . . . . . . . . . . . . . . . . . . . . . . . . . . . . . . . . . . . . . . . . . . . . . . . . . . *Paracricotopus* **Thienemann & Harnisch** (p. 135)
9. Mandible with spines on inner margin (Fig. 305) . . . . . . . . . . . . . . . . . . . . . . . . . . . . . . . . . . . . . . . . . . . . . . . . *Eukiefferiella* **Thienemann** (in part) (p. 122)
   Mandible without spines on inner margin (Figs. 330, 379) . . . . . . . . . . . 10
10. Abdominal segments with some setae plumose (Fig. 287) . . . . . . . . . . . . . . . . . . . . . . . . . . . . . . . . . . . . . . . . . . . . . . . . *Cricotopus* **Wulp** (in part ) (p. 117)
    Abdominal segments with only simple setae . . . . . . . . . . . . . . . . . . . . . . . . 11
11. Hypostoma with broad toothless median area and 13 pairs of lateral teeth (Figs. 377–378). Paralabial plates fused medially with 3 teeth in a median concavity (Fig. 377). Mandible without seta interna . . . . . . . . . . . . . . . . . . . . . . . . . . . . . . . . . . . . . . . . . . . . . . . . . . . . . . . . . . . . . . *Phycoidella* **Sæther** (p. 139)
    Hypostoma with single, partly double, or 2 median teeth and 5 or 6 pairs of lateral teeth (Fig. 336). Paralabial plates not fused medially. Mandible with seta interna (Fig. 331) . . . . . . . . . . . . . . . . . . . . . . . . . . . . . . . . . . . 12
12. Hypostoma with 2 deeply recessed median teeth. Antenna about one-quarter length of mandible . . . . . . . . . . . *Metriocnemus* **Wulp** (in part) (p. 130)
    Hypostoma with median teeth not recessed, longer or as long as first lateral tooth (Figs. 265, 336). Antenna at least one-half length of mandible . . . . . . . . . . . . . . . . . . . . . . . . . . . . . . . . . . . . . . . . . . . . . . . . . . . . . . . . 13
13. Procercus without posterior spurs; 2 anal setae longer and thicker than remaining setae (Fig. 269). Apical tooth of mandible shorter than combined width of inner teeth (Fig. 266). . . . . . . . . . . . . . . . . . . . . . . . . . . . . . . . . . . . . . . . . . . . . . . . . . . . . . . . . . . . . . . . *Cardiocladius* **Kieffer** (p. 115)
    Procercus with 1 or 2 posterior spurs; anal setae all more or less of equal length. Apical tooth of mandible longer than combined width of inner teeth (Fig. 339) . . . . . . . . . . . . . . . . . . . . *Nanocladius* **Kieffer** (p. 131)
14. SI bifid (Figs. 8, 354, 361) . . . . . . . . . . . . . . . . . . . . . . . . . . . . . . . . . . . . 15
    SI plumose (Figs. 296, 322) or, rarely, palmate (Fig. 392) . . . . . . . . . . . . 24
15. Paralabial plate with setae (Figs. 258, 358) . . . . . . . . . . . . . . . . . . . . . . . . 16
    Paralabial plate without setae . . . . . . . . . . . . . . . . . . . . . . . . . . . . . . . . . 19
16. Mandible without seta interna . . . . . . . . . . . . . . *Acricotopus* **Kieffer** (p. 111)
    Mandible with seta interna . . . . . . . . . . . . . . . . . . . . . . . . . . . . . . . . . . 17
17. Hypostoma with 2 median and 5 or 6 pairs of lateral teeth (Figs. 398–399). Procercus usually with a posterior spur . . . . . . . . . . . . . . . . . . . . . . . . . . . . . . . . . . . . . . . . . . *Rheocricotopus* **Thienemann & Harnisch** (p. 143)
    Hypostoma with 1 median and 6–9 pairs of lateral teeth (Figs. 356–357). Procercus without posterior spur . . . . . . . . . . . . . . . . . . . . . . . . . . . . . 18
18. Anal tubules present. Lauterborn's organs large, at least three-quarters length of third antennal segment. Median tooth of hypostoma more than twice as wide as first lateral tooth (Fig. 357) . . . . . . . . . . . . . . . . . . . . . . . . . . . . . . . . . . . . . . . . . . . . . . . . . . . . . *Paracladius* **Hirvenoja** (p. 134)
    Anal tubules absent. Lauterborn's organs indistinct or vestigial. Median tooth of hypostoma less than twice as wide as first lateral tooth . . . . . . . . . . . . . . . . . . . . . . . . . . . . . . . . . . . . . . . . . . . . *Halocladius* **Hirvenoja** (p. 124)
19. Hypostoma with 2 median teeth (Fig. 325) . . . . . . . . . . . . . . . . . . . . . . 20
    Hypostoma with 1 median tooth (Figs. 275, 281) . . . . . . . . . . . . . . . . . . 21
20. Hypostoma with 20 teeth. Antennal blade shorter than length of terminal segments. . . . . . . . . . . . . . . . . . . . . . . . . *Baeoctenus* **Sæther** (p. 112)

Hypostoma with 12 teeth. Antennal blade longer than length of terminal segments .................... ***Limnophyes* Eaton** (in part) (p. 129)
21. Hypostoma with more than 6 pairs of lateral teeth (Figs. 350–351) ...... ............................ ***Orthocladius* Wulp** (in part) (p. 133)
    Hypostoma with 6 pairs of lateral teeth (Figs. 278, 281) ............ 22
22. Abdominal segments with plumose or tufts of setae in addition to simple setae ............................. ***Cricotopus* Wulp** (in part) (p. 117)
    Abdominal segments with only simple setae ....................... 23
23. First lateral tooth of hypostoma bulbous, with greatest width at middle of tooth ................... ***Paratrichocladius* Santos Abreu** (p. 138)
    First lateral tooth of hypostoma triangular or apically rounded, with greatest width at base of tooth ........... ***Cricotopus* Wulp** (in part) (p. 117) ........................... or ***Orthocladius* Wulp** (in part) (p. 133)
24. Labral lamellae present, consisting of 2 broad pectinate lobes (Fig. 334) ......................................................................... 25
    Labral lamellae usually absent, if present, then lobes smooth or unevenly toothed or single ............................................. 28
25. Paralabial plate with setae (Fig. 294) ....... ***Diplocladius* Kieffer** (p. 121)
    Paralabial plate without setae ..................................... 26
26. Hypostoma with 2 median teeth slightly shorter than first lateral teeth (Fig. 328), or with both median and first lateral teeth shorter than second lateral teeth (Fig. 329). Labral lamellae coarsely pectinate, lying above or between bases of SI (Fig. 334) ........................................ ........................... ***Metriocnemus* Wulp** (in part) (p. 130)
    Hypostoma with 2 median teeth much longer than first lateral teeth (Fig. 262), or first lateral teeth much longer than single median and second lateral teeth (Fig. 307). Labral lamellae finely pectinate, lying between tormae and SI (Fig. 263) ..................................... 27
27. Antenna 5-segmented; second segment shorter than third .............. ......................................... ***Brillia* Kieffer** (p. 112)
    Antenna 4-segmented; second segment longer than third .............. ........................................ ***Euryhapsis* Oliver** (p. 123)
28. Paralabial plate with setae (Figs. 388, 423) ......................... 29
    Paralabial plate without setae ..................................... 30
29. SI palmate, with 3–9 branches more or less equal in size (Fig. 392). Premandible with 1 apical tooth (Fig. 391) .... ***Psectrocladius* Kieffer** (p. 140)
    SI plumose, with branches unequal in size. Premandible with 2 apical teeth .......................................... ***Zalutschia* Lipina** (p. 147)
30. Procercus with 1 or 2 long anal setae; setae about one-quarter of body length or longer; and with or without shorter anal setae (Fig. 299) ...... 31
    Procercus with all anal setae short (always more than 1 seta present); setae more or less same length, with none as long as one-quarter of body length ......................................................................... 33
31. Procercus with 7 anal setae. Body thickly covered with long, simple setae. One pair of anal tubules present .......... ***Epoicocladius* Zavřel** (p. 121)
    Procercus with 1–4 anal setae. Body sparsely covered with setae. Two pairs of anal tubules present ................................... 32
32. Hypostoma without nipplelike projection on median tooth and with 4 pairs of lateral teeth (Fig. 395). Apical tooth of mandible shorter than combined length of inner teeth. Anterior parapods fused ................... ........................... ***Pseudorthocladius* Goetghebuer** (p. 141)

       Hypostoma with nipplelike projection on median tooth and with 6 pairs of lateral teeth (Fig. 323). Apical tooth of mandible longer than combined width of inner teeth (Fig. 324). Anterior parapods separate .........
       .................... **Krenosmittia Thienemann & Krüger** (p. 128)
33. Antenna 6- or 7-segmented, sixth or seventh segment sometimes minute (Fig. 318) ...................................................... 34
    Antenna usually 5-segmented, rarely 4-segmented .................... 38
34. Second antennal segment much shorter than third (Fig. 312); Lauterborn's organs on apex of third segment ......... **Heleniella Gowin** (p. 125)
    Second antennal segment longer than third; Lauterborn's organs on apex of second segment ............................................... 35
35. Hypostoma with 5 pairs of lateral teeth (Fig. 316). Antenna 7-segmented; third segment much shorter than fourth (Fig. 318) .................
    ............................... **Heterotrissocladius Spärck** (p. 126)
    Hypostoma with 6 pairs of lateral teeth (Figs. 366–368). Antenna 6-segmented; third segment longer than fourth ................... 36
36. Premandible with 1 apical tooth ... **Parakiefferiella Thienemann** (p. 136)
    Premandible with 2 or 3 apical teeth (Figs. 321, 342) ................ 37
37. SI with less than 20 branches (Fig. 322). Hypostoma with 2 median teeth (Fig. 319) ................................... **Hydrobaenus Fries** (p. 127)
    SI with more than 30 branches. Hypostoma with 1 broad median tooth (Fig. 340) ................................... **Oliveridia Sæther** (p. 132)
38. Inner margin of mandible with spines (as in Fig. 305) ..................
    .................... **Eukiefferiella Thienemann** (in part) (p. 122)
    Inner margin of mandible without spines ........................ʻ. 39
39. Anal end bent, directed ventrally; procerci directed posteriorly (Fig. 376)
    .......................... **Paraphaenocladius Thienemann** (p. 138)
    Anal end not bent, directed posteriorly; procerci directed dorsally ... 40
40. Hypostoma with 1 median tooth, tooth entire or weakly notched medially
    .................................................................. 41
    Hypostoma with 2 distinct median teeth (Figs. 325, 370) ............ 42
41. Hypostoma streaked longitudinally; median tooth with weak median notch (Fig. 300) ............. **Eukiefferiella Thienemann** (in part) (p. 122)
    Hypostoma uniform in color; median tooth entire ....................
    .......................... **Chaetocladius Kieffer** (in part) (p. 116)
42. Labral lamellae present, usually consisting of a pair of unevenly toothed lobes
    .......................... **Chaetocladius Kieffer** (in part) (p. 116)
    Labral lamellae usually absent, if present, then consisting of a single weakly sclerotized lobe between bases of SI ............................ 43
43. Paralabial plate well developed, with distolateral corner rounded (Fig. 370). Lauterborn's organs longer than length of third antennal segment, sometimes as long as combined length of third and fourth segments
    .......................... **Parametriocnemus Goetghebuer** (p. 137)
    Paralabial plate small, with distolateral corner angular (Fig. 327). Lauterborn's organs as long as length of third antennal segment, never longer
    ................................... **Limnophyes Eaton** (in part) (p. 129)
44. Anal end bent, directed ventrally (as in Fig. 376). Hypostoma with 2 median teeth (Fig. 309) ............................................... 45
    Anal end not bent, directed posteriorly. Hypostoma with 1 median tooth (Figs. 396, 412), or without median teeth (Fig. 405) ............. 46
45. Anal tubules absent. Posterior parapod apically divided, with anterior part bearing spines .......... **Gymnometriocnemus Goetghebuer** (p. 124)

Anal tubules present. Posterior parapod entire, with claws scattered on apex.
................................ ***Bryophaenocladius* Thienemann** (p. 113)
46. Hypostoma with median concavity flanked by low protuberance; lateral teeth peglike, grouped on both sides and well spaced from low protuberances (Fig. 406). Mandible with peglike inner teeth (Fig. 407) ............
................................ ***Symbiocladius* Kieffer** (p. 144)
Hypostoma with 1 broad median tooth; lateral teeth more or less conical and contiguous with median tooth (Figs. 396, 412). Mandible with broadly rounded or subtriangular inner teeth (Figs. 404, 413) ........... 47
47. Mandible without seta interna. SI bifid; SII bifid ................... 48
Mandible with seta interna. SI plumose; SII simple ................ 49
48. Posterior parapods absent. Terminal antennal segment shorter than preceding segment ...................... ***Camptocladius* Wulp** (p. 114)
Posterior parapods usually present, if absent, then terminal antennal segment longer than preceding segment ............................
............................ ***Pseudosmittia* Goetghebuer** (p. 142)
49. Hypostoma with 5 pairs of lateral teeth (Fig. 403). Second antennal segment longer than first segment .............. ***Smittia* Holmgren** (p. 143)
Hypostoma with 4 pairs of lateral teeth (Fig. 412). Second antennal segment shorter than first segment (Fig. 414) .............................
.................... ***Thalassosmittia* Strenzke & Remmert** (p. 146)

## Clé des genres de Orthocladiinæ

1. Extrémité anale avec procerques portant des soies anales (fig. 299, 306). Parapodes antérieurs séparés, rarement soudés ................. 2
Extrémité anale sans procerques; soie anale, si présente, s'élevant directement de la paroi du corps. Parapodes antérieurs soudés (fig. 3) .. 44
2. Antenne plus longue qu'une demie de la longueur de la capsule céphalique, habituellement plus longue que cette dernière .................. 3
Antenne plus courte qu'une demie de la longueur de la capsule céphalique, généralement pas plus longue que la mandibule ................. 5
3. Parapodes antérieurs avec groupes apical et préapical de griffes (comme à la figure 256). Hypostome avec dents médianes et premières dents latérales fortement renfoncées (fig. 313) .... ***Heterotanytarsus* Spärck** (p. 126)
Parapodes antérieurs avec seulement un groupe apical de griffes. Hypostome avec dents médianes et premières dents latérales non renfoncées (fig. 270, 415, 416) ............................................. 4
4. Antenne à 4 articles, plus longue que la capsule céphalique (fig. 273) ....
............................ ***Corynoneura* Winnertz** (p. 116)
Antenne à 5 articles, plus courte que la capsule céphalique (fig. 419) ...
............................ ***Thienemanniella* Kieffer** (p. 146)
5. Tubercule antennaire présent avec un appendice apicomésal (fig. 255). Parapodes antérieurs avec groupes apical et préapical de griffes (fig. 256)
............................ ***Abiskomyia* Edwards** (p. 110)
Tubercule antennaire absent. Parapodes antérieurs avec groupe apical de griffes, seulement ................................................ 6
6. SI simple (fig. 268, 291) ......................................... 7
SI bifide (fig. 8, 345, 361), plumeuse (fig. 296, 322) ou palmée (fig. 392)
............................................................. 14

7. Plaque paralabiale avec soies ................................... 8
   Plaque paralabiale sans soies[5] ................................... 9
8. Mandibule avec une longue épine sur le bord interne (fig. 410); soie interne absente. Segments abdominaux avec soies simples et soies plumosées (fig. 411) .................. ***Synorthocladius* Thienemann** (p. 145)
   Mandibule avec bord interne lisse; soie interne présente. Segments abdominaux avec soies simples, seulement ................. ***Paracricotopus* Thienemann & Harnisch** (p. 135)
9. Mandibule avec épines sur le bord interne (fig. 305) ................... ***Eukiefferiella* Thienemann** (en partie) (p. 122)
   Mandibule sans épines sur le bord interne (fig. 330, 379) ............ 10
10. Segments abdominaux avec quelques soies plumeuses (fig. 287) ......... ***Cricotopus* Wulp** (en partie) (p. 117)
    Segments abdominaux avec soies simples, seulement ................ 11
11. Hypostome avec région médiane large sans dents et 13 paires de dents latérales (fig. 377, 378). Plaques paralabiales soudées au milieu avec 3 dents disposées dans une cavité médiane (fig. 377). Mandibule sans soie interne .............................. ***Phycoidella* Sæther** (p. 139)
    Hypostome avec dent simple ou partiellement dédoublée ou encore 2 dents médianes et 5 ou 6 paires de dents latérales (fig. 336). Plaques paralabiales non soudées au milieu. Mandibule avec soie interne (fig. 331) ................................................. 12
12. Hypostome avec 2 dents médianes profondément renfoncées. Antenne égale à environ un quart de la longueur de la mandibule ................ ***Metriocnemus* Wulp** (en partie) (p. 130)
    Hypostome avec dents médianes non renfoncées, plus longues ou aussi longues que la première dent latérale (fig. 265, 336). Antenne au moins égale à la demi-longueur de la mandibule ..................... 13
13. Procerque sans éperons postérieurs; 2 soies anales plus longues et plus épaisses que les autres (fig. 269). Dent apicale de la mandibule plus courte que la largeur combinée des dents internes (fig. 266) ........ ***Cardiocladius* Kieffer** (p. 115)
    Procerque avec 1 ou 2 éperons postérieurs; soies anales toutes plus ou moins d'égale longueur. Dent apicale de la mandibule plus longue que la largeur combinée des dents internes (fig. 339) ..................... ***Nanocladius* Kieffer** (p. 131)
14. SI bifide (fig. 8, 354, 361) ....................................... 15
    SI plumeuse (fig. 296, 322) ou, rarement, palmée (fig. 392) .......... 24
15. Plaque paralabiale avec soies (fig. 258, 358) ...................... 16
    Plaque paralabiale sans soies .................................... 19
16. Mandibule sans soie interne ............... ***Acricotopus* Kieffer** (p. 111)
    Mandibule avec soie interne ..................................... 17
17. Hypostome avec 2 dents médianes et 5 ou 6 paires de dents latérales (fig. 398, 399). Procerque avec, habituellement, un éperon postérieur ........ ***Rheocricotopus* Thienemann & Harnisch** (p. 143)
    Hypostome avec 1 dent médiane et de 6 à 9 paires de dents latérales (fig. 356, 357). Procerque sans éperon postérieur ....................... 18
18. Tubules anaux présents. Organes de Lauterborn gros, d'une longueur au moins égale aux trois quarts du troisième article antennaire. Dent

---

[5]Une espèce de *Nanocladius* présente une petite plaque située latéralement par rapport à l'hypostome et dorsalement par rapport à la plaque paralabiale et qui porte une touffe de soies. À ne pas confondre avec les soies qui s'élèvent de la surface ventrale de la plaque paralabiale.

médiane de l'hypostome d'une largeur supérieure au double de la première dent latérale (fig. 357) ....... ***Paracladius*** **Hirvenoja** (p. 134)
Tubules anaux absents. Organes de Lauterborn indistincts ou vestigiaux. Dent médiane de l'hypostome d'une largeur inférieure au double de la première dent latérale ............. ***Halocladius*** **Hirvenoja** (p. 124)
19. Hypostome avec 2 dents médianes (fig. 325) ........................ 20
    Hypostome avec 1 dent médiane (fig. 275, 281) .................... 21
20. Hypostome avec 20 dents. Lame antennaire plus courte que les articles terminaux .................. ***Bæoctenus*** **Sæther** (p. 112)
    Hypostome avec 12 dents. Lame antennaire plus longue que les articles terminaux ................. ***Lymnophyes*** **Eaton** (en partie) (p. 129)
21. Hypostome avec plus de 6 paires de dents latérales (fig. 350, 351) .......
    ............................ ***Orthocladius*** **Wulp** (en partie) (p. 133)
    Hypostome avec 6 paires de dents latérales (fig. 278, 281) .......... 22
22. Segments abdominaux avec soies plumeuses ou touffes de soies en plus des soies simples ................. ***Cricotopus*** **Wulp** (en partie) (p. 117)
    Segments abdominaux avec soies simples seulement ................. 23
23. Première dent latérale de l'hypostome bulbeuse, de largeur maximale au milieu ................. ***Paratrichocladius*** **Santos Abreu** (p. 138)
    Première dent latérale de l'hypostome triangulaire ou arrondie à l'extrémité, de largeur maximale à la base... ***Cricotopus*** **Wulp** (en partie) (p. 117)
    ........................ ou ***Orthocladius*** **Wulp** (en partie) (p. 133)
24. Lamelles labrales présentes, consistant en 2 larges lobes pectinés (fig. 334) ................................................................ 25
    Lamelles labrales habituellement absentes et, si présentes, alors lobes lisses ou inégalement dentés ou simples ............................... 28
25. Plaque paralabiale avec soies (fig. 294) ..... ***Diplocladius*** **Kieffer** (p. 121)
    Plaque paralabiale sans soies ..................................... 26
26. Hypostome avec 2 dents médianes légèrement plus courtes que les premières dents latérales (fig. 328) ou avec dents médianes et premières dents latérales plus courtes que les deuxièmes dents latérales (fig. 329). Lamelles labrales grossièrement pectinées, s'étalant au-dessus ou entre les bases de SI (fig. 334) ............. ***Metriocnemus*** **Wulp** (en partie) (p. 130)
    Hypostome avec 2 dents médianes beaucoup plus longues que les premières dents latérales (fig. 262) ou première dent latérale beaucoup plus longue que la seule dent médiane et les deuxièmes dents latérales (fig. 307). Lamelles labrales finement pectinées, s'étalant entre les tormæ et SI (fig. 263) ..................................................... 27
27. Antenne à 5 articles; deuxième article plus court que le troisième .......
    ............................................ ***Brillia*** **Kieffer** (p. 112)
    Antenne à 4 articles; deuxième article plus long que le troisième ........
    ............................................ ***Euryhapsis*** **Oliver** (p. 123)
28. Plaque paralabiale avec soies (fig. 388, 423) ....................... 29
    Plaque paralabiale sans soies ..................................... 30
29. SI palmé, avec de 3 à 9 ramifications de taille plus ou moins égale (fig. 392). Prémandibule avec 1 dent apicale (fig. 391) ........................
    ............................................ ***Psectrocladius*** **Kieffer** (p. 140)
    SI plumeuse, avec ramifications de taille inégale. Prémandibule avec 2 dents apicales ................. ***Zalutschia*** **Lipina** (p. 147)
30. Procerque avec 1 ou 2 longues soies anales; soies d'une longueur égale à environ un quart de celle du corps ou plus longues; et avec ou sans soies anales plus courtes (fig. 299) ................................... 31

Procerque avec toutes les soies anales courtes (toujours plus que 1 soie présente); soies plus ou moins de la même longueur, aucune n'étant aussi longue que le quart du corps .................................. 33
31. Procerque avec 7 soies anales. Corps abondamment couvert de soies simples et longues. Une paire de tubules anaux présente ................... ................................. ........... ***Epoicocladius* Zavřel** (p. 121)
Procerque avec de 1 à 4 soies anales. Corps couvert de soies peu abondantes. Deux paires de tubules anaux présentes ....................... 32
32. Hypostome sans appendice en forme de mamelon sur la dent médiane et avec 4 paires de dents latérales (fig. 395). Dent apicale de la mandibule plus courte que la longueur combinée des dents internes. Parapodes antérieurs soudés ............. ***Pseudorthocladius* Gœtghebuer** (p. 141)
Hypostome avec appendice en forme de mamelon sur la dent médiane et avec 6 paires de dents latérales (fig. 323). Dent apicale de la mandibule plus longue que la largeur combinée des dents internes (fig. 324). Parapodes antérieurs séparés .... ***Krenosmittia* Thienemann & Krüger** (p. 128)
33. Antenne à 6 ou 7 articles, le sixième ou le septième étant parfois minuscule (fig. 318) ................................................. 34
Antenne habituellement à 5 articles, rarement à 4 .................. 38
34. Deuxième article antennaire beaucoup plus court que le troisième (fig. 312); organes de Lauterborn sur l'apex du troisième article .............. ....................................... ***Heleniella* Gowin** (p. 125)
Deuxième article antennaire plus long que le troisième; organes de Lauterborn sur l'apex du deuxième article .......................... 35
35. Hypostome avec 5 paires de dents latérales (fig. 316). Antenne à 7 articles; troisième article beaucoup plus court que le quatrième (fig. 318) .... ............................... ***Heterotrissocladius* Spärck** (p. 126)
Hypostome avec 6 paires de dents latérales (fig. 366 à 368). Antenne à 6 articles; troisième article plus long que le quatrième .......... 36
36. Prémandibule avec 1 dent apicale .. ***Parakiefferiella* Thienemann** (p. 136)
Prémandibule avec 2 ou 3 dents apicales (fig. 321, 342) .............. 37
37. SI avec moins de 20 ramifications (fig. 322). Hypostome avec 2 dents médianes (fig. 319) .................... ***Hydrobænus* Fries** (p. 127)
SI avec plus de 30 ramifications. Hypostome avec 1 dent médiane large (fig. 340) .......................... ***Oliveridia* Sæther** (p. 132)
38. Bord interne de la mandibule avec épines (comme à la figure 305) ...... .................... ***Eukiefferiella* Thienemann** (en partie) (p. 122)
Bord interne de la mandibule sans épines .......................... 39
39. Extrémité anale courbée, dirigée ventralement; procerques dirigés vers l'arrière (fig. 376) ............ ***Paraphænocladius* Thienemann** (p. 138)
Extrémité anale non courbée, dirigée vers l'arrière; procerques dirigés dorsalement ................................................ 40
40. Hypostome avec 1 dent médiane, dent entière ou faiblement entaillée au milieu .................................................... 41
Hypostome avec 2 dents médianes distinctes (fig. 325, 370) .......... 42
41. Hypostome strié longitudinalement; dent médiane avec légère entaille médiane (fig. 300) ... ***Eukiefferiella* Thienemann** (en partie) (p. 122)
Hypostome de couleur uniforme; dent médiane entière ................ ........................ ***Chætocladius* Kieffer** (en partie) (p. 116)
42. Lamelles labrales présentes, consistant habituellement en un paire de lobes dentés inégalement ........ ***Chætocladius* Kieffer** (en partie) (p. 116)
Lamelles labrales habituellement absentes et, si présentes, consistant en un lobe unique faiblement sclérotisé entre les bases de SI .......... 43

43. Plaque paralabiale bien développée, avec coin distolatéral arrondi (fig. 370). Organes de Lauterborn plus longs que le troisième article antennaire, parfois égaux à la longueur combinée du troisième et du quatrième articles .................. ***Parametriocnemus* Gœtghebuer** (p. 137)
    Plaque paralabiale petite, avec coin distolatéral angulaire (fig. 327). Organes de Lauterborn égaux au troisième article antennaire, jamais plus longs ............................. ***Limnophyes* Eaton** (en partie) (p. 129)
44. Extrémité anale courbée, dirigée ventralement (comme à la figure 376). Hypostome avec 2 dents médianes (fig. 309) .................... 45
    Extrémité anale non courbée, dirigée postérieurement. Hypostome avec 1 dent médiane (fig. 396, 412) ou sans dents médianes (fig. 405) .. 46
45. Tubules anaux absents. Parapode postérieur divisé à l'extrémité, la partie antérieure portant des épines ........................................ ............................ ***Gymnometriocnemus* Gœtghebuer** (p. 124)
    Tubules anaux présents. Parapode postérieur entier, avec griffes dispersées sur l'apex .................. ***Bryophænocladius* Thienemann** (p. 113)
46. Hypostome avec concavité médiane flanquée d'une protubérance basse; dents latérales en forme de cheville, groupées sur les deux côtés et bien écartées des protubérances basses (fig. 406). Mandibule avec dents internes en forme de chevilles (fig. 407) ........................................ ............................... ***Symbiocladius* Kieffer** (p. 144)
    Hypostome avec 1 dent médiane large; dents latérales plus ou moins coniques et contiguës à la dent médiane (fig. 396, 412). Mandibule avec dents internes largement arrondies ou subtriangulaires (fig. 404, 413) .. 47
47. Mandibule sans soie interne. SI bifide; SII bifide ................... 48
    Mandibule avec soie interne. SI plumeuse; SII simple .............. 49
48. Parapodes postérieurs absents. Article antennaire terminal plus court que l'article précédent .................. ***Camptocladius* Wulp** (p. 114)
    Parapodes postérieurs habituellement présents; si absents, alors article antennaire terminal plus long que l'article précédent ................. .............................. ***Pseudosmittia* Gœtghebuer** (p. 142)
49. Hypostome avec 5 paires de dents latérales (fig. 403). Deuxième article antennaire plus long que le premier ..... ***Smittia* Holmgren** (p. 143)
    Hypostome avec 4 paires de dents latérales (fig. 412). Deuxième article antennaire plus court que le premier (fig. 414) ........................ ....................... ***Thalassosmittia* Strenzke & Remmert** (p. 146)

## Genus *Abiskomyia* Edwards

Figs. 253–256

*Abiskomyia* Edwards, 1937:140; Thienemann 1941:205; Brundin 1956:67.

Antenna 5-segmented, with second and third segments subequal in length, and with fourth and fifth small; borne on a prominent tubercle with an apicomesal projection; blade shorter than or equal to combined length of terminal segments; one Lauterborn's organ on apex of second segment, other on side of third segment. Hypostoma with 2 median and 5 pairs of lateral teeth; median and first lateral teeth lighter in color than remaining lateral teeth; second lateral tooth reduced. Paralabial plate well developed, with short and sparse setae. Mandible with 3 or 4 inner teeth; apical tooth shorter than combined width of inner teeth; seta interna present. Premandible with 2 apical teeth. SI plumose. Labral lamella consisting of 1 single lobe. Pecten epipharyngis consisting of bifid scale. Anterior parapods separate, each with 1 apical and 1 preapical group of claws. Abdominal segments with some plumose setae. Procercus bearing 7 anal setae, without spur. Two pairs of ovoid anal tubules present; tubules shorter than posterior parapods. Posterior parapod claws arranged in horseshoe-shaped rows.

**Remarks.** Among larvae of Orthocladiinae, larvae of *Abiskomyia* are unique in that the antennae are mounted on a tubercle having an apicomesal projection. As in *Heterotanytarsus* the anterior parapods each have two groups of claws, but the shape of the hypostoma will easily separate the two genera. Larvae of both genera build cases similar to those of *Stempellina* and related genera of the Tanytarsini. In structure of the antennae and arrangement of claws on the posterior parapods they resemble the larvae of Tanytarsini but lack striated paralabial plates. Larvae are described by Thienemann (1941) and Pankratova (1970).

**Distribution.** Arctic region.

**Habitat.** Larvae of *Abiskomyia* are cool-adapted, inhabiting the shallow region of arctic lakes and small bodies of flowing water (Thienemann 1937a, 1941; Brundin 1956).

## Genus *Acricotopus* Kieffer

Figs. 257–261

*Acricotopus* Kieffer, 1921a:90; Hirvenoja 1973:81.

Antenna 5-segmented, with segments consecutively smaller; Lauterborn's organs about three-quarters as long as third segment; blade shorter than combined length of terminal segments. Hypostoma with 1 median and 6 pairs of lateral teeth; median tooth broad, with anterior margin weakly divided into 4 parts. Paralabial plate well developed, with long setae. Mandible with 3 inner teeth; apical tooth slightly longer than combined width of inner teeth; seta interna absent. Premandible with 1 apical tooth. SI bifid, with branches finely feathered. Pecten epipharyngis consisting of 3 pointed scales. Anterior parapods separate, each with apical group of claws. Abdomen with long simple setae. Procercus strongly sclerotized, with posterior spur and 6 anal setae. Two pairs of anal tubules present; tubules shorter than length of posterior parapods.

**Remarks.** The combination of bifid SI, paralabial plate with setae, and mandible without seta interna is distinctive. The SI is intermediate between the bifid and plumose types, having two main branches and each with a number of finer subdivisions. Hirvenoja (1973) reported that the SI may also be single. If this type is found in Canada it would key to the couplet with *Synorthocladius* and *Paracricotopus*, but the absence of a seta interna on the mandible will distinguish the larvae of *Acricotopus*. Larvae are described by Johannsen (1937a; as *Spaniotoma* (*Trichocladius*)) and Hirvenoja (1973).

**Distribution.** Alberta and District of Mackenzie to New Brunswick.

**Habitat.** Larvae of *Acricotopus* live in standing water of different types (Brundin 1956).

## Genus *Baeoctenus* Sæther

*Baeoctenus* Sæther, 1977c:2354.

Antenna 5-segmented, short, and stubby, with segments consecutively smaller; Lauterborn's organs (2 present ?) large; blade shorter than combined length of terminal segments. Hypostoma with 2 median and 9 pairs of lateral teeth, with apex truncated. Paralabial plate elongate, narrow, without setae. Mandible with 3 inner teeth; apical tooth longer than combined width of inner teeth; seta interna present. Premandible apically broad, with 5 pointed teeth. SI bifid. Pecten epipharyngis consisting of 3 elongate scales. Anterior parapods separate, each with an apical group of claws. Procercus short, bearing about 7 anal setae. Anal tubules apparently absent.

**Remarks.** The larvae of *Baeoctenus* with a bifid SI, apically truncated hypostoma, and no anal tubules are distinctive. The foregoing description is based on that given by Sæther (1977c).

**Distribution.** Southern Manitoba; New Brunswick.

**Habitat.** Larvae of *Baeoctenus* live within the gills of the freshwater clam, *Anodonta* (Sæther 1977c).

## Genus *Brillia* Kieffer

Figs. 260–264

*Brillia* Kieffer, 1913c:34; Brundin 1956:68.

Antenna 5-segmented, with first segment slightly curved, and with second segment usually much shorter than third; Lauterborn's organs distinct; either on apex of second or of third segment; blade length variable. Hypostoma with 2 median and 4 or 5 pairs of lateral teeth; median teeth large, well separated, sometimes with a small tooth between them. Paralabial plate small, without setae. Mandible with 3 inner teeth; apical tooth shorter than combined width of inner teeth; seta interna present. Premandible broad, with 2 apical teeth. SI plumose. Labral lamellae consisting of 2 transverse elongate pectinate lobes; lobes almost contiguous medially, situated between bases of SI and tormae. Pecten epipharyngis consisting of 3 blunt scales. Anterior parapods separate, each with apical crown of claws. Abdomen with simple scattered setae or with strong lateral fringe of plumose setae. Procercus elongate, usually about twice as long as wide, bearing 8 anal setae, without spur. Two pairs of anal tubules present; tubules slender, shorter than posterior parapods.

**Remarks.** The larvae of *Brillia* and *Euryhapsis* are similar, differing primarily in antennal structure (see key, couplet 27). Also the procerci are longer and the premandible usually more clearly bifid in *Brillia* than in *Euryhapsis*. Larvae of both genera have distinctive labral lamellae that will distinguish them from those of other Orthocladiinae. These lobes are transverse elongate pectinate structures occupying the area between the bases of the SI and the pecten epipharyngis. Being almost contiguous they superficially resemble the labral lamellae of the Chironomini. The labral lamellae of *Diplocladius* and some *Metriocnemus* are similar, but in these genera the lamellae are situated closer to the bases of the SI.

There are two larval types of *Brillia*. One has a lateral setal fringe on the abdomen, and the antenna has the Lauterborn's organs on the apex of the second segment, which is slightly shorter than the third segment. The other, which is more common, lacks the setal fringe and has an antenna with the Lauterborn's organs on the apex of the third segment. Also the second segment is distinctly shorter than the third. Only *Heleniella* has an antenna similar to that of the second type, but this genus lacks broad pectinate labral lamellae and has a six-segmented antenna. Larvae are described by Johannsen (1937a), Roback (1957), and Sæther (1969).

**Distribution.** Widespread south of treeline.

**Habitat.** Larvae of the common type of *Brillia* inhabit flowing water (Brundin 1956) usually associated with plant debris. The type with a lateral abdominal setal fringe mines in submerged rotten wood (Johannsen 1937a).

## Genus *Bryophaenocladius* Thienemann

*Bryophaenocladius* Thienemann 1934:36; Brundin 1956:128.

Antenna 5-segmented, with second segment about three-quarters as long as first and much longer than third segment, and with third and fourth segments subequal in length; blade shorter than combined length of terminal segments. Lauterborn's organs on apex of second segment. Hypostoma with 2 large median and 4 pairs of smaller lateral teeth;

median teeth separated by wide V-shaped notch. Paralabial plate well developed, without setae. Mandible with 3 inner teeth; apical tooth shorter than combined width of inner teeth; seta interna absent. Premandible simple. SI simple, sometimes broad. Anterior parapods completely fused, apically with short bristlelike claws. Preanal segment rounded, without procerci, and with or without short anal setae; anal segment retractile, directed ventrally, bearing 2 pairs of short anal tubules and 2 posterior parapods with small claws.

**Remarks.** The bent posterior end with parapods directed ventrally will distinguish the larvae of *Bryophaenocladius*, except from those of *Gymnometriocnemus* and *Paraphaenocladius*. Larvae of *Paraphaenocladius* have procerci, which are absent in the larvae of the other two genera. Larvae of *Bryophaenocladius* and *Gymnometriocnemus* are similar, especially in the structure of the head, differing primarily in the proportion of the antennal segments—the third segment is much shorter than the fourth in *Gymnometriocnemus*. However, the main diagnostic characters are found in the anal end. Larvae of *Bryophaenocladius* have two pairs of anal tubules and one pair of posterior parapods, although these are very short. In *Gymnometriocnemus*, the anal tubules are absent and each of the posterior parapods is divided.

Although the adults of several species of *Bryophaenocladius* have been recorded from Canada (Sæther 1973a) no larvae have been found as yet. The foregoing description is based on those given by Johannsen (1937a; as *Spaniotoma* (*Orthocladius*), Group *Bryophaenocladius*), Thienemann (1944), and Pankratova (1970). Figures and keys may also be found in these publications.

**Distribution.** British Columbia; southeastern Ontario to New Brunswick.

**Habitat.** Larvae of *Bryophaenocladius* are terrestrial (Brundin 1956).

## Genus *Camptocladius* Wulp

*Camptocladius* Wulp, 1874:133.

Antenna very reduced, less than one-half as long as mandible, with 3 segments, and with third segment shorter than second; blade longer than combined length of terminal segments; Lauterborn's organs vestigial. Hypostoma with 1 broad median and 4 pairs of lateral teeth. Paralabial plate small, without setae. Mandible with 4 inner teeth; apical tooth shorter than combined width of inner teeth; seta interna absent. Premandible with 3 apical teeth. SI and SII bifid. Anterior parapods reduced, fused, and with an apical crown of fine claws. Procerci, anal setae, anal tubules, and posterior parapods absent.

**Remarks.** The absence of posterior parapods will distinguish the larvae of *Camptocladius* from those of all Orthocladiinae except from a few larvae of *Pseudosmittia*. In these larvae of *Pseudosmittia* without posterior parapods, the terminal antennal segment is longer than the preceding segment. In *Camptocladius* the terminal antennal segment is shorter than the preceding segment. The foregoing description is based on that given by Strenzke (1940).

**Distribution.** Widespread south of treeline.

**Habitat.** Larvae of *Camptocladius* inhabit cow dung.

## Genus *Cardiocladius* Kieffer

Figs. 265–269

*Cardiocladius* Kieffer, 1912:22; Brundin 1956:66.

Antenna 5-segmented, with third and fourth segments subequal in length; blade shorter than combined length of terminal segments. Lauterborn's organs about as long as third segment. Hypostoma strongly sclerotized, with 1 broad median and 5 pairs of lateral teeth. Paralabial plate well developed, without setae. Mandible with 3 inner teeth; apical tooth short, not much longer than first inner tooth; seta interna present. Premandible stout, with 1 blunt or apically notched tooth. SI simple. Pecten epipharyngis consisting of 3 scales. Anterior parapods partly fused, each with apical crown of claws. Procercus short, bearing 2 long thick and 2–5 shorter thinner anal setae, without spur. Two pairs of anal tubules present.

**Remarks.** Larvae of *Cardiocladius*, among genera with a simple SI, are distinctive, in that the premandibles and hypostoma are strongly sclerotized, and the procercus bears two anal setae much longer and thicker than the remaining setae. According to Johannsen (1937*a*) the anterior parapods are fused bearing two crowns of claws. The parapods appear to arise from a common base but are not fused, for example, to the extent found in the larvae of *Smittia*, *Bryophaenocladius*, and *Gymnometriocnemus*. The hypostoma of one of the species described by Saunders (1924) has no definite teeth; this type of larva has not been observed in Canadian material. Larvae are described by Johannsen (1937*a*).

**Distribution.** Yukon Territory; Ontario; Quebec; Newfoundland.

**Habitat.** Larvae of *Cardiocladius* live in swift flowing bodies of water (Brundin 1956).

## Genus *Chaetocladius* Kieffer

*Chaetocladius* Kieffer, 1911*b*:182; Brundin 1956:121.

Antenna 5-segmented, with segments consecutively smaller or with second and third segments subequal in length; blade length variable; Lauterborn's organs distinct, usually about as long as third segment. Hypostoma with 1 or 2 median and 5 pairs of lateral teeth. Paralabial plate distinct, without setae. Mandible with 3 or 4 inner teeth; apical tooth shorter than combined width of inner teeth; seta interna present. Premandible with 1–3 apical teeth. SI plumose. Labral lamellae distinct, pectinate or simple. Pecten epipharyngis consisting of 3 subequal scales. Anterior parapods separated, each with apical crown of claws. Procercus bearing 6 or 7 anal setae without spur. Two pairs of anal tubules present.

**Remarks.** No figures are given for this genus as only one poorly mounted larva definitely associated with an adult *Chaetocladius* was available for study. The foregoing description is based mainly on those provided by Thienemann (1944) and Pankratova (1970). The generic limits are uncertain and the larvae assigned to this genus are quite diverse in structure. Probably all have a pair of labral lamellae located between the bases of the SI. These lamellae are pectinate or simple. Based on this character they could only be confused with some *Metriocnemus*, but in this genus the lamellae appear to be finer pectinate. They are also located above or between the bases of the SI.

**Distribution.** Yukon Territory to District of Franklin (occurs throughout Arctic Archipelago); Ontario to New Brunswick.

**Habitat.** Larvae of *Chaetocladius* inhabit flowing water and cooler bodies of still water (Brundin 1956).

## Genus *Corynoneura* Winnertz

Figs. 270–274

*Corynoneura* Winnertz, 1846:12; Brundin 1956:172; Schlee 1968:1.

Antenna 4-segmented, longer than head, with third (and usually second) segment brown, with second and third segments long, with third slightly longer than second, and with fourth short; blade shorter than second segment; Lauterborn's organs indistinct or absent. Hypostoma triangular, with 2 or 3 median and 5 pairs of lateral teeth; middle median tooth small to absent and outer pair enlarged; first lateral tooth usually smaller than either outer median or second lateral teeth. Paralabial plate vestigial, without setae. Mandible with 4 or 5 inner teeth; apical tooth about same size to smaller than first inner tooth; seta interna present. Premandible with no distinct apical teeth, but serrated apically. SI simple.

Pecten epipharyngis consisting of 5 scales. Second and third thoracic segments partly to completely fused. Anterior parapods separate, each with apical crown of claws. Procercus bearing 4 anal setae, without spur. Two pairs of anal tubules present; pointed, about one-half as long as posterior parapods. Posterior parapod with plumose brown seta arising from basal half.

**Remarks.** The larvae of *Corynoneura* are easily recognized by their narrow elongate head with long straight antennae. In *Heterotanytarsus*, the only other genus of Orthocladiinae with antennae longer than the head capsule, the antennae are strongly curved. These two genera differ in most other characters, especially the shape of the hypostoma.

The larvae of *Corynoneura* and *Thienemanniella* differ primarily in antennal characters (see key, couplet 4). They share a number of characters that distinguish them from other Orthocladiinae, namely, a strong brown seta on the basal half of posterior parapods; pecten epipharyngis with five scales; and their general elongate appearance. The adults are also distinctive and this was reflected in older classifications where they were placed in a separate subfamily, Corynoneurinae (see Goetghebuer 1939*a*). Brundin (1956) placed them with the European genus *Corynoneurella*, in a group (*Corynoneura* group) in the Orthocladiinae, a system followed by Schlee (1968) in his revision of the European species. Larvae are described by Johannsen (1937*a*) and Roback (1957).

**Distribution.** Widespread, occurring north of treeline.

**Habitat.** Larvae of *Corynoneura* live in standing water (Brundin 1956), and are sometimes found on submerged or floating aquatic plants (Johannsen 1937*a*).

## Genus *Cricotopus* Wulp

Figs. 275–292

*Cricotopus* Wulp, 1874:132; Brundin 1956:109; Hirvenoja 1973:131.
*Eucricotopus* Thienemann, 1936:200.
*Isocladius* Kieffer, 1909:44.
*Trichocladius* Kieffer, 1906*b*:356.

Antenna usually 5-segmented, sometimes 4-segmented, with segments usually consecutively smaller; blade usually shorter than combined length of terminal segments; Lauterborn's organs distinct to absent. Hypostoma with 1 median and 5 or 6 pairs of lateral teeth. Paralabial plate small, without setae. Mandible with 3 inner teeth; apical tooth shorter than combined width of inner teeth; seta interna present; inner margin usually smooth, rarely with spines; outer margin often strongly crenulated. Premandible usually with 1, sometimes with 2, apical teeth. SI usually bifid, rarely simple. Pecten epipharyngis consisting of either 3 subequal scales or broad conical plate composed of fused scales. Abdominal segments usually with some plumose setae or tufts of setae, all setae sometimes simple. Anterior parapods separate, each with apical crown of claws. Procercus present, bearing 6 or 7 anal setae, without spur. Two pairs of anal tubules present; tubules usually shorter than posterior parapods.

**Remarks.** The combination of a bifid SI and plumose setae or setal tufts on at least the first six abdominal segments will distinguish most larvae of *Cricotopus*. Those with a simple SI are also easily recognized as they have plumose setae. However, larvae with only simple setae on the abdomen (some *Cricotopus* (*Cricotopus*)) cannot be separated from those of *Orthocladius* (*Orthocladius*) with confidence. Despite the distinctness of the adults of these two genera, no good character or combination of characters has been demonstrated to separate the larvae. One character worth further investigation is the anteromesal extent of the paralabial plates. In *Cricotopus* (*Cricotopus*) it ends between the first and second lateral teeth of the hypostoma, usually at the posterior margin of the groove separating the two teeth, whereas in *Orthocladius* (*Orthocladius*) it ends about the level of the middle of the second lateral tooth. The tip is generally not curved anteriorly but may even curve posteriorly.

Hirvenoja (1973) recognized two subgenera, *Cricotopus* (*Cricotopus*) and *Cricotopus* (*Isocladius*), with a number of species groups within each subgenus. The two subgenera are primarily distinguished by the structure of the pecten epipharyngis. In *Cricotopus* (*Cricotopus*) it consists of three subequal scales. The three scales in *Cricotopus* (*Isocladius*) are fused to form a single conical plate; sometimes the two outer scales are discernible as two small structures appressed to the base of the enlarged median scale. In this subgenus the chaetulae laterales are enlarged and more or less fused basally, one on each side, to the pecten epipharyngis. In preparations where the labrum is not well laid out the pecten epipharyngis may appear to be composed of three broad conical scales because of the enlarged chaetulae laterales.

The following key to subgenera and species groups known to occur in Canada is adapted from Hirvenoja (1973). It is a simplified version as several characters employed by Hirvenoja are not used here. These characters such as parts of the maxilla and prementohypopharyngeal complex are very difficult to see even with high-quality microscopes. The key should serve to separate most larvae of *Cricotopus* into species groups; however, whenever possible Hirvenoja (1973) should be consulted. Also not all the larvae known in Canada will fit into the groups defined by Hirvenoja.

# Key to the subgenera and species groups of Cricotopus

1. Pecten epipharyngis consisting of 3 separate scales (as in Fig. 280). Abdominal segments with or without plumose setae ........................
 ................... *Cricotopus (Cricotopus)* .................... 2
 Pecten epipharyngis consisting of a single conical scale (Fig. 288). Abdominal segments with plumose setae (Fig. 287) ..........................
 ................... *Cricotopus (Isocladius)* .................... 7
2. Median tooth of hypostoma less than two and one-half times as wide as a first lateral tooth (Fig. 275) ........................................... 3
 Median tooth of hypostoma more than three times as wide as a first lateral tooth (Figs. 277–279, 281–282) ................................... 5
3. Premandible with 1 apical tooth (Fig. 276) ............. ***tremulus*** group
 Premandible with 2 apical teeth ................................... 4
4. Abdominal segments I–VIII with plumose setae .......... ***tibialis*** group
 Abdominal segments I–VI with plumose setae or with only simple strong setae ............................................... ***fuscus*** group
5. First and second lateral teeth of hypostoma small and partly fused with median tooth (Fig. 277) ........................... ***trifascia*** group
 All lateral teeth of hypostoma uniform in length or second lateral tooth somewhat smaller than other teeth (when smaller, first and second lateral teeth not fused to median tooth) ............................ 6
6. Plumose setae indistinct, shorter than one-quarter as long as an abdominal segment. Inner margin of mandible with spines (Fig. 283) or smooth
 ................................................ ***bicinctus*** group
 Plumose setae distinct, more than one-quarter as long as an abdominal segment. Inner margin of mandible smooth .......................
 ............................ ***cylindraceus*** group, ***festivellus*** group
7. Premandible with 2 apical teeth (Fig. 286) ............. ***sylvestris*** group
 Premandible with 1 apical tooth .................................. 8
8. SI simple (Fig. 291) .................................. ***obnixus*** group
 SI bifid .......................................................... 9
9. Plumose seta about twice as long as an abdominal segment, with about 10 branches on all abdominal segments ............. ***laricomalis*** group
 Plumose seta about as long as an abdominal segment, with about 20–50 branches on abdominal segments I–VI and about 10 branches on abdominal segment VII ............................................. 10
10. Plumose setae on abdominal segments I–VII with 20–50 branches each ..
 ............................................... ***reversus*** group
 Plumose setae on abdominal segments I–VI with about 20–30 branches each and on segment VII with about 10 branches each .. ***intersectus*** group

# Clé des sous-genres et des groupes d'espèces de Cricotopus

1. Peigne épipharyngéal consistant en 3 écailles séparées (comme à la figure 280). Segments abdominaux avec ou sans soies plumeuses ..........
 ................... *Cricotopus (Cricotopus)* .................... 2
 Peigne épipharyngéal consistant en une seule écaille conique (fig. 288). Segments abdominaux avec soies plumeuses (fig. 287) ................
 ................... *Cricotopus (Isocladius)* .................... 7

2. Dent médiane de l'hypostome d'une largeur inférieure à deux fois et demi celle d'une première dent latérale (fig. 275) .................... 3
   Dent médiane de l'hypostome d'une largeur supérieure à trois fois celle d'une première dent latérale (fig. 277 à 279, 281, 282) ................ 5
3. Prémandibule avec 1 dent apicale (fig. 276) ......... groupe des *tremulus*
   Prémandibule avec 2 dents apicales ............................... 4
4. Segments abdominaux I–VIII avec soies plumeuses .. groupe des *tibialis*
   Segments abdominaux I–VI avec soies plumeuses ou avec seulement des soies simples et fortes .......................... groupe des *fuscus*
5. Premières et deuxièmes dents latérales de l'hypostome petites et partiellement soudées à la dent médiane (fig. 277) ...... groupe des *trifascia*
   Toutes les dents latérales de l'hypostome de longueur uniforme ou deuxième dent latérale légèrement plus petite que les autres (lorsqu'elle est plus petite, première et deuxième dents latérales non soudées à la dent médiane) ...................................................... 6
6. Soies plumeuses indistinctes, d'une longueur inférieure au quart de celle d'un segment abdominal. Bord interne de la mandibule garni d'épines (fig. 283) ou lisse .......................... groupe des *bicinctus*
   Soies plumeuses distinctes, d'une longueur supérieure au quart d'un segment abdominal. Bord interne de la mandibule lisse .....................
   ................... groupe des *cylindraceus*, groupe des *festivellus*
7. Prémandibule avec 2 dents apicales (fig. 286) ...... groupe des *sylvestris*
   Prémandibule avec 1 dent apicale ............................... 8
8. SI simple (fig. 291) ............................. groupe des *obnixus*
   SI bifide ......................................................... 9
9. Soie plumeuse environ deux fois plus longue qu'un segment abdominal, avec environ 10 ramifications sur tous les segments abdominaux .........
   ......................................... groupe des *laricomalis*
   Soie plumeuse à peu près aussi longue qu'un segment abdominal, avec environ de 20 à 50 ramifications sur les segments abdominaux I–VI et environ 10 ramifications sur le segment abdominal VII .......... 10
10. Soies plumeuses sur les segments abdominaux I–VII avec de 20 à 50 ramifications chacune ............................... groupe des *reversus*
    Soies plumeuses sur les segments abdominaux I–VI avec environ de 20 à 30 ramifications chacune et, sur le segment VII, avec environ 10 ramifications chacune .................... groupe des *intersectus*

**Distribution.** Widespread, common north of treeline.

**Habitat.** Larvae of *Cricotopus* inhabit all types of freshwater and occur to a lesser extent in brackish water. A few species live in bottom mud, but most are associated with submerged vegetation, including moss, or live on stones (Hirvenoja 1973). Some mine living plant tissue (Berg 1950; Hirvenoja 1973). Larvae of the subgenus *Isocladius* are generally closely associated with plants in still water, whereas those of the subgenus *Cricotopus* are commoner in flowing water (Hirvenoja 1973).

## Genus *Diplocladius* Kieffer

Figs. 293–296

*Diplocladius* Kieffer, 1908:6; Brundin 1956:70.

Antenna 5-segmented, with segments consecutively smaller, and with fifth segment minute; blade shorter than combined length of terminal segments; Lauterborn's organs small. Hypostoma with 2 median and 6 pairs of lateral teeth, all teeth subequal in size. Paralabial plate elongate, with long setae. Mandible with 4 inner teeth; apical tooth shorter than combined width of inner teeth; seta interna present, with some branches pectinate. Premandible broad, with 2 apical teeth. SI plumose. Labral lamellae broad, pectinate. Pecten epipharyngis consisting of 3 subtriangular scales. Anterior parapods separate, each with apical crown of claws. Procercus usually longer than wide, bearing 6 or 7 anal setae, without spur. Two pairs of ar'al tubules present; tubules shorter than posterior parapods.

**Remarks.** The presence of a thick patch of long setae on each paralabial plate is distinctive among the larvae of the Orthocladiinae. Only larvae of *Synorthocladius* have comparatively long setae on the paralabial plates but the hypostoma is triangular in this genus. Larvae of *Diplocladius* might be confused with larvae of *Prodiamesa* which have equally long dense paralabial plate setae, but the central part of the hypostoma is deeply recessed in the latter. The larvae are described by Johannsen (1937a; as *Spaniotoma (Diplocladius)*).

**Distribution.** Yukon Territory, Northwest Territories, and Manitoba to New Brunswick.

**Habitat.** Larvae of *Diplocladius* inhabit small cool bodies of still water (Brundin 1956) and pools in small flowing waters.

## Genus *Epoicocladius* Zavřel

Figs. 297–299

*Epoicocladius* Zavřel *in* Sulc and Zavřel, 1924:368; Brundin 1956:147.

Antenna 4-segmented, with segments consecutively smaller or with segments 3 and 4 very small; blade longer than combined length of terminal segments; Lauterborn's organs vestigial. Hypostoma truncated, with 6 or 9 teeth across apex and 5 pairs of lateral teeth along sloping sides; lateral and outer pair of apical teeth on apex brown, remainder of apical teeth lighter in color. Paralabial plate small, without setae. Mandible hooked, with 3 inner teeth; apical tooth about as long as combined width of inner teeth; first inner tooth broad, truncated; seta interna present. Premandible with 2 apical teeth. SI simple, with finely pectinate apex. Pecten epipharyngis consisting of at least 7 pointed scales. Anterior parapods separate, each with apical crown of claws. Abdominal segments with numerous long simple setae. Procercus about twice as long as wide, bearing 7 anal setae of unequal length; 2 longest setae about one-quarter as long as body, without spur. One pair of anal tubules present; tubules about one-half to one-third as long as posterior parapods.

**Remarks.** Larvae of *Epoicocladius* are distinctive with long setae on the abdomen, only one pair of anal tubules, and an apically truncated hypostoma. According to Johannsen (1937a) the body viewed from above distinctly tapers posteriorly from the fifth segment. Although the SI has a finely pectinate apex, it is called simple here because in most preparations the serrations are difficult to see and it appears simple. Larvae are described by Sulc and Zavřel (1924), Johannsen (1937a; as *Spaniotoma* (*Smittia*) sp. E, Group *Epoicocladius*), and Roback (1953).

**Distribution.** District of Mackenzie; Manitoba to Quebec.

**Habitat.** Larvae of *Epoicocladius* are phoretic on mayfly nymphs (Ephemeroptera) (Sulc and Zavřel 1924; Johannsen 1937a).

## Genus *Eukiefferiella* Thienemann

Figs. 300–306

*Eukiefferiella* Thienemann, 1926a:325; Brundin 1956:82; Lehmann 1972:347.

Antenna 4- or 5-segmented, with fourth segment subequal to or longer than third, rarely shorter; blade length variable; Lauterborn's organs usually large, sometimes longer than third segment. Hypostoma with 1 or 2 median and 4 or 5 pairs of lateral teeth; median tooth sometimes notched or with apical protuberance; sclerotization of plate not always uniform, usually with longitudinal or oblique lighter streaks. Paralabial plate vestigial, without setae. Mandible with 3 or 4 inner teeth; apical tooth shorter than combined width of inner teeth; seta interna present; inner margin usually with spines, rarely smooth. Premandible with 1 apical tooth, sometimes very broad. SI simple or apically plumose. Pecten epipharyngis consisting of 3 pointed scales. Abdominal segments often with long simple setae. Anterior parapods separate, each with apical crown of claws. Procercus variable in length bearing 7 anal setae, and with posterior margin frequently strongly sclerotized and sometimes with posterior spur. Two pairs of anal tubules present; tubules shorter than posterior parapods.

**Remarks.** The larvae of *Eukiefferiella* are diverse in structure and a revision involving larvae is needed. Most can be distinguished by the presence of spines on the inner margin of the mandible and these can only be confused with some members of the *bicinctus* group of the genus *Cricotopus*. However, the latter have a bifid SI, plumose setae on the abdomen, and six pairs of lateral teeth on the hypostoma. The absence of spines is usually due to wear, although they are lacking in a few species. These can be distinguished by the combinations of long simple abdominal setae and streaked hypostoma with a weakly notched median tooth. Larvae are described by Zavřel (1939), Johannsen (1937a; as *Spaniotoma* (*Eukiefferiella*)), and Sæther (1969).

**Distribution.** Widespread, occurring north of treeline.

**Habitat.** Larvae of *Eukiefferiella* inhabit flowing water of different types (Brundin 1956).

## Genus *Euryhapsis* Oliver

Figs. 307–308

*Euryhapsis* Oliver, 1981:711.

Antenna 4-segmented, with segments consecutively smaller; blade longer than combined length of terminal segments; Lauterborn's organs small. Hypostoma with 1 small median and 6 pairs of lateral teeth; first lateral teeth large; outer lateral teeth enlarged. Paralabial plate vestigial, with very short setae. Mandible with 3 inner teeth; apical tooth shorter than combined length of inner teeth; seta interna present. Premandible broad, weakly notched apically. SI plumose. Labral lamellae elongate, pectinate, almost contiguous medially, situated between bases of SI and tormae. Pecten epipharyngis consisting of 3 broad scales. Anterior parapods separate, each with apical crown of claws. Procercus shorter than wide bearing 6 anal setae, without spur. Two pairs of anal tubules present; tubules shorter than posterior parapods.

**Remarks.** The larvae of *Euryhapsis* and *Brillia* are similar (see remarks under *Brillia*). Except at very high magnifications the area of the paralabial plate bearing setae appears rugose. Because the individual setae are difficult to see, *Euryhapsis* is treated in the key as having no paralabial setae.

**Distribution.** Yukon Territory; District of Mackenzie; Alberta; Manitoba.

**Habitat.** Larvae of *Euryhapsis* inhabit bodies of flowing water of moderate size.

# Genus *Gymnometriocnemus* Goetghebuer

Figs. 309–310

*Gymnometriocnemus* Goetghebuer, 1932:23; Brundin 1956:140.

Antenna 5-segmented, with second segment longer than first, and with fourth segment longer than third; blade about as long as length of terminal segments; Lauterborn's organs distinct. Hypostoma with 2 large median and 4 pairs of smaller lateral teeth; median teeth separated by wide V-shaped notch. Paralabial plates well developed, without setae. Mandible with 3 inner teeth; apical tooth shorter than combined width of inner teeth; seta interna absent. Premandible simple. SI simple. Pecten epipharyngis consisting of 3 blunt scales. Anterior parapods completely fused, apex with short fine spinelike claws. Preanal segment rounded, without procerci, and with 1 anal seta on each side; anal segment retractile, directed ventrally, with anterior parts of divided parapods bearing row of 7 or 8 small claws, posterior part bare. Anal tubules absent.

**Remarks.** The larvae of *Gymnometriocnemus* and *Bryophaenocladius* are similar (see remarks under *Bryophaenocladius*). Larvae are described by Krüger and Thienemann (1941) and Strenzke (1950) and keyed by Thienemann (1944).

**Distribution.** Yukon Territory and British Columbia; Ontario; New Brunswick.

**Habitat.** Larvae of *Gymnometriocnemus* are terrestrial (Brundin 1956).

# Genus *Halocladius* Hirvenoja

*Halocladius* Hirvenoja, 1973:106.

Antenna 5-segmented, with third and fourth segments subequal; blade shorter than combined length of terminal segments; Lauterborn's organs vestigial. Hypostoma convex or sometimes concave, with 1 median and 6 pairs of lateral teeth. Paralabial plate small, with setae. Mandible with 3 inner teeth; apical tooth shorter than combined width of inner teeth; seta interna present. Premandible with 1–4 teeth. SI bifid or plumose. Pecten epipharyngis consisting of 3 pointed scales. Anterior parapods separate, each with apical crown of claws. Abdominal segments with short simple setae. Procercus bearing 5 or 6 anal setae, without spur. Anal tubules absent.

**Remarks.** The foregoing description is based on that of Hirvenoja (1973). He recognized two subgenera, *Halocladius* (*Halocladius*) and *Halocladius* (*Psammocladius*), distinguished primarily by the shape of SI and hypostoma. The SI is bifid and the hypostoma convex in the subgenus *Halocladius* and plumose and concave, respectively, in the subgenus *Psammocladius*.

Adults belonging to *Halocladius (Halocladius)* are known from one locality in Canada. Larvae have not been collected, hence no microphotographs are given and only this subgenus is keyed out. If *Halocladius (Psammocladius)* occurs in Canada, the larvae will probably key to *Chaetocladius* (couplet 41). However, the absence of anal tubules would afford easy separation from *Chaetocladius*.

**Distribution.** Northern Manitoba.

**Habitat.** Larvae of *Halocladius* inhabit saline water, either inland or seacoasts (Hirvenoja 1973).

## Genus *Heleniella* Gowin

Figs. 311–312

*Heleniella* Gowin, 1943:114; Brundin 1956:144.

Antenna 6-segmented, with second segment much shorter than third, and with sixth segment minute; blade about twice length of terminal segments; Lauterborn's organs on apex of third segment and about as long as fourth segment. Hypostoma with 2 broad median and 5 pairs of lateral teeth; lateral teeth narrow, with sides almost straight; fourth lateral tooth shorter than third or fifth lateral tooth. Paralabial plate vestigial, without setae. Mandible with 3 inner teeth; seta interna present. Premandible with 3 teeth. SI plumose. Labral lamellae consisting of 2 small simple lobes. Pecten epipharyngis consisting of 3 broad scales. Anterior parapods separate, each with apical crown of claws. Procercus bearing 6 anal setae of unequal length, without spur. Two pairs of anal tubules present; tubules shorter than length of posterior parapods.

**Remarks.** The antenna of *Heleniella* is unusual in that the second segment is shorter than the third and the Lauterborn's organs are borne on the apex of the third segment. Only some members of *Brillia* have a similar antenna but these have pronounced pectinate labral lamellae. The sixth antennal segment is not always clearly differentiated from the fifth, and these might be confused with *Limnophyes*, which has a similar hypostoma. However, the other antennal characters of *Limnophyes* are distinctly different, except for the "*Limnophyes karelicus*" type (see remarks under *Limnophyes*). Larvae are described in Gowin (1943) and Thienemann (1944).

**Distribution.** Southern British Columbia and Alberta; southern Ontario and Quebec.

**Habitat.** Larvae of *Heleniella* live in flowing water (Brundin 1956).

## Genus *Heterotanytarsus* Spärck

Figs. 313–315

*Heterotanytarsus* Spärck, 1922:92; Brundin 1956:80; Sæther 1975a:259.

Antenna 4-segmented, curved, longer than head, with second segment longer than first; blade shorter than combined length of terminal segments. Lauterborn's organs long, alternate on sides of second segment. Hypostoma with 1 pair of median and 6 pairs of lateral teeth; median and first lateral teeth deeply recessed. Paralabial plate well developed, without setae. Mandible with 3 inner teeth; apical tooth about as long as combined width of inner teeth; seta interna present. Premandible with 4 teeth; apical 3 teeth long and slender. SI plumose. Labral lamella consisting of 1 single lobe. Pecten epipharyngis consisting of 3 fine pointed scales. Anterior prolegs separate, each with apical and preapical groups of claws. Procercus longer than wide, bearing 6–8 anal setae, without spur. Two pairs of anal tubules present; tubules shorter than posterior parapods.

**Remarks.** The deeply emarginated hypostoma and the long curved antennae are distinctive. Only the larvae of *Heterotanytarsus* and *Abiskomyia* have a group of claws on the shaft of the anterior parapod in addition to the usual apical group, but the presence of an apicomesal projection on the antennal tubercle will distinguish *Abiskomyia*. Larvae are described by Thienemann (1941) and Sæther (1975a).

**Distribution.** Southern British Columbia; eastern Ontario.

**Habitat.** Larvae of *Heterotanytarsus* primarily inhabit small lakes, but one species occurs in flowing water (Sæther 1975a).

## Genus *Heterotrissocladius* Spärck

Figs. 316–318

*Heterotrissocladius* Spärck, 1922:94; Brundin 1956:80; Sæther 1975c:1.

Antenna 7-segmented, with third segment much shorter than fourth, and with seventh segment minute; blade length variable; Lauterborn's organs vestigial or absent. Hypostoma with 1 or 2 median and 5 pairs of lateral teeth. Paralabial plates well developed, without setae. Mandible with 3 inner teeth; apical tooth shorter than combined width of inner teeth; seta interna present. Premandible bifid. SI plumose. Labral lamellae consisting of 2 simple lobes. Pecten epipharyngis consisting of 3 weakly sclerotized serrated scales. Anterior parapods separate, each with apical crown of claws. Procercus usually longer than wide, bearing 7 anal setae, without spur. Two pairs of anal tubules present; tubules shorter than posterior prolegs.

**Remarks.** This genus resembles *Hydrobaenus* in the general structure of the head capsule, but differs in antennal and hypostoma characters (see key, couplet 37). Also the scales of the pecten epipharyngis are serrate not smooth as in *Hydrobaenus*.

Sæther (1975c) in a recent revision of *Heterotrissocladius* recognized three species groups: *subpilosus, marcidus,* and *maeaeri.* The *subpilosus* and *marcidus* groups can be separated by the characters given in the following key. Both groups have a hypostoma with two median teeth, which distinguishes them from the *maeaeri* group which has a single median tooth. The last group is not known from Canada. Larvae are described by Sæther (1975c).

## Key to species groups of *Heterotrissocladius*

1. Antennal blade longer than combined length of terminal segments; A.R. greater than 1.5 .................................. **subpilosus** group
   Antennal blade shorter than combined length of terminal segments; A.R. less than 1.3 ........................................... **marcidus** group

## Clé des groupes d'espèces d'*Heterotrissocladius*

1. Lame antennaire plus longue que la longueur combinée des articles terminaux; R.A. supérieur à 1,5 ................ groupe des **subpilosus**
   Lame antennaire plus courte que la longueur combinée des segments terminaux; R.A. inférieur à 1,3 ................. groupe des **marcidus**

**Distribution.** Widespread, occurring north of treeline.

**Habitat.** Larvae of *Heterotrissocladius* generally inhabit cool and large bodies of still water, but some occur in smaller habitats and in flowing water (Sæther 1975c).

## Genus *Hydrobaenus* Fries

Figs. 319–322

*Hydrobaenus* Fries, 1830:177; Sæther 1976:54.
*Trissocladius*; Brundin 1956:75 (in part).

Antenna 6-segmented, with segments consecutively smaller, and with sixth segment minute; blade shorter than combined length of terminal segments; Lauterborn's organs distinct, about as long as third segment. Hypostoma with 2 median and 6 pairs of lateral teeth. Paralabial plate well developed, without setae. Mandible with 3 or 4 inner teeth; apical tooth shorter than combined width of inner teeth; seta interna present. Premandible bifid. SI plumose, with less than 20 branches. Labral lamella consisting of 1 simple triangular scale. Pecten epipharyngis consisting of 3 smooth scales. Anterior parapods separate, each with apical crown of claws. Procercus usually shorter than wide, bearing 7 anal setae, without spur. Two pairs of anal tubules present, tubules shorter to slightly longer than posterior parapods.

**Remarks.** This genus resembles *Heterotrissocladius* (see remarks under *Heterotrissocladius*) and *Oliveridia* in general structure. In addition to the characters given in the key (couplet 37), the uniformly dark hypostoma will distinguish it from *Oliveridia*.

Sæther (1976) placed the North American species assigned to *Trissocladius* in three genera—*Hydrobaenus*, *Oliveridia*, and *Zalutschia*. *Trissocladius* as defined by Sæther (1976) does not occur in North America. In *Hydrobaenus* he recognized five species groups: *conformis*, *distylus*, *lapponicus*, *pilipes*, and *lugubris*. Only the first four groups occur in North America. These groups are not well defined in the larval stage, and their recognition often involves the use of difficult hard-to-see characters not used here. Sæther (1976) should be consulted for definition of the groups as well as for keys and descriptions of larvae of species of *Hydrobaenus* occurring in Canada.

**Distribution.** Widespread, occurring north of treeline.

**Habitat.** Larvae of *Hydrobaenus* inhabit shallow regions of both still and flowing water (Sæther 1976).

### Genus *Krenosmittia* Thienemann & Krüger

Figs. 323–324

*Krenosmittia* Thienemann and Krüger, 1939:253; Brundin 1956:154.

Antenna 5-segmented, with second segment much shorter than third, and with fourth shorter than fifth; blade shorter than combined length of terminal segments; Lauterborn's organ distinct, arising from side of third segment. Hypostoma with 1 median and 6 pairs of narrow lateral teeth; median tooth broad, with medial nipplelike projection. Paralabial plate small, without setae. Mandible with 3 inner teeth; apical tooth longer than combined width of inner teeth; seta interna present. Premandible with 2 apical teeth. SI plumose. Pecten epipharyngis consisting of 3 fine scales. Anterior parapods separate, each with apical group of claws. Procercus bearing 4 anal setae of unequal length, with longest seta about one-quarter to one-half length of body, without spur. Two pairs of anal tubules present; tubules shorter than posterior parapods.

**Remarks.** The larvae of *Krenosmittia* resemble those of *Pseudorthocladius* and *Epoicocladius* in having long anal setae on each procercus. Otherwise they are quite distinct and can easily be separated by the characters given in the key (couplets 31 and 32).

The antenna of *Krenosmittia* is unusual in that a Lauterborn's organ arises from about the midpoint of the third segment, rather than from the apex of the second segment. We are unable to determine if a second organ is present. Also, in some specimens a fine small sixth segment appears to be present, similar to that found in *Hydrobaenus*. Larvae are described by Thienemann and Krüger (1939) and Sæther (1969).

**Distribution.** Yukon Territory; District of Mackenzie; British Columbia; New Brunswick.

**Habitat.** The larvae of *Krenosmittia* inhabit small mountain streams (Sæther 1969).

## Genus *Limnophyes* Eaton

Figs. 325–327

*Limnophyes* Eaton, 1875:60; Brundin 1956:131.

Antenna 5-segmented, with fourth segment usually longer than third, at most subequal; blade length variable; Lauterborn's organs small to absent. Hypostoma with 2 median and 5 pairs of lateral teeth; median teeth broader than first lateral teeth. Paralabial plate indistinct to well developed, without setae. Mandible with 3 inner teeth; apical tooth shorter than combined width of inner teeth; seta interna present. Premandible bifid. SI usually plumose, rarely bifid. Setae on abdomen usually simple, sometimes plumose. Pecten epipharyngis consisting of 3 simple scales, often difficult to distinguish from chaetulae laterales. Anterior parapods separate, each with apical crown of claws. Procercus bearing 7 anal setae, without spur. Two pairs of anal tubules present; tubules variable in length, usually shorter than posterior parapods; apices usually pointed, sometimes blunt.

**Remarks.** *Limnophyes* is a species-rich genus (Brundin 1956), but larvae of very few species have been described and the characteristics of the genus based on larvae are not clearly established. Several larval types have been described by various authors (Thienemann 1921, 1933, 1944; Strenzke 1950; Pankratrova 1970). The *longiseta*-type included in *Limnophyes* by some of these authors belongs to *Paralimnophyes* (Brundin 1956).

The type of larvae commonly collected in Canada is whitish or some shade of purple. Preserved larvae sometimes retain the purple color. The abdomen has simple setae and the apices of the anal tubules are pointed.

Also, some of them have short antennae, less than one-half the length of the mandible and two circular organs in addition to the ring organ on the first segment. These larvae all have a plumose SI, but sometimes the branches are restricted to the apex of the seta, and they can be separated from other larvae of Orthocladiinae with plumose SI by the characters given in the key.

Another larval type, which closely resembles the larvae of *Limnophyes karelicus* (Chernovski) (Pankratova 1970) occurs in the deep zone of the Great Lakes. Except for a bifid SI this larva is similar to that of *Heleniella*. It has an unusually long antennal blade, fourth antennal segment shorter than third, and broad median teeth with sloping shoulders. Placement of this larval type in *Limnophyes* must be confirmed by association with adult specimens.

**Distribution.** Widespread, occurring north of treeline.

**Habitat.** Larvae of *Limnophyes* inhabit the margins and shallow regions of all types of freshwater. Some live in damp terrestrial habitats (Brundin 1956). They are frequently associated with aquatic plants.

## Genus *Metriocnemus* Wulp

Figs. 328–335

*Metriocnemus* Wulp, 1874:136; Brundin 1956:132.

Antenna 5-segmented (sometimes 4-segmented?), with segments consecutively smaller or with fourth segment longer than third segment; sometimes less than one-half as long as mandible; number of terminal segments difficult to discern, but usually 4 present; blade length variable, much longer than combined length of terminal segments in reduced form; Lauterborn's organs often large. Hypostoma with 1 or 2 median and 5 (usually) or 6 pairs of lateral teeth; median and first lateral teeth sometimes recessed. Paralabial plate small, without setae. Mandible with 4 inner teeth; apical tooth shorter than combined width of inner teeth; seta interna present. Premandible with 2–4 apical teeth. Labral lamellae usually consisting of 2 pectinate lobes. Pecten epipharyngis consisting of 3 scales. Anterior parapods separate, each with apical crown of claws. Procercus often longer than wide, bearing 6 or 7 anal setae, sometimes with posterior spur present. Two pairs of anal tubules present; tubules variable in length, but usually shorter than posterior parapods.

**Remarks.** A variety of larval types has been assigned to the genus *Metriocnemus* (see Thienemann 1944; Strenzke 1950; Johannsen 1937a; Zavřel 1941a). The genus is in need of revision.

The larval type commonly encountered in Canada has normal antennae, convex hypostoma with a broad median tooth or a pair of median teeth, and long procerci bearing anal setae of unequal length or all setae of equal length but short. The mandible has four inner teeth. The last character will aid in separating the larvae of *Metriocnemus* from that of *Chaetocladius* with three inner teeth. Larvae are described by Potthast (1915).

**Distribution.** Widespread, occurring north of treeline.

**Habitat.** Larvae of *Metriocnemus* live in a wide variety of habitats ranging from streams, ponds, shallow regions of larger bodies of still water to damp terrestrial and semiaquatic conditions (Strenzke 1950; Brundin 1956). One species lives in the water contained by pitcher plants (Johannsen 1937a).

## Genus *Nanocladius* Kieffer

Figs. 336–339

*Nanocladius* Kieffer, 1913d:31; Sæther 1977a:2.
*Microcricotopus* Thienemann and Harnisch, 1932:137.
*Plecopteracoluthus* Steffan, 1965:1323.

Antenna 4- or 5-segmented, with segments consecutively smaller; blade shorter than combined length of terminal segments; Lauterborn's organs distinct. Hypostoma with 1 broad to weakly double median tooth and 5 or 6 pairs of lateral teeth; each half of median tooth with an apicomedial nipplelike projection and broad shoulders. Paralabial plate large, often long, without setae, but with tuft of setae on small plate lateral to hypostoma in one species. Mandible with 3 inner teeth; apical tooth longer than combined width of inner teeth; seta interna present. Premandible with 1–5 apical teeth. SI simple. Pecten epipharyngis consisting of 3 scales. Anterior parapods separate, each with apical crown of claws. Procercus bearing 5 or 6 anal setae, strongly sclerotized posteriorly, with 2 or 3 small spurs. Two pairs of anal tubules present; tubules variable in length.

**Remarks.** Among larvae of Orthocladiinae possessing a simple SI, the shape of the hypostoma of *Nanocladius* is distinctive. In this character and the long apical tooth of the mandible it resembles some larvae of *Psectrocladius*, but the single SI and bare paralabial plate will distinguish it.

Sæther (1977a) recognized two subgenera, *Nanocladius (Nanocladius)* and *Nanocladius (Plecopteracoluthus)*, which can be separated by the characters given in the key below. One species, *Nanocladius (Plecopteracoluthus) downesi* (Steffan), has a small plate bearing setae located lateral to the hypostoma. In poorly prepared specimens these setae may appear to arise from the region of the paralabial plate. Larvae are described by Steffan (1965) and Sæther (1977a).

## Key to the subgenera of Nanocladius

1.  All anal tubules subequal in length and slightly shorter than posterior parapods. Posterolateral apex of paralabial plate straight (Fig. 337) ......
    .................................. **Nanocladius (Plecopteracoluthus)**
    One pair of anal tubules longer than other pair and usually longer than posterior parapods. Posterolateral apex of paralabial plate usually rounded (Fig. 338), rarely straight ....... **Nanocladius (Nanocladius)**

## Clé des sous-genres de Nanocladius

1.  Tous les tubules anaux à peu près de la même longueur et légèrement plus courts que les parapodes postérieurs. Apex postérolatéral de la plaque paralabiale droit (fig. 337) ........ **Nanocladius (Plecopteracoluthus)**
    Une paire de tubules anaux plus longue que l'autre paire et habituellement plus longue que les parapodes postérieurs. Apex postérolatéral de la plaque paralabiale habituellement arrondi (fig. 338), rarement droit .
    .................................................. **Nanocladius (Nanocladius)**

**Distribution.** Widespread, occurring north of treeline.

**Habitat.** Larvae of *Nanocladius* inhabit shallow regions of still water and small bodies of flowing water (Sæther 1977a). Members of the subgenus *Plecopteracoluthus* live in streams in phoretic association with stonefly nymphs (Plecoptera) (Steffan 1965; Sæther 1977a).

## Genus *Oliveridia* Sæther

Figs. 340–342

*Oliveridia* Sæther, 1980:399.
*Oliveria* Sæther, 1976:48 (nec *Oliveria* Sutherland, 1965:48).

Antenna 6-segmented, with segments consecutively smaller, and with sixth segment very small; blade shorter than combined length of terminal segments; Lauterborn's organs distinct. Hypostoma with 1 median and 6 pairs of lateral teeth; median and first lateral teeth lighter in color than remaining teeth. Paralabial plate relatively well developed, without setae. Mandible with 3 inner teeth; apical tooth shorter than combined width of inner teeth; seta interna present. Premandible bifid. SI finely plumose, with over 30 branches. Labral lamella consisting of 1 single triangular lobe. Pecten epipharyngis consisting of 3 scales. Anterior parapod separate, each with apical crown of claws. Procercus bearing 6 or 7 anal setae, without spur. Two pairs of anal tubules present; tubules shorter than posterior parapods.

**Remarks.** The larvae of this genus are similar to those of *Hydrobaenus*. The main character separating the larvae of the two genera is the number of branches on the SI, i.e., over 30 in *Oliveridia* and less than 20 in *Hydrobaenus* (Sæther 1976). The hypostomal character used in the generic key is also of doubtful status, although it will work for most specimens. It is possible that the hypostoma of the specimens described by Oliver (1976) and Sæther (1976) were worn. Since preparing the description we have seen a few specimens with a weakly incised median tooth, which otherwise assigns them to *Oliveridia*. The incision is very shallow and could easily be worn away.

**Distribution.** District of Franklin.

**Habitat.** Larvae of *Oliveridia* are known only from medium to large arctic lakes (Oliver 1976; Welch 1973; both as *Trissocladius*).

## Genus *Orthocladius* Wulp

Figs. 343–355

*Orthocladius* Wulp, 1874:132; Brundin 1956:103; Soponis 1977:13.
*Rheorthocladius* Thienemann, 1935:205.
*Euorthocladius* Thienemann, 1935:201.
*Eudactylocladius* Thienemann, 1935:206.

Antenna usually 5-segmented, rarely 4-segmented, with segments consecutively smaller or third and fourth subequal; blade usually shorter than combined length of terminal segments; Lauterborn's organs small to distinct. Hypostoma with 1 median and 6, 7, 9, or 10 pairs of lateral teeth. Paralabial plate small, without setae. Mandible with 3 or 4 inner teeth; apical tooth shorter than combined width of inner teeth; seta interna present; inner margin smooth; outer margin smooth or crenulated. Premandible usually simple, rarely bifid. SI bifid. Labral lamellae absent. Pecten epipharyngis consisting of 3 scales. Anterior parapods separate, each with apical crown of claws. Abdominal segments with only simple setae. Procercus bearing 6 or 7 anal setae, without spur. Two pairs of anal tubules present; tubules shorter than posterior parapods.

**Remarks.** The larvae of *Orthocladius* can be distinguished from all other larvae of the Orthocladiinae except some of *Cricotopus (Cricotopus)* by the characters given in the key. See remarks under *Cricotopus* for a discussion of the problem of distinguishing some of the larvae of these two genera.

Brundin (1956) included the subgenera *Orthocladius (Orthocladius)*, *O. (Eudactylocladius)*, *O. (Euorthocladius)*, and *O. (Pogonocladius)*, in the genus *Orthocladius*. This classification was followed by Soponis (1977) in her revision of the Nearctic *Orthocladius (Orthocladius)*. The four subgenera can be distinguished by the characters given in the following key, which is based on that of Soponis. In the subgenus *Orthocladius (Euorthocladius)* Soponis recognized three larval types: Type I is similar to that of *Orthocladius (Orthocladius)* but has a light brown head capsule; Type II has a hypostoma with 7 pairs of lateral teeth and a dark brown head capsule; and Type III also has a light brown head capsule but the mandible has 4 inner teeth. The larval type with 9 or 10 pairs of lateral teeth on the hypostoma probably would be grouped with Type II larvae.

## Key to the subgenera of *Orthocladius*

1. Hypostoma with 7, 9, or 10 pairs of lateral teeth (Figs. 350, 351) ........ .......................... ***Orthocladius* (*Euorthocladius*)** (in part)
   Hypostoma with 6 pairs of lateral teeth (Figs. 343, 348) .............. 2
2. Head capsule yellow ...................... ***Orthocladius* (*Orthocladius*)**
   Head capsule brown ............................................... 3
3. Antenna 4-segmented ................... ***Orthocladius* (*Pogonocladius*)**
   Antenna 5-segmented (Fig. 355) .................................. 4
4. Head capsule dark brown ............... ***Orthocladius* (*Eudactylocladius*)**
   Head capsule light brown ........ ***Orthocladius* (*Euorthocladius*)** (in part)

## Clé des sous-genres de *Orthocladius*

1. Hypostome avec 7, 9 ou 10 paires de dents latérales (fig. 350, 351) ...... ........................ ***Orthocladius (Euorthocladius)*** (en partie)
   Hypostome avec 6 paires de dents latérales (fig. 343, 348) ............ 2
2. Capsule céphalique jaune .................. ***Orthocladius (Orthocladius)***
   Capsule céphalique brune ......................................... 3
3. Antenne à 4 articles ..................... ***Orthocladius (Pogonocladius)***
   Antenne à 5 articles (fig. 355) .................................... 4
4. Capsule céphalique brun foncé .......... ***Orthocladius (Eudactylocladius)***
   Capsule céphalique brun pâle .. ***Orthocladius (Euorthocladius)*** (en partie)

**Distribution.** Widespread, occurring north of treeline.

**Habitat.** Larvae of *Orthocladius* generally inhabit cool bodies of still or flowing water, often associated with aquatic plants (Soponis 1977). Larvae of the subgenus *Euorthocladius* occur in flowing water, subgenus *Pogonocladius* in small bodies of still water, subgenus *Orthocladius* in both still and flowing water, and subgenus *Eudactylocladius* in small bodies of still water including marshes and bogs.

## Genus *Paracladius* Hirvenoja

Figs. 356–361

*Paracladius* Hirvenoja, 1973:91.
*Paratrichocladius* Thienemann, 1942:314, not Santos Abreu 1918:204.

Antenna 5-segmented, with segments consecutively smaller; blade shorter than combined length of terminal segments; Lauterborn's organs distinct, at least two-thirds as long as third segment. Hypostoma with 1 median and 6 pairs of lateral teeth; median tooth broad, dome-shaped, translucent to lightly pigmented, lighter in color than lateral teeth. Paralabial plate well developed, with setae. Mandible with 3 inner teeth; apical tooth as long as or longer than combined width of inner teeth; seta interna present. Premandible with 1 or 2 apical teeth. SI bifid. Labral lamellae absent. Pecten epipharyngis consisting of 3 scales. Anterior parapods separate, each bearing apical crown of claws. Abdominal segments with simple setae or sometimes also with plumose setae. Procercus bearing 7 anal setae, without spur. Two pairs of anal tubules present; tubules shorter than posterior parapods.

**Remarks.** Larvae of *Paracladius* are similar to those of *Cricotopus* (*Cricotopus*) and *Halocladius* but the broad, dome-shaped and light-colored median tooth of the hypostoma is distinctive. According to Hirvenoja (1973) the body setae are simple, but we have seen larvae from the Great Lakes that have plumose setae. Larvae are described by Hirvenoja (1973) and Oliver (1976).

**Distribution.** Districts of Franklin and Mackenzie south to Labrador, Quebec, and southern Ontario.

**Habitat.** Larvae of *Paracladius* inhabit cold lakes and slow reaches of flowing water (Hirvenoja 1973; Welch 1973, as *Cricotopus*; Oliver 1976).

## Genus *Paracricotopus* Thienemann & Harnisch

Figs. 362–365

*Paracricotopus* Thienemann and Harnisch, 1932:136; Brundin 1956:119.

Antenna 5-segmented, with segments consecutively smaller or third segment slightly longer than fourth; blade shorter than combined length of terminal segments; Lauterborn's organs as long as or longer than third segment. Hypostoma with 1 median and 6 pairs of lateral teeth; median tooth broad, and with apical notch or 2 low tubercles. Paralabial plate well developed, with short scattered setae. Mandible with 4 inner teeth; apical tooth longer than combined width of inner teeth; seta interna present. Premandible with 1 weakly bifid apical tooth. SI simple. Pecten epipharyngis consisting of 3 long, pointed, smooth scales. Anterior parapods separate, each with apical crown of claws. Abdominal segments each with 1 long simple seta and several shorter simple setae on each side. Procercus bearing 5 anal setae, with posterior half rugose or with spurs. Two pairs of tubules present; tubules shorter than posterior parapods.

**Remarks.** The larvae of *Paracricotopus* resemble those of some *Psectrocladius* in the shape of the hypostoma when two small nipplelike projections or tubercles are present on the median tooth. Both genera have paralabial plates with setae and posterior spurs or a rugose area on procerci. However, the presence of a pair of long dark setae on each abdominal segment and a simple SI will separate larvae of *Paracricotopus* from those of *Psectrocladius*. Larvae are described by Thienemann and Harnisch (1932) and keyed by Thienemann (1944).

**Distribution.** Districts of Franklin and Mackenzie; southeastern Ontario; southern Quebec.

**Habitat.** Larvae of *Paracricotopus* live in the moss zone of flowing water (Brundin 1956).

## Genus *Parakiefferiella* Thienemann

Figs. 366–369

*Parakiefferiella* Thienemann, 1936:195; Brundin 1956:148.

Antenna 6-segmented, sometimes apparently 4-segmented, with segments usually consecutively smaller; first segment variable, usually shorter than combined length of terminal segments; blade shorter than combined length of terminal segments; Lauterborn's organs vestigial to absent. Hypostoma with 1 median and 6 pairs of lateral teeth; median tooth broad, usually entire, sometimes with apical notch or nipplelike median projection; first lateral teeth sometimes appressed to median tooth giving appearance of a tripartite median tooth. Paralabial plate well developed, without setae. Mandible with 3 inner teeth; apical tooth length variable; seta interna present. Premandible simple, rarely bifid. SI plumose, sometimes bifid. Labral lamellae present. Pecten epipharyngis consisting of 3 scales. Anterior parapods separate, each with apical crown of claws. Procercus bearing 7 anal setae, without spur. Two pairs of anal tubules present; tubules shorter than length of posterior parapods.

**Remarks.** Larvae of *Parakiefferiella* are difficult to characterize as no character or combination of characters will invariably distinguish them from other Orthocladiinae. The type commonly collected in Canada has an A.R. greater than 1, plumose SI, and the apical tooth of the mandible shorter than the combined width of the inner teeth. All apparently have a six-segmented antenna; however, sometimes the sixth segment is minute and easily overlooked. In this case it would key to the couplet with *Chaetocladius* and *Eukiefferiella*. All have a hypostoma with 1 median and 6 pairs of lateral teeth, although sometimes the first lateral teeth are appressed to the median tooth which could be regarded as tripartite (Sæther 1969).

Several other larval types have been assigned to *Parakiefferiella*. One has an A.R. of 1 or less, a bifid SI, and the apical tooth of the mandible longer than the combined width of the inner teeth (Thienemann 1933, 1944). Except for the low A.R. this type could be confused with some *Cricotopus* and *Orthocladius*. Another type described by Wülker (1957) has 5 distinct pairs of lateral teeth on the hypostoma. Also Hamilton (1965) tentatively assigned larvae with a four-segmented antenna to *Parakiefferiella*. One of these larvae has two minute median teeth and no lateral teeth. The first two larval types have not been collected in Canada.

**Distribution.**   Widespread, occurring north of treeline.

**Habitat.**   Larvae of *Parakiefferiella* usually inhabit still water (Brundin 1956), although some live in flowing water (Thienemann 1944).

## Genus *Parametriocnemus* Goetghebuer

Figs. 370–372

*Parametriocnemus* Goetghebuer, 1932:22; Brundin 1956:135.

Antenna 5-segmented, with segments consecutively smaller or with third segment shorter than fourth; blade shorter than or subequal to combined length of terminal segments; Lauterborn's organs large, as long as third segment or as long as third and fourth segments combined. Hypostoma with 2 median and 5 pairs of lateral teeth; median teeth usually broader and higher than first lateral teeth. Paralabial plate well developed, with rounded distolateral apex, without setae. Mandible with 3 inner teeth; apical tooth shorter than combined width of inner teeth; seta interna present. Premandible with 2–6 teeth. SI plumose. Labral lamellae consisting of 2 weak lobes. Pecten epipharyngis consisting of 3 scales. Anterior parapods separate, each with apical crown of claws. Procercus bearing 5–7 anal setae, without spur. Two pairs of anal tubules present; tubules shorter than posterior parapods.

**Remarks.**   The mouth parts of the larvae of *Parametriocnemus* and of some *Paraphaenocladius* are similar. The hypostoma of *Parametriocnemus* has two distinct median teeth, whereas the median tooth of *Paraphaenocladius* is notched medially sometimes giving the appearance of two teeth. The anal end of *Paraphaenocladius* is bent ventrally, which usually affords easy separation of the two genera, but this character is often lost in microscope slide preparations. However, in *Parametriocnemus* the anal tubules are distinctly shorter than the length of the posterior parapods, whereas in *Paraphaenocladius* they are as long as or longer than the posterior parapods. The hypostoma and paralabial plates of *Paraphaenocladius* are also similar to those of some *Hydrobaenus* and *Heterotrissocladius*, but the antenna is either six- or seven-segmented in these two genera. Larvae are described by Zavřel (1941a) and Sæther (1969).

**Distribution.** Yukon Territory; District of Mackenzie; southeastern Ontario to New Brunswick.

**Habitat.** Larvae of *Parametriocnemus* inhabit flowing water (Brundin 1956), usually swift flowing streams.

## Genus *Paraphaenocladius* Thienemann

Figs. 373–376

*Paraphaenocladius* Thienemann, 1926b:9; Brundin 1956:136.

Antenna 5-segmented, with fourth segment usually longer than third; blade length variable; Lauterborn's organs distinct. Hypostoma with 1 median and 5 pairs of lateral teeth; median tooth very broad, sometimes with 1 or 2 weak notches on apex. Paralabial plate well developed, without setae, sometimes a second pair of plates lying medial to main pair. Mandible with 3 or 4 inner teeth; apical tooth shorter than combined width of inner teeth; seta interna present. Premandible with 2 or 3 apical teeth. SI plumose. Labral lamella consisting of 1 single lobe. Pecten epipharyngis consisting of 3 short simple scales. Anterior parapods separate, each with apical crown of claws. Procercus bearing 2–4 anal setae, without spur. Two pairs of anal tubules present; tubules subequal in length to posterior parapods. Anal end bent ventrally.

**Remarks.** Larvae of *Paraphaenocladius* are variable in many mouthpart characters but all have a ventrally bent anal end. This character is best observed in specimens preserved in fluid, as there is often a tendency for the body to straighten during microscope slide preparation. Larvae of *Bryophaenocladius* and *Gymnometriocnemus* also have a bent anal end but both of these genera lack procerci. Larvae are described by Strenzke (1950).

**Distribution.** Widespread, occurring north of treeline.

**Habitat.** Larvae of *Paraphaenocladius* are usually terrestrial (Strenzke 1950; Brundin 1956), but in Canada several species are aquatic inhabiting small bodies of flowing water.

## Genus *Paratrichocladius* Santos Abreu

*Paratrichocladius* Santos Abreu, 1918:204; Hirvenoja 1973:88.
*Syncricotopus* Brundin, 1956:106.

Antenna 5-segmented, with segments consecutively smaller; blade shorter than combined length of terminal segments; Lauterborn's organs distinct. Hypostoma evenly convex, with 1 median and 6 pairs of lateral teeth. Paralabial plate small, without setae. Mandible with 3 inner teeth; apical tooth shorter than combined width of inner teeth; seta interna present. Premandible with 1 or 2 apical teeth. SI bifid. Weak labral lamellae present. Pecten epipharyngis consisting of 3 narrow scales. Anterior parapods separate, each with apical crown of claws. Procercus bearing 5–7 anal setae, without spur. Two pairs of anal tubules present; tubules shorter than posterior parapods.

**Remarks.** The above diagnosis is based on Fittkau (1954; as *Trichocladius*) and Hirvenoja (1973). No Canadian larvae assignable to *Paratrichocladius* as defined by Hirvenoja (1973) are known, although Sæther (1973*a*; as *Syncricotopus*) recorded one species, based on adult specimens, from Canada. Apparently the only character that will distinguish larvae of *Paratrichocladius* from those of *Cricotopus* without abdominal setal tufts and from *Orthocladius* is the shape of the first lateral tooth of the hypostoma. The greatest width is at the middle of the tooth in *Paratrichocladius* and at the base of the tooth in *Cricotopus* and *Orthocladius*. Contrary to that stated by Hirvenoja (1973) the seta interna is present on the mandible.*

Prior to Hirvenoja (1973) larvae assigned to *Paratrichocladius* were based on the description of Thienemann (1942). For this larval type Hirvenoja (1973) erected a new genus (*Paracladius*). In material from the Canadian Northwest we have seen larvae similar to that of *Paratrichocladius triquetra* Chernovskii (see Pankratova 1970). Hirvenoja (1973) is uncertain of the generic placement of this species, but the larva has more of the attributes of *Paracladius* than of *Paratrichocladius*.

**Distribution.** Manitoba.

**Habitat.** Larvae of *Paratrichocladius* inhabit cool lakes, springs, flowing water, and brackish water (Hirvenoja 1973).

## Genus *Phycoidella* Sæther

Figs. 377–379

*Phycoidella* Sæther, 1971*c*:1810.

Antenna 5-segmented, with segments consecutively smaller; blade shorter than combined length of terminal segments; Lauterborn's organs indistinct. Hypostoma with very broad dome-shaped median area and 13 pairs of small sharp lateral teeth. Paralabial plates very large, extending to area beneath lateral teeth of hypostoma and broadly fused medially with 3 teeth in median concavity, without setae. Mandible with 3 inner teeth; apical tooth longer than combined width of inner teeth; seta interna absent. Premandible with 3 teeth. SI simple. Pecten epipharyngis consisting of 3 pointed scales. Anterior parapods separate, each with apical

---
*Peter Cranston: personal communication.

crown of claws. Abdominal segments with setae long and simple. Procercus about twice as long as wide, bearing 7 anal setae, without spur. Two pairs of anal tubules present; tubules subequal in length to posterior parapods.

**Remarks.** The greatly enlarged, medially fused paralabial plates with three small teeth set in a median concavity will distinguish the larvae of *Phycoidella* from those of all other Orthocladiinae. Larvae are described by Sæther (1971c).

**Distribution.** Northwestern Ontario.

**Habitat.** Larvae of *Phycoidella* live in colonies of blue green algae in shallow sheltered regions of lakes (Sæther 1971c).

## Genus *Psectrocladius* Kieffer

Figs. 380–394

*Psectrocladius* Kieffer, 1906b:356; Brundin 1956:116.

Antenna 5-segmented, with segments consecutively smaller; blade shorter than combined length of terminal segments; Lauterborn's organs vestigal to small. Hypostoma with 1 or 2 median and 5 pairs of lateral teeth; median tooth or teeth broad, usually lighter in color than lateral teeth, and often with 1 or more nipplelike median projections. Paralabial plate well developed, with setae. Mandible with 3 inner teeth; apical tooth length variable; seta interna present or absent. Premandible with 1 apical tooth. SI palmate, with 3–9 subequal branches or outer branches smaller. Labral lamellae absent. Pecten epipharyngis consisting of 3 subequal scales. Anterior parapods separate, each with apical crown of claws. Procercus bearing 6 or 7 anal setae, with posterior spur or spurs. Two pairs of anal tubules present; tubules shorter than posterior parapods.

**Remarks.** The palmate SI of the larvae of *Psectrocladius* is unique. In its typical form the branches are arranged like prongs on a fork. Sometimes the branches are not parallel but radiate from a common base, but they are always deeply divided and generally subequal. This character plus procerci with posterior spurs and paralabial plate with setae will separate larvae of *Psectrocladius* from those of other Orthocladiinae.

There are two different larval types, one consisting of the subgenera *Psectrocladius* (*Monopsectrocladius*) and *Psectrocladius* (*Psectrocladius*) and the other of the subgenera *Psectrocladius* (*Allopsectrocladius*) and *Psectrocladius* (*Mesopsectrocladius*) (see following key). Larvae are described by Johannsen (1937a; as *Spaniotoma* (*Psectrocladius*)), Thienemann (1944), Roback (1957), Sæther (1969), and Pankratova (1970).

## Key to the subgenera of *Psectrocladius*

1.  Mandible with seta interna; apical tooth less than 1.75 as long as combined width of inner teeth ........................................... 2
    Mandible without seta interna; apical tooth 1.75 as long as or longer than combined width of inner teeth ................................... 3
2.  Hypostoma with 1 median tooth with 1 median projection ..............
    ............................. ***Psectrocladius (Monopsectrocladius)***
    Hypostoma with 2 median teeth or with 1 median tooth with 2 median nipplelike projections ................ ***Psectrocladius (Psectrocladius)***
3.  SI with 3 or 4 branches .............. ***Psectrocladius (Allopsectrocladius)***
    SI with 6–9 branches ................ ***Psectrocladius (Mesopsectrocladius)***

## Clé des sous-genres de *Psectrocladius*

1.  Mandibule avec soie interne; dent apicale d'une longueur inférieure à 1,75 de la largeur combinée des dents internes .......................... 2
    Mandibule sans soie interne; dent apicale d'une longueur égale ou supérieure à 1,75 de la largeur combinée des dents internes .......... 3
2.  Hypostome avec 1 dent médiane et 1 appendice médian ...............
    ............................. ***Psectrocladius (Monopsectrocladius)***
    Hypostome avec 2 dents médianes ou avec 1 dent médiane portant 2 appendices médians en forme de mamelon . ***Psectrocladius (Psectrocladius)***
3.  SI avec 3 ou 4 ramifications .......... ***Psectrocladius (Allopsectrocladius)***
    SI avec de 6 à 9 ramifications ....... ***Psectrocladius (Mesopsectrocladius)***

**Distribution.** Widespread, occurring north of treeline.

**Habitat.** Larvae of *Psectrocladius* inhabit still water (Brundin 1956).

## Genus *Pseudorthocladius* Goetghebuer

Fig. 395

*Pseudorthocladius* Goetghebuer, 1932:93; Brundin 1956:137.

Antenna 5-segmented, with third segment much shorter than fourth; blade slightly shorter to longer than combined length of terminal segments; Lauterborn's organs distinct, about as long as third segment. Hypostoma with 1 median and 4 pairs of lateral teeth; median tooth broad, truncated, sometimes weakly notched apically. Paralabial plate small, without setae. Mandible with 3 inner teeth; apical tooth shorter than combined width of inner teeth; seta interna present. Premandible broad, with 1 apical tooth; sometimes apical tooth notched. SI plumose. Weak labral lamellae present. Pecten epipharyngis consisting of 3 fine scales. Anterior parapods reduced, fused, with apical crown of claws weakly divided into two groups. Procercus small, bearing 2 anal setae, one as long as one-third length of body, without spur. Two pairs of anal tubules present; tubules subequal to posterior parapods.

**Remarks.** Larvae of *Pseudorthocladius* resemble those of *Krenosmittia* in having a long anal seta on each procercus, but the anterior parapods are separate in the latter. The larvae of *Pseudorthocladius* are the only larvae of Orthocladiinae treated here that have the combination of procerci bearing anal setae and fused anterior parapods. Larvae are described by Thienemann and Krüger (1939) and Strenzke (1950).

**Distribution.** British Columbia; District of Mackenzie; Ontario; New Brunswick.

**Habitat.** Larvae of *Pseudorthocladius* inhabit terrestrial or semiterrestrial habitats and sometimes very shallow water at the margins of still water (Brundin 1956).

## Genus *Pseudosmittia* Goetghebuer

Figs. 396–397

*Pseudosmittia* Goetghebuer, 1932:126; Brundin 1956:165.

Antenna reduced, less than one-half as long as mandible, with 4 (or fewer?) segments, number often difficult to discern, and with terminal segment usually longer than preceding segment; blade long, broad, often three to four times length of first segment; Lauterborn's organs vestigial. Hypostoma with 1 median and 4 pairs of lateral teeth; median tooth sometimes with weak apical notch or projection. Paralabial plate well developed, without setae; second pair of plates if present located median to main pair. Mandible with 3 inner teeth; apical tooth shorter than combined width of inner teeth; seta interna absent. Premandible with 2–4 apical teeth. SI and SII bifid. Labral lamellae absent. Pecten epipharyngis consisting of 3 simple or serrated scales. Anterior parapods reduced, fused, with apical crown of claws. Procerci absent; 1 short anal seta present. Two pairs of anal tubules present, with length variable. Posterior parapods absent or weakly developed, with few claws.

**Remarks.** Larvae of *Pseudosmittia*, *Camptocladius*, *Thalassosmittia*, and *Smittia* are similar. All have fused anterior parapods, a similar type of hypostoma, and no procerci. The anal end is directed posteriorly, not ventrally as in larvae of *Bryophaenocladius* and *Gymnometriocnemus*. However, among larvae without procerci only those of *Pseudosmittia* and *Camptocladius* have a bifid SII. Most larvae of *Pseudosmittia* have posterior parapods, which will distinguish them from those of *Camptocladius*. Larvae of *Pseudosmittia* without posterior parapods can only be separated from those of *Camptocladius* by an antennal character (see remarks under *Camptocladius*). Larvae are described by Strenzke (1950).

**Distribution.** British Columbia; District of Mackenzie; southeastern Ontario; New Brunswick.

**Habitat.** Typically the larvae of *Pseudosmittia* inhabit terrestrial and semiaquatic habitats often associated with wetland areas. Several species are aquatic and live in still water (Thienemann 1954; Brundin 1956).

## Genus *Rheocricotopus* Thienemann & Harnisch

Figs. 398–401

*Rheocricotopus* Thienemann and Harnisch, 1932:135; Brundin 1956:118.

Antenna 5-segmented, with segments usually consecutively smaller, sometimes with fifth segment subequal to or slightly longer than fourth; blade shorter than combined length of terminal segments, Lauterborn's organs usually as long as third segment, sometimes shorter. Hypostoma with 2 median and 5 or 6 pairs of lateral teeth; each median tooth broader than first lateral tooth. Paralabial plate well developed, with setae. Mandible with 3 inner teeth; apical tooth shorter than combined width of inner teeth; seta interna present. Premandible simple. SI bifid. Labral lamellae absent. Pecten epipharyngis consisting of 3 unequal scales. Anterior parapods separate, each with apical crown of claws. Procercus bearing 3 or 5 anal setae, with median posterior spur. Two pairs of anal tubules present; tubules shorter than posterior parapods.

**Remarks.** The larvae of *Rheocricotopus* with bifid SI and paralabial plates bearing setae are similar in these characters to the larvae of *Paracladius* and *Halocladius*. They may be distinguished by the presence of two median teeth on the hypostoma; *Paracladius* and *Halocladius* have a single median tooth. As in some other genera of Orthocladiinae, the number of lateral teeth on the hypostoma varies. In some there are clearly 5 pairs of lateral teeth and in others there are small first lateral teeth appressed to the median tooth in addition to five distinct pairs. Larvae are described by Roback (1957; as *Trichocladius*); Beck and Beck (1964; as *Trichocladius*); and Sæther (1969; 1970).

**Distribution.** Widespread, occurring north of treeline.

**Habitat.** Larvae of *Rheocricotopus* inhabit small bodies of flowing water (Brundin 1956).

## Genus *Smittia* Holmgren

Figs. 402–404

*Smittia* Holmgren, 1869:47; Brundin 1956:146.
*Euphaenocladius* Thienemann, 1934:32.

Antenna 5-segmented, sometimes 4-segmented, with second segment longer than first, and with remaining segments small; blade stout, shorter than combined length of terminal segments; Lauterborn's organs usually indistinct. Hypostoma with 1 median and 5 pairs of lateral teeth; median tooth broad, dome-shaped, weakly tripartite or with small median nipplelike projection. Paralabial plate small, without setae. Mandible with 3 inner teeth; apical tooth shorter than combined width of inner teeth; seta interna present. Premandible with 1–3 teeth. SI plumose; SII simple. Labral lamellae absent. Anterior parapods reduced, fused, with apical groups of claws separate. Procercus and anal setae absent. Two pairs of reduced anal tubules present. Posterior parapods present but reduced.

**Remarks.** Larvae of *Smittia* are similar to those of *Camptocladius*, *Pseudosmittia*, and *Thalassosmittia* (see remarks under *Pseudosmittia*), but can be separated by the characters given in the key (couplets 47–49).

*Smittia* is a species-rich genus (Brundin 1956), and larvae of very few species are known. Among those assigned to *Smittia* there is considerable variation. Some of the larvae described by Strenzke (1950; as *Euphaenocladius*) differ from the type on which the above diagnosis was based, i.e., plumose SII and well developed anal tubules sometimes longer than posterior parapods. We have not seen this larval type in Canadian material, but it would be distinguished from those of *Pseudosmittia* and *Thalassosmittia* by the number of lateral teeth on the hypostoma—five in *Smittia* and four in *Pseudosmittia* and *Thalassosmittia*. Larvae are described by Johannsen (1937a; as *Spaniotoma* (*Smittia*)), Strenkze (1950; as *Euphaenocladius*), and Thienemann (1944; as *Euphaenocladius*).

**Distribution.** Widespread, occurring north of treeline.

**Habitat.** Larvae of *Smittia* are terrestrial (Brundin 1956).

## Genus *Symbiocladius* Kieffer

Figs. 405–407

*Symbiocladius* Kieffer, 1925:565.

Antenna 4-segmented, reduced, about one-quarter as long as mandible; blade shorter than combined length of terminal segments; Lauterborn's organs apparently absent. Hypostoma with median toothless concavity flanked on each side by conical projection and by group of 4 or 5 peglike lateral teeth. Paralabial plates large, without setae. Mandible narrowed apically, with 4 spinelike inner teeth; apical tooth subequal to combined width of inner teeth; fourth inner tooth larger than any of first 3 inner teeth. Premandible small, narrow, and simple. SI simple. Anterior parapods reduced, fused, with apical crown of claws. Procercus absent; tuft of short anal setae present. Anal tubules absent. Posterior parapods reduced, each with apical crown of claws.

**Remarks.** Structurally the larvae of *Symbiocladius* are modified in a number of characters. The antenna, premandible, labral setae, anterior and posterior parapods are reduced, and the proterci and anal tubules are absent. Because of the reduction exhibited in the anal end and the fusion of the anterior parapods this larva keys out with larvae belonging to genera usually inhabiting terrestrial or brackish environments. However, it is easily distinguished by the structure of the hypostoma. The larvae is described by Codreanu (1939).

**Distribution.** District of Mackenzie.

**Habitat.** Larvae of *Symbiocladius* are parasitic on mayfly nymphs (Ephemeroptera) (Codreanu 1939).

## Genus *Synorthocladius* Thienemann

Figs. 408–411

*Synorthocladius* Thienemann, 1935:211; Brundin 1956:91.

Antenna 4- or 5-segmented, with third and fourth segments usually clearly defined, but sometimes with division between these segments consisting of a narrow band of weak sclerotization giving appearance of a 4-segmented antenna; second segment expanded apically; blade length variable; Lauterborn's organs large. Hypostoma with 2 long median and 4 pairs of lateral teeth. Paralabial plate well developed, with group of long setae radiating from common base. Mandible with 3 inner teeth; apical tooth shorter than combined width of inner teeth; seta interna absent. Premandible simple. SI simple. Labral lamellae absent. Pecten epipharyngis consisting of 3 short simple scales. Abdominal segments with 2 long simple setae and 2 plumose setae on each side. Anterior parapods separate, each with apical crown of claws. Procercus bearing 5 anal setae, without spur. Two pairs of anal tubules present; tubules subequal to or slightly longer than posterior parapods.

**Remarks.** The larvae of *Synorthocladius* are distinctive in several respects, notably in the combination of both well developed plumose and simple setae on abdominal segments, expanded apex of second antennal segment, and the arrangement of setae on paralabial plates. These setae are long and relatively thick, and radiate out from a small area; they have been referred to as "star-shaped." The larvae of *Parorthocladius*, which also have "star-shaped" setae on the paralabial plate, are distinguished from those of *Synorthocladius* by having three teeth on the apex of the hypostoma. *Parorthocladius* has not been reported from Canada. Thienemann (1944) reports that the antenna is four-segmented, but all Canadian specimens have five distinct segments. Larvae are described by Johannsen (1937a; as *Spaniotoma* (*Dactylocladius*), group *Synorthocladius*).

**Distribution.** Yukon Territory; District of Mackenzie; British Columbia; Alberta; southeastern Ontario to New Brunswick.

**Habitat.** Larvae of *Synorthocladius* inhabit flowing water (Brundin 1956).

## Genus *Thalassosmittia* Strenzke & Remmert

Figs. 412–414

*Thalassosmittia* Strenzke and Remmert, 1957:270.
*Saunderia* Sublette, 1967:318.

Antenna 5-segmented, reduced, about one-third as long as mandible, with fourth segment longer than third; blade about as long as or slightly shorter than combined length of terminal segments; Lauterborn's organs small. Hypostoma with 1 median and 4 pairs of lateral teeth; median tooth broad, truncated to dome-shaped. Paralabial plate small, without setae. Mandible with 3 inner teeth; apical tooth shorter than combined width of inner teeth; seta interna present. Premandible simple. SI plumose. Labral lamellae consisting of 2 pectinate or simple lobes. Pecten epipharyngis consisting of 3 smooth blunt scales. Anterior parapods partly fused, each part with apical crown of claws. Procercus absent; several short anal setae present. Anal tubules absent. Posterior parapods normal.

**Remarks.** Larvae of *Thalassosmittia* are similar to *Pseudosmittia* and *Smittia* (see remarks under these genera). Larvae are described by Saunders (1928*b*; as *Camptocladius*), and Strenzke and Remmert (1957).

**Distribution.** Coastal British Columbia.

**Habitat.** Larvae of *Thalassosmittia* inhabit intertidal rock pools (Saunders 1928*b*; Strenzke and Remmert 1957).

## Genus *Thienemanniella* Kieffer

Figs. 415–419

*Thienemanniella* Kieffer, 1911*b*:201; Brundin 1956:171; Schlee 1968:10.

Antenna 5-segmented, about one-half as long as head, with second and sometimes with first segment darkened; second and third segments long, subequal in length, with fourth and fifth short; blade shorter than combined length of second segment; Lauterborn's organs small. Hypostoma triangular, with 3 median teeth and 5 pairs of lateral teeth; middle median tooth variable in length but usually as long as or longer than outer pair. Paralabial plate small, without setae. Mandible with 4 inner teeth; sometimes apical and first inner teeth subequal and both distinctly larger than remaining inner teeth; seta interna present. Premandible simple. SI simple. Labral lamellae absent. Pecten epipharyngis consisting of 5 narrow scales. Second and third thoracic segments partly to completely fused. Anterior parapods separate, each bearing apical crown of claws. Procercus bearing 5 anal setae, without spur. Two pairs of anal tubules present; tubules shorter than posterior parapods. Seta arising from basal half of posterior parapod simple, brown, and thickened.

**Remarks.** Larvae of *Thienemanniella* and *Corynoneura* are similar (see remarks under *Corynoneura*). Larvae are described by Lenz (1939) and by Roback (1957; as *Corynoneura* (*Thienemanniella*)).

**Distribution.** British Columbia; District of Mackenzie; Manitoba to Quebec.

**Habitat.** Larvae of *Thienemanniella* inhabit flowing water (Brundin 1956).

## Genus *Zalutschia* Lipina

Figs. 420–424

*Zalutschia* Lipina, 1939:106; Sæther 1976:173.
*Trissocladius* Kieffer, 1908:3 (in part).

Antenna 6-segmented, with segments consecutively smaller or third and fourth segments subequal, and with sixth segment minute; blade shorter than combined length of terminal segments; Lauterborn's organs distinct. Hypostoma with 2 median and 6 pairs of lateral teeth; median teeth sometimes light colored and/or with small tooth between them; first lateral teeth small. Paralabial plate well developed, with short setae. Mandible with 3 inner teeth; apical tooth shorter than combined width of inner teeth; seta interna present, some branches pectinate. Premandible with 2 apical teeth. SI plumose. Labral lamella consisting of 1 simple triangular lobe. Pecten epipharyngis consisting of 3 simple scales. Anterior parapods separate, each with apical crown of claws. Procercus bearing anal setae, without spur. Two pairs of anal tubules present; tubules as long as to much longer than posterior parapods.

**Remarks.** The larvae of *Zalutschia* and *Diplocladius* are the only Orthocladiinae larvae occurring in Canada with plumose SI and setae on the paralabial plates. The paralabial plate of *Diplocladius* is elongate and the setae are relatively long and distinct, whereas the setae in *Zalutschia* are often difficult to see. Larvae are described by Sæther (1976).

**Distribution.** Northwest Territories; British Columbia; Manitoba to New Brunswick; occurs north of treeline.

**Habitat.** Larvae of *Zalutschia* primarily live in lakes, but also are found in smaller bodies of still water and occasionally in flowing water (Sæther 1976).

## Subfamily Prodiamesinae

Figs. 15, 454–460

One or two eye spots present, if 2, then dorsal eye spot lying posterior to ventral eye spot. Antenna 4-segmented, with segments consecutively smaller, and with third and fourth segments much smaller than either first or second segment and with third segment non-annulate; ring organ on basal third of first segment; blade arising from apex of first segment, with length variable; Lauterborn's organs on apex of second segment. Hypostoma with 13–18 teeth. Paralabial plates long, without striations, with setae. Mandible with 1 apical tooth and 3 or 4 inner teeth; seta subdentalis and seta interna present; apicodorsal tooth and pecten mandibularis absent. Premandible simple or bifid. SI plumose or simple, with weak inner serrations; SII simple. Labral lamellae consisting of either 2 pectinate lobes or 2 tufts of setae. Pecten epipharyngis consisting of 3 smooth scales or vestigial. Maxillary palpus about twice as long as wide. Anterior parapods well developed, separate, and each with apical crown of claws. Abdomen with simple setae. Proverci well developed, each bearing 7 or 8 anal setae. Two pairs of anal tubules present; tubules shorter than posterior parapods. Posterior parapods well developed, each with apical crown of claws.

**Remarks.** In general, the larvae of Prodiamesinae resemble those of the Orthocladiinae but only the larvae of *Diplocladius* are likely to be mistaken for Prodiamesinae. Larvae of this genus have large paralabial plates with strong setae but they have a five-segmented antennae. Also, the hypostoma is unlike those of the Prodiamesinae (compare Fig. 287 with the hypostomal figures in this section). The larvae of the Prodiamesinae also resemble those of the Diamesinae, but can be distinguished by the combination of paralabial plates with setae and a non-annulated third antennal segment. *Protanypus*, the only genus of Diamesinae without an annulated third segment, lacks setae on the paralabial plates.

Three genera, *Monodiamesa*, *Odontomesa*, and *Prodiamesa*, are placed in the subfamily Prodiamesinae by Sæther (1977*b*). These genera have been previously placed in the Diamesinae (Goetghebuer 1939*b*; Pagast 1947; Sublette and Sublette 1965) or in the Orthocladiinae (Brundin 1956).

**Distribution.** The Prodiamesinae are reported only from the northern hemisphere and southern South America.

**Biology.** Larvae of Prodiamesinae are generally cool-adapted, living in both still and flowing water. They are often associated with sandy substrates but occur in mud. They are probably free-living, but Zavřel (1926) reported the larvae of *Odontomesa* living in small fine tubes. However, Shilova (1966) observed that only a few larvae in the laboratory built poorly constructed tubes. She also observed that they fed on diatoms and green or blue green algae by filtration.

## Key to the genera of Prodiamesinae

1. Mandible circular, with seta interna composed of row of 30–40 setae (Fig. 458). Hypostoma with convex pale median tooth (Fig. 457) .........
   ................................... *Odontomesa* **Pagast** (p. 150)
   Mandible not circular, with seta interna composed of a tuft of about 7 setae (Fig. 455). Hypostoma unicolorous with recessed median teeth or with median pair separated by U-shaped depression (Figs. 454, 459) ... 2
2. Apical tooth of mandible much longer than combined width of inner teeth. Hypostoma with median pair of teeth longer than first laterals (Fig. 454)
   ................................... *Monodiamesa* **Kieffer** (p. 149)
   Apical tooth of mandible shorter than combined width of inner teeth. Hypostoma with recessed median teeth (Fig. 459) ........................
   ................................... *Prodiamesa* **Kieffer** (p. 150)

## Clé des genres de Prodiamesinæ

1. Mandibule circulaire avec soie interne composée d'une rangée de 30 à 40 soies (fig. 458). Hypostome avec dent médiane convexe et pâle (fig. 457) ............................. *Odontomesa* Pagast (p. 150)
   Mandibule non circulaire, avec soie interne composée d'une touffe d'environ 7 soies (fig. 455). Hypostome unicolore avec dents médianes renfoncées ou avec une paire de dents médianes séparées par une dépression en forme de U (fig. 454, 459) ..................................... 2
2. Dent apicale de la mandibule beaucoup plus longue que la largeur combinée des dents internes. Hypostome avec paire médiane de dents plus longues que les premières dents latérales (fig. 454) ........................
   ................................... *Monodiamesa* **Kieffer** (p. 149)
   Dent apicale de la mandibule plus courte que la largeur combinée des dents internes. Hypostome avec dents médianes renfoncées (fig. 459) .....
   ................................... *Prodiamesa* **Kieffer** (p. 150)

## Genus *Monodiamesa* Kieffer

Figs. 454–456

*Monodiamesa* Kieffer, 1921*c*:287; Sæther 1973*b*:665.

Antenna with blade longer than combined length of terminal segments. Hypostoma unicolorous, with 2 median and 6 pairs of lateral teeth, and with median teeth broad, shallowly separated by shallow U-shaped notch. Paralabial plate with short fine setae. Mandible with apical tooth longer than combined width of 2 inner teeth; seta interna composed of a tuft of about 7 setae. Premandible simple. SI plumose. Labral lamellae pectinate. Pecten epipharyngis consisting of 3 smooth scales.

**Remarks.** The hypostoma with a pair of low median teeth separated by a shallow concavity distinguishes the larvae of *Monodiamesa* from those of the other Prodiamesinae. Also, the setae on the paralabial plate are short and sometimes difficult to see. In the other two genera the paralabial setae are usually long and easy to see. Larvae are described by Johannsen (1937*a*) and Sæther (1973*b*).

**Distribution.** Southern British Columbia; District of Mackenzie; Manitoba to southern Ontario.

**Habitat.** Larvae of *Monodiamesa* inhabit large bodies of still water and occasionally slow flowing waters (Sæther 1973*b*).

## Genus *Odontomesa* Pagast

Figs. 457–459

*Odontomesa* Pagast, 1947:502.

Antenna with blade shorter than combined length of terminal segments. Hypostoma with 1 broad, light-colored, dome-shaped median tooth and 7 pairs of lateral teeth. Paralabial plate with strong setae. Mandible circular in outline, with 3 or 4 inner teeth, and with first inner tooth broader than apical tooth and remaining inner teeth; seta interna consisting of a row of 30–40 setae. Premandible bifid, with teeth long, narrow, and widely separated. SI mounted on elongate socle, simple, with weak serrations along apical half of inner margin. Labral lamellae consisting of 2 tufts of setae. Pecten epipharyngis vestigial.

**Remarks.** The larvae of *Odontomesa* with circular mandibles are distinctive. Larvae are described by Shilova (1966).

**Distribution.** Southern Alberta; southeastern Ontario.

**Habitat.** Larvae of *Odontomesa* inhabit small, slow flowing water (Brundin 1956; Shilova 1966).

## Genus *Prodiamesa* Kieffer

Figs. 15, 459–460

*Prodiamesa* Kieffer, 1906*c*:37.

Antenna with blade longer than combined length of terminal segments. Hypostoma with 2 median and 7 pairs of lateral teeth, and with median and first lateral teeth recessed. Paralabial plate usually with strong setae, sometimes setae weak. Mandible with apical tooth much shorter than combined width of 4 inner teeth; seta interna consisting of tuft of about 7 setae. Premandible bifid. SI plumose. Labral lamellae pectinate. Pecten epipharyngis consisting of 3 smooth scales.

**Remarks.** The hypostoma with recessed median and first lateral teeth distinguishes *Prodiamesa* from the other Prodiamesinae. Larvae are described by Johannsen (1937*a*) and Pankratova (1970).

**Distribution.** Yukon Territory; southeastern Ontario; southern Quebec.

**Habitat.** Larvae of *Prodiamesa* inhabit still and running water but unlike *Monodiamesa* they are commoner in flowing water (Brundin 1956).

## Subfamily Diamesinae

Figs. 9, 10, 425–453

Body usually brownish or whitish; head capsule brownish. Usually 1 eye spot present, if 2 present, then dorsal eye spot lying posterior to ventral eye spot. Antenna nonretractile, mounted directly on head capsule or on low antennal tubercle, shorter than length of head capsule, 4- or 5-segmented; ring organ on basal half of first antennal segment; blade arising from apex of first segment; Lauterborn's organs on apex of second segment; third segment usually annulate. Hypostoma usually with teeth, rarely without teeth; anterior margin usually convex, rarely straight. Paralabial plate vestigial to large, without striations or setae, and with anterior margin smooth. Mandible with 1 apical and 4 or 5 inner teeth; apicodorsal tooth and pecten mandibularis absent; seta subdentalis present; seta interna present or absent. Premandible with 1–16 teeth. SI and SII simple. Labral lamellae usually present. Pecten epipharyngis consisting of 3, 5, or 7 scales. Maxillary palpus short. Prementohypopharyngeal complex usually with 3 tufts of long setae, sometimes with only scales or modified setae. Abdominal setae simple. One pair of procerci usually present, and bearing 4–8 anal setae. Two pairs of anal tubules present; usually 1 tubule directed distoventrally between posterior parapods. Anterior and posterior parapods well developed, usually separate, and each bearing apical crown of claws.

**Remarks.** The combination of an annulated third antennal segment plus the prementohypopharyngeal complex bearing long setae will distinguish most Diamesinae from other chironomids. Only those of *Protanypus* lack these characters but the head capsule with numerous setae and a pair of long posteriorly directed projections on the occipital margin make these larvae easy to recognize.

Edwards (1929), Goetghebuer (1939b), and Johannsen (1937a) recognized the subfamily Diamesinae, but Thienemann (1944), Pagast (1947), and Brundin (1956) considered the species to be part of the subfamily Orthocladiinae. Later Brundin (1966) and Serra-Tosio (1971) reestablished the subfamily and recognized seven tribes comprising three groups: Heptagyini, Lobodiamesini, and Boreoheptagyini comprising one group, Harrisonini, Diamesini, and Protanypodini a second, and Prodiamesini a third. Subsequently Sæther (1977b) gave the Prodiamesini subfamily status.

Only the Diamesini, Protanypodini, and Boreoheptagyini occur in Canada. The remaining tribes are restricted to the southern hemisphere (Brundin 1966). *Pagastia* is the only genus occurring in Canada with undescribed larvae. The occurrence of *Sympotthastia* in Canada is problematical and the generic limits of the larva are unclear. According to Pagast (1947), the pupa of *Diamesa (Psilodiamesa) fulva* Johannsen (1937a) belongs to *Sympotthastia*, but Sæther (1970) believes the larva of *fulva* to be "nearly identical" to *Potthastia gaedii* (Meigen).

**Distribution.** The subfamily Diamesinae is worldwide in distribution, occurring in the cooler parts of all major geographical regions except Antarctica. It occurs throughout Canada, and only the genus *Potthastia* has not been recorded from the arctic region, but it occurs close to treeline in the Northwest. As in the Orthocladiinae, there is a reduction in the number of species from north to south and from cool to warm habitats.

**Biology.** Larvae of Diamesinae are primarily cool-adapted, living in flowing water. They are a major component of the bottom fauna of torrential streams, especially those originating from glacial melt-water. Some also occur in cool lakes and in seepage zones. A number of species are adapted to completing larval development below 4°C.

Most Diamesinae are apparently free-living in that few build cases of silk but anchor themselves by attaching their posterior parapod claws to a glob of silk deposited on substrate. Others that live in crevices and small spaces between parts of substrate do build loose cocoons (Serra-Tosio 1972). Silk is also used in locomotion, although most species appear capable of swimming. Foreward movement is achieved by anchoring the anterior parapod claws to silk attached to substrate, after which the posterior parapod claws are anchored to the silk behind the anterior parapods. Anchored by only the posterior parapods the body is extended forward reanchoring the anterior parapods in another bit of silk (Serra-Tosio 1972). By repeating this procedure a larva can change its position and not be swept away by strong currents.

Most larvae of Diamesini feed on algae, primarily diatoms. The larvae of *Protanypus*, the *longimanus* type of *Potthastia*, and *Pseudodiamesa* are predaceous (Serra-Tosio 1972), but those of *Pseudodiamesa* are also algal eaters and scavengers, feeding on carrion such as dead fish.

## Key to the tribes of Diamesinae

1. Hypostoma without teeth (Fig. 432) or with only extreme lateral teeth (Fig. 446). Mandible without seta interna ............................ 2
   Hypostoma toothed (Figs. 425, 438). Mandible with seta interna ...... 3
2. Hypostoma with extreme lateral teeth (Fig. 446). Occipital margin of head capsule with a posteriorly directed ventrolateral projection on each side (Fig. 8) .......... Tribe **Protanypodini**; *Protanypus* **Kieffer** (p. 158)
   Hypostoma without teeth (Fig. 432). Occipital margin of head capsule without or with weak posteriorly directed ventrolateral projections ...... ................................. Tribe **Diamesini** (in part) (p. 154)
3. Head capsule with prominent dorsal tubercles (Fig. 450). Body with multipointed spines (Fig. 453) ........................................... .......... Tribe **Boreoheptagyini**; *Boreoheptagyia* **Brundin** (p. 153)
   Head capsule without tubercles. Body without multipointed spines ...... .............................. Tribe **Diamesini** (in part) (p. 154)

## Clé des tribus de Diamesinæ

1. Hypostome sans dents (fig. 432) ou avec des dents latérales à l'extrémité seulement (fig. 446). Mandibule sans soie interne ................ 2
   Hypostome denté (fig. 425, 438). Mandibule avec soie interne ........ 3
2. Hypostome avec dents latérales à l'extrémité (fig. 446). Bord occipital de la capsule céphalique avec, de chaque côté, un appendice ventrolatéral dirigé vers l'arrière (fig. 8) ........................................ ............... tribu des **Protanypodini**; *Protanypus* **Kieffer** (p. 158)
   Hypostome sans dents (fig. 432). Bord occipital de la capsule céphalique avec ou sans appendices ventrolatéraux faibles dirigés vers l'arrière ...... ........................... tribu des **Diamesini** (en partie) (p. 154)
3. Capsule céphalique avec tubercules dorsaux proéminents (fig. 450). Corps avec épines à pointes multiples (fig. 453) ........................... ....... tribu des **Boreoheptagyini**; *Boreoheptagyia* **Brundin** (p. 153)
   Capsule céphalique sans tubercules. Corps sans épines à pointes multiples ........................... tribu des **Diamesini** (en partie) (p. 154)

## Tribe Boreoheptagyini

Figs. 448–453

This tribe consists of the single genus *Boreoheptagyia*, and the description given for the genus is sufficient to define the characters of the tribe.

## Genus *Boreoheptagyia* Brundin

Figs. 448–453

*Boreoheptagyia* Brundin, 1966:420.

Head capsule with normal number of setae, with elongate dorsal tubercles, and with occipital margin without strongly developed, posteriorly directed ventrolateral projections. Antenna 4-segmented, with third segment annulate and longer than second segment; blade shorter than combined length of terminal segments. Lauterborn's organs distinct. Hypostoma with 1 or 2 median and 5–7 pairs of lateral teeth. Paralabial plate distinct. Mandible with 4 inner teeth; seta interna present. Labral lamellae consisting of 2 broad pectinate lobes. Pecten epipharyngis consisting of 3 blunt scales. Prementohypopharyngeal complex with 3 groups of long setae. Procercus very short to absent, if present, then on lateral margin of preanal segment, bearing 6–8 short anal setae. Anterior parapods fused, apically divided into 2 lobes, each lobe bearing circle of claws. Posterior parapods partly fused basally, each bearing circle of claws. Body dorsally with areas of multipointed spines; abdominal segments each sometimes with 2 pairs of dorsolateral tubercles.

**Remarks.** The combination of the head capsule with dorsal tubercles and the body with areas of multipointed spines is distinctive. Larvae are described by Saunders (1928a; 1930; as *Diamesa* and *Heptagyia*).

**Distribution.** British Columbia and Yukon Territory.

**Habitat.** Larvae of *Boreoheptagyia* inhabit cool mountain or glacial-fed streams, clinging to rocks and boulders at the splash line (Saunders 1928a, 1930).

## Tribe Diamesini

Figs. 10, 425–445

Head capsule with normal number of setae, without dorsal tubercles; occipital margin without posteriorly directed ventrolateral projections, or if present, then very weak. Antenna 5-segmented; Lauterborn's organs vestigial to small; third segment annulate. Hypostoma with or without teeth. Mandible usually with seta interna. Premandible curved medially. Labral lamellae usually present. Pecten epipharyngis consisting of 3, 5, or 7 scales. Prementohypopharyngeal complex with 3 groups of long setae. Abdominal segments usually with thin light-colored setae, sometimes with strong dark setae. Procercus usually present. Anterior and posterior parapods separate, each bearing apical group of irregularly placed claws.

**Remarks.** The presence of long setae on the prementohypopharyngeal complex and an annulated third antennal segment will distinguish the larvae of Diamesini from those of the Protanypodini. Besides the Diamesinae, only the Telmatogetoninae, a marine group of chironomids, have a prementohypopharyngeal complex bearing long setae, but this group lacks an annulated third antennal segment.

## Key to the genera of Diamesini

1. Hypostoma without teeth (Fig. 432) .. ***Potthastia* Kieffer**(in part) (p. 156)
   Hypostoma with teeth ............................................. 2
2. Abdominal segments with strong dark setae (Fig. 445) .................
   ........................................ ***Pseudokiefferiella* Zavřel** (p. 157)
   Abdominal segments with thin light-colored setae .................... 3
3. Premandible with less than 3 teeth (Fig. 437). Pecten epipharyngis consisting of 3 scales .................... ***Potthastia* Kieffer** (in part) (p. 156)
   Premandible with 5 or more teeth (Fig. 430). Pecten epipharyngis consisting of 5 or 7 scales ............................................... 4
4. Hypostoma with median tooth triangular, distinctly separated from first laterals by deep V-shaped notches (Fig. 438) ......................
   ................................... ***Pseudodiamesa* Goetghebuer** (p. 157)
   Hypostoma with median tooth rounded, truncated, or notched, not distinctly separated from first laterals (Figs. 425–427) ......................
   ........................................ ***Diamesa* Meigen** (p. 155)

## Clé des genres de Diamesini

1. Hypostome sans dents (fig. 432) .. ***Potthastia* Kieffer** (en partie) (p. 156)
   Hypostome avec dents ............................................. 2
2. Segments abdominaux avec soies foncées et fortes (fig. 445) ............
   ........................................ ***Pseudokiefferiella* Zavřel** (p. 157)
   Segments abdominaux avec soies fines et pâles ...................... 3
3. Prémandibule avec moins de 3 dents (fig. 437). Peigne épipharyngéal consistant en 3 écailles .......... ***Potthastia* Kieffer** (en partie) (p. 156)
   Prémandibule avec 5 dents ou plus (fig. 430). Peigne épipharyngéal consistant en 5 ou 7 écailles ........................................ 4
4. Hypostome avec dent médiane triangulaire, nettement séparée des premières dents latérales par de profondes entailles en forme de V (fig. 438)
   .................................. ***Pseudodiamesa* Gœtghebuer** (p. 157)
   Hypostome avec dent médiane arrondie, tronquée ou entaillée, non nettement séparée des premières dents latérales (fig. 425 à 427) .........
   ........................................ ***Diamesa* Meigen** (p. 155)

## Genus *Diamesa* Meigen

Figs. 425–431

*Diamesa* Meigen *in* Waltl, 1835:66; Pagast 1947:462; Hansen and Cook 1976:1.

Antennal segments consecutively smaller; blade shorter than length of terminal segments. Hypostoma with 1 or 2 median and 6–10 pairs of lateral teeth. Paralabial plate small. Mandible with 4 inner teeth, with first lateral tooth often as long as apical tooth; seta interna present. Premandible with 5–8 apical teeth. Labral lamellae numerous, consisting of narrow elongate scales with pointed or serrated apices. Pecten epipharyngis consisting of 5 contiguous or overlapping scales. Abdominal segments with thin light-colored setae. Procercus usually reduced, sometimes longer than wide, often absent, bearing 4(usually)–7 anal setae.

**Remarks.** The combination of five scales comprising the pecten epipharyngis, multiple labral lamellae, a reduced to absent procercus, and usually only four anal setae will distinguish most larvae of *Diamesa*. Those larvae with an elongate procercus bearing 5–7 anal setae can be separated from *Pseudokiefferiella* by having pale abdominal setae. Larvae are described by Johannsen (1937a), Roback (1957), Pankratova (1970), and Sæther (1968, 1970).

**Distribution.** Widespread, occurring north of treeline.

**Habitat.** Larvae of *Diamesa* inhabit flowing water (Brundin 1956; Serra-Tosio 1972).

## Genus *Potthastia* Kieffer

Figs. 432–437

*Potthastia* Kieffer, 1922b:362; Pagast 1947:489; Serra-Tosio 1971:281.

Second antennal segment usually longer than combined length of third to fifth segments, sometimes longer than first segment; blade length variable. Hypostoma without teeth or with 1 broad, light-colored median and 7 pairs of lateral teeth. Paralabial plate vestigial or large. Mandible with 3 minute inner teeth, with hooked projection distal to seta subdentalis, and without seta interna, or of normal shape with 4 inner teeth and a seta interna. Premandible broad, with 11–16 teeth or narrow, with 1–3 teeth. Labral lamellae consisting of pair of bifid or trifid lobes or pair of transparent lobes. Pecten epipharyngis consisting of 3 scales. Abdominal segments with thin, light-colored setae. Procercus at most as long as wide, bearing 5–7 anal setae.

**Remarks.** There are two distinct larval types. One, the *longimana* type, has a toothless hypostoma, a mandible with a hooked process distal to the seta subdentalis, and a premandible with 11–16 teeth. The other, the *gaedii* type, has a toothed hypostoma, a normal mandible, and a premandible with less than three teeth. Larvae are described by Pagast (1933; as *Diamesa*) and Pankratova (1970).

**Distribution.** Yukon Territory; District of Mackenzie; Ontario to Newfoundland.

**Habitat.** Larvae of the *longimana* type inhabit still and flowing water and that of the *gaedii* type only flowing water (Serra-Tosio 1972).

## Genus *Pseudodiamesa* Goetghebuer

Figs. 438–441

*Syndiamesa* (*Pseudodiamesa*) Goetghebuer, 1939*b*:9.
*Pseudodiamesa*; Pagast, 1947:451; Serra-Tosio 1971:60.

Second antennal segment longer than combined length of third and fourth segments; blade shorter than combined length of terminal segments. Hypostoma with 1 median and 7 pairs of lateral teeth; median tooth triangular, clearly set off from first lateral teeth by deep V-shaped notches; second lateral teeth appressed to first lateral; sixth lateral tooth longer than fifth or seventh lateral. Paralabial plate vestigial. Mandible with 4 inner teeth; seta interna present. Premandible with 6–11 teeth. Labral lamellae consisting of 2 pectinate lobes or absent. Pecten epipharyngis consisting of 7 overlapping scales. Abdominal segments with thin, light-colored setae. Procercus longer than wide, bearing 6–8 anal setae.

**Remarks.** The mentum with the triangular median tooth separated from the first lateral teeth by deep V-shaped notches is distinctive. Larvae are described by Johannsen (1937*a*; as *Syndiamesa*) and Oliver(1976).

**Distribution.** Alberta; Northwest Territories.

**Habitat.** Larvae of *Pseudodiamesa* inhabit flowing water and deep still water (Serra-Tosio 1972).

## Genus Pseudokiefferiella Zavřel

Figs. 442–445

*Pseudokiefferiella* Zavřel, 1941*b*:9; Serra-Tosio 1971:258.
*Diplomesa* Pagast, 1947:539.

Antennal segments consecutively smaller; blade shorter than combined length of terminal segments. Hypostoma with 1 median and 6 or 7 pairs of lateral teeth. Paralabial plate vestigial. Mandible with 4 inner teeth; seta interna present. Premandible with 5 or 6 teeth. Labral lamellae consisting of 2 pectinate lobes. Pecten epipharyngis consisting of 5 scales. Abdominal segments with strong dark setae. Procercus longer than wide, bearing 6–8 anal setae.

**Remarks.** Larvae of *Pseudokiefferiella* are similar to those of *Diamesa*, but may be distinguished by the presence of strong dark setae on the abdomen. In *Diamesa* the abdominal setae are thinner and light colored. *Pseudokiefferiella* has been placed as a subgenus of *Diamesa* (e.g., Sæther 1970) but now is recognized as a separate genus. Larvae are described by Zavřel (1941*b*) and Sæther (1970).

**Distribution.** Yukon Territory; Districts of Franklin and Mackenzie.

**Habitat.** Larvae of *Pseudokiefferiella* inhabit small bodies of flowing water (Serra-Tosio 1972) and seepage areas.

## Tribe Protanypodini

Figs. 8, 446–447

This tribe consists of the single genus *Protanypus*, and the description given for the genus is sufficient to define the characters of the tribe.

## Genus *Protanypus* Kieffer

Figs. 8, 446–447

*Protanypus* Kieffer, 1906a:318; Sæther 1975b:367.

Head capsule covered with numerous setae, without dorsal tubercles; occipital margin with distinct posteriorly directed ventrolateral projection on each side. Antenna 4-segmented, with second segment longer than combined length of third and fourth segments, and with third segment non-annulate; blade shorter than combined length of terminal segments; Lauterborn's organs larger. Hypostoma with toothless anterior margin except for extreme lateral teeth. Paralabial plates small. Mandible with 1 apical and 5 inner teeth; seta interna absent. Premandible straight, simple, or with 3 or 4 weak teeth. Labral lamellae consisting of row of overlapping scales, often serrated apically. Pecten epipharyngis consisting of 3 simple, bifid, or toothed scales. Prementohypopharyngeal complex bearing flat scales. Abdominal setae thin, light colored. Procercus longer than wide, bearing 6 anal setae. Anterior and posterior parapods separate, each bearing apical group of irregularly placed claws.

**Remarks.** The larvae of *Protanypus* are easily recognized by the presence of numerous setae on the head capsule and by the hypostoma having only widely separated lateral teeth. Larvae are described by Sæther (1975b).

**Distribution.** British Columbia and Alberta to Northwest Territories; Ontario; Quebec.

**Habitat.** Larvae of *Protanypus* inhabit deep cool lakes (Sæther 1975b).

## Subfamily Telmatogetoninae

Figs. 461–462

Only one genus, *Paraclunio*, is reported from Canada. Therefore only this genus is described here. Larvae of the Telmatogetoninae may be distinguished from all other Chironominae by the presence of long setae on the prementohypopharyngeal complex and by a non-annulate third antennal segment. Wirth (1949) gave a description of the larvae and a key to genera.

Larvae of Telmatogetoninae are generally restricted to warm seashores, often in waters of low salinity. One genus, *Telmatogeton*, has freshwater representatives living in streams in Hawaii. Little is known about their biology.

## Genus *Paraclunio* Kieffer

Figs. 461–462

*Paraclunio* Kieffer, 1911c:103; Wirth 1949:174.

Body yellowish; head capsule dark brown to black. One small eye spot present. Antenna 4-segmented, reduced, with segments consecutively smaller, and with total length about one-quarter as long as mandible; ring organ on basal quarter of first segment; blade slightly longer than combined length of terminal segments; Lauterborn's organs small, on apex of second segment. Hypostoma with 1 triangular tripartite median and 5 pairs of lateral teeth. Paralabial plate very small, without setae. Mandible with 1 apical and 5 inner teeth; seta subdentalis very long, extending to level of apical tooth; seta interna present; apicodorsal tooth and pecten mandibularis absent. Premandible broad apically, with 3 rounded teeth. SI and SII simple. Pecten epipharyngis consisting of 3 scales. Anterior parapods partly fused, with claws on each part. Abdominal segments devoid of setae, except last 2 segments, with simple setae. Procercus absent, replaced by 1 anal seta. Anal tubules absent. Posterior parapods well developed, each with apical irregularly placed claws.

**Remarks.** Two genera, *Thalassomyia* and *Telmatogeton* occur on the Pacific and southern Atlantic seashores of the United States. Larvae of *Thalassomyia* may occur in southern British Columbia waters. They can be distinguished from larvae of *Paraclunio* by having six inner teeth on the mandible. The premandible of both *Thalassomyia* and *Telmatogeton* is broad, toothless, and with an apical fringe of setae. Larvae of *Paraclunio* are described by Saunders (1928*b*) and Wirth (1949).

**Distribution.** Pacific coast.

**Habitat.** Larvae of *Paraclunio* inhabit the intertidal zone of seashores, generally in filamentous algae (Saunders 1928*b*).

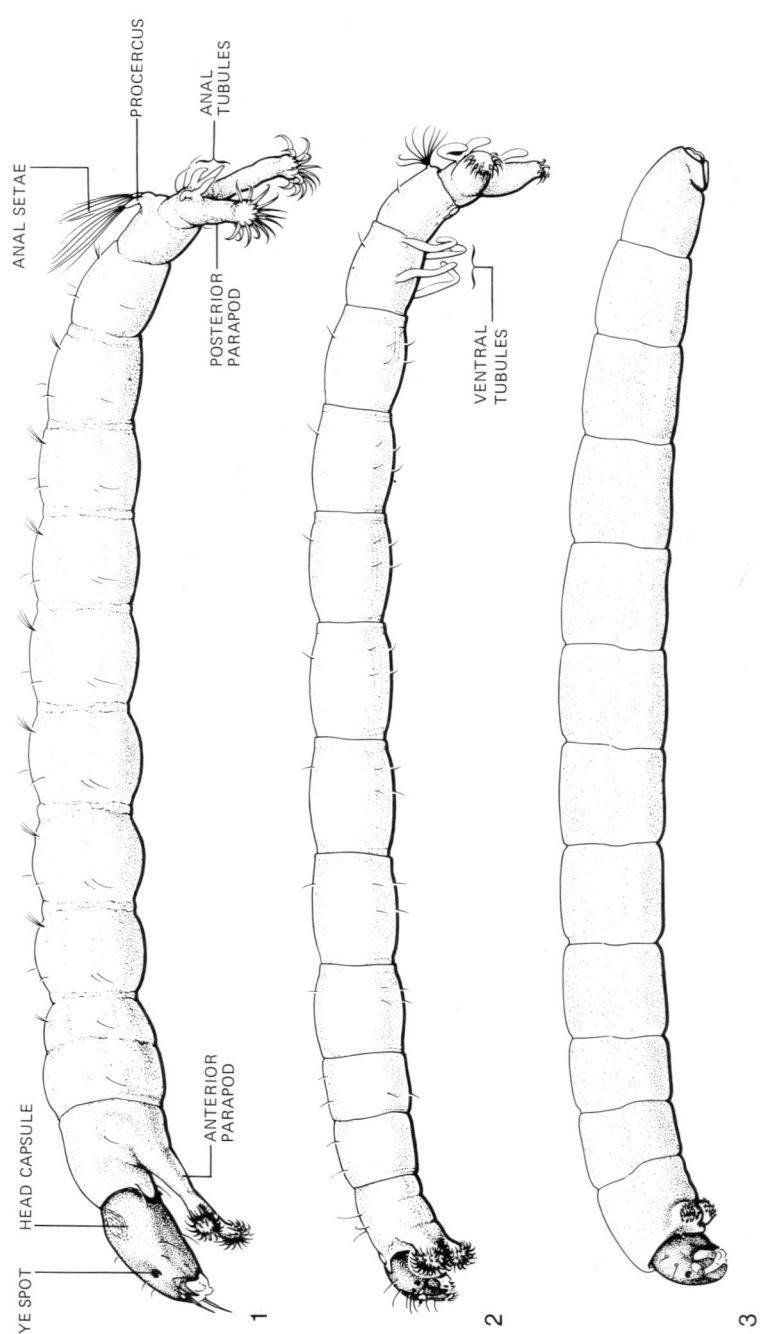

Figs. 1–3. Whole larva, lateral view. 1, *Ablabesmyia*; 2, *Chironomus*; 3, *Pseudosmittia*.

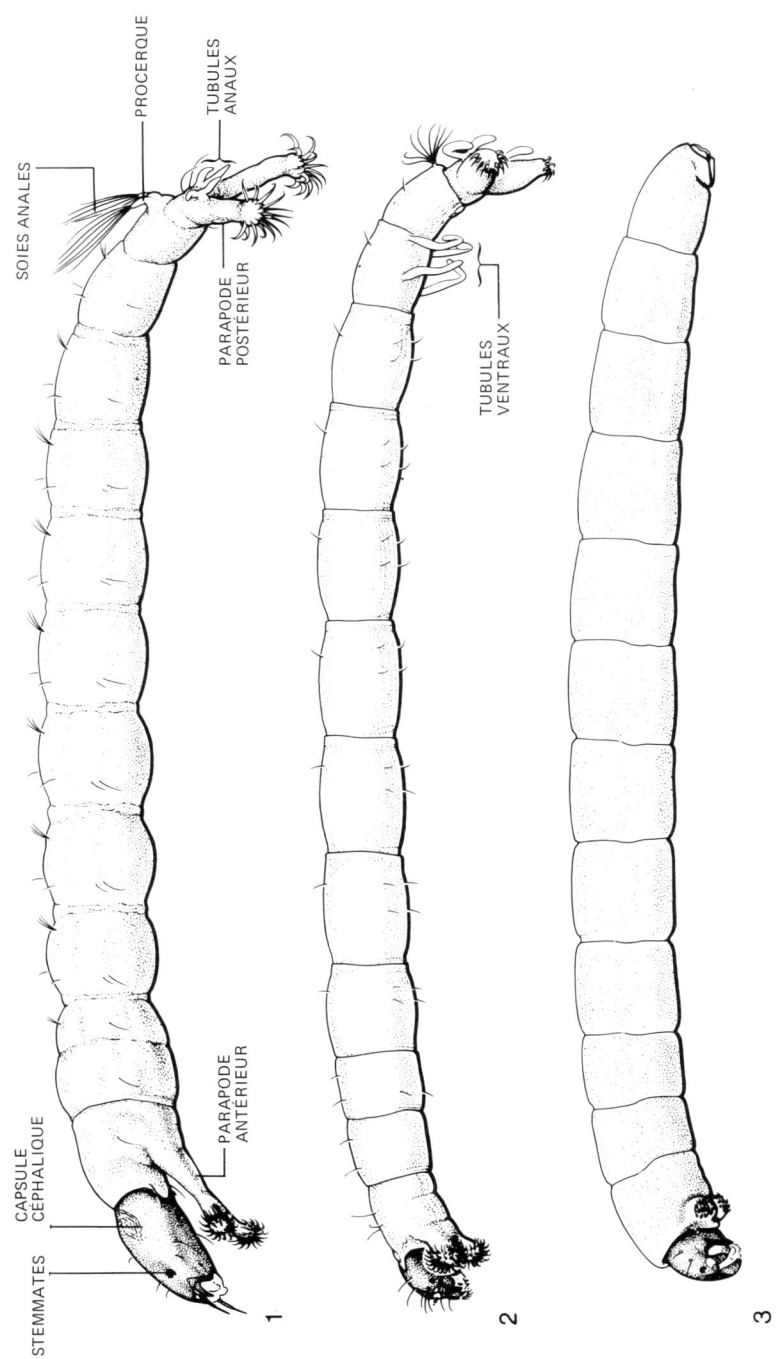

Fig. 1–3. Larve entière, vue latérale. 1, *Ablabesmyia*; 2, *Chironomus*; 3, *Pseudosmittia*.

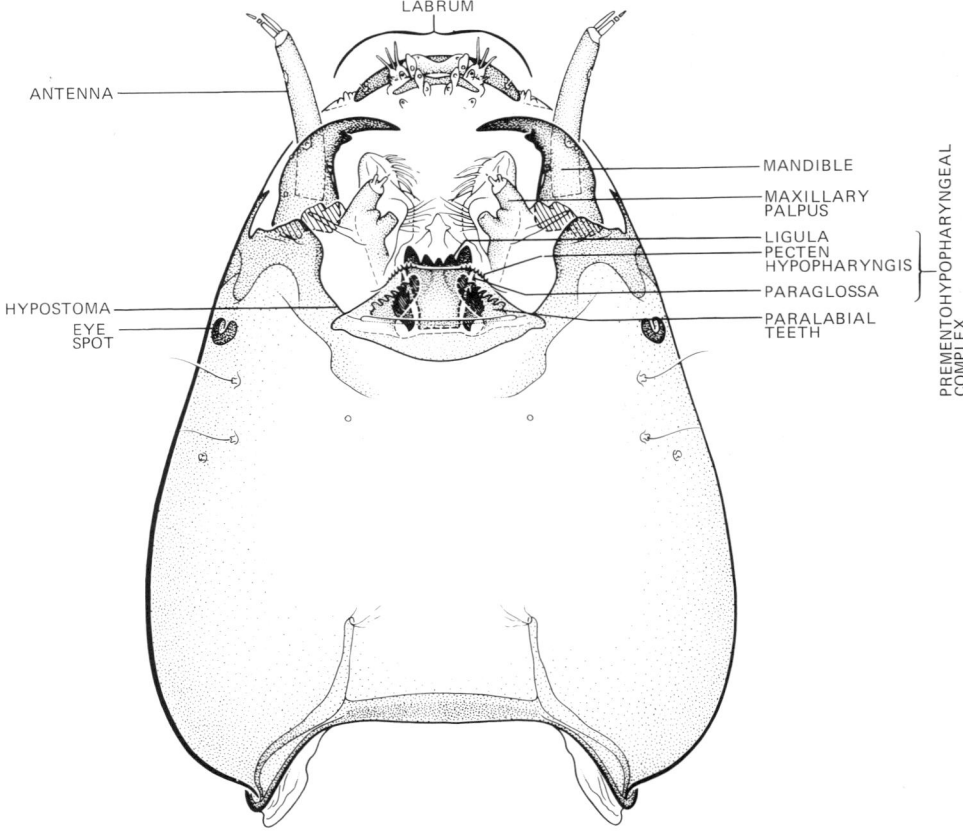

Fig. 4. *Procladius* – head capsule, ventral view.

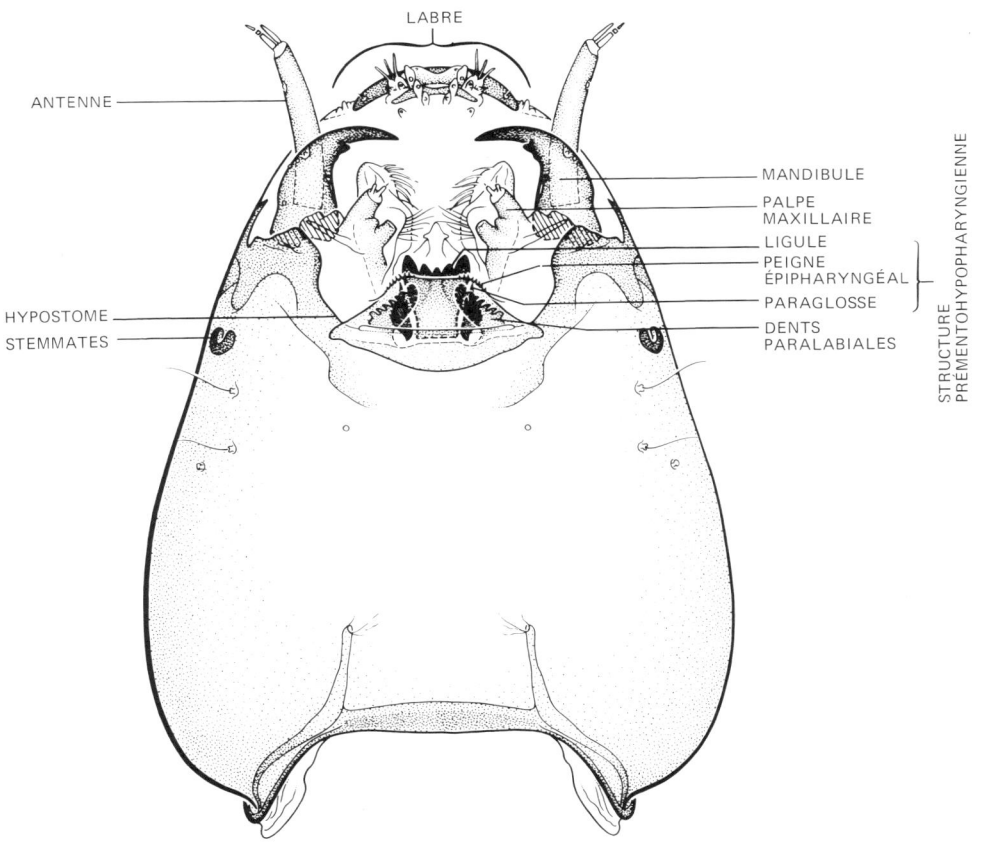

Fig. 4. *Procladius* – capsule céphalique, vue ventrale.

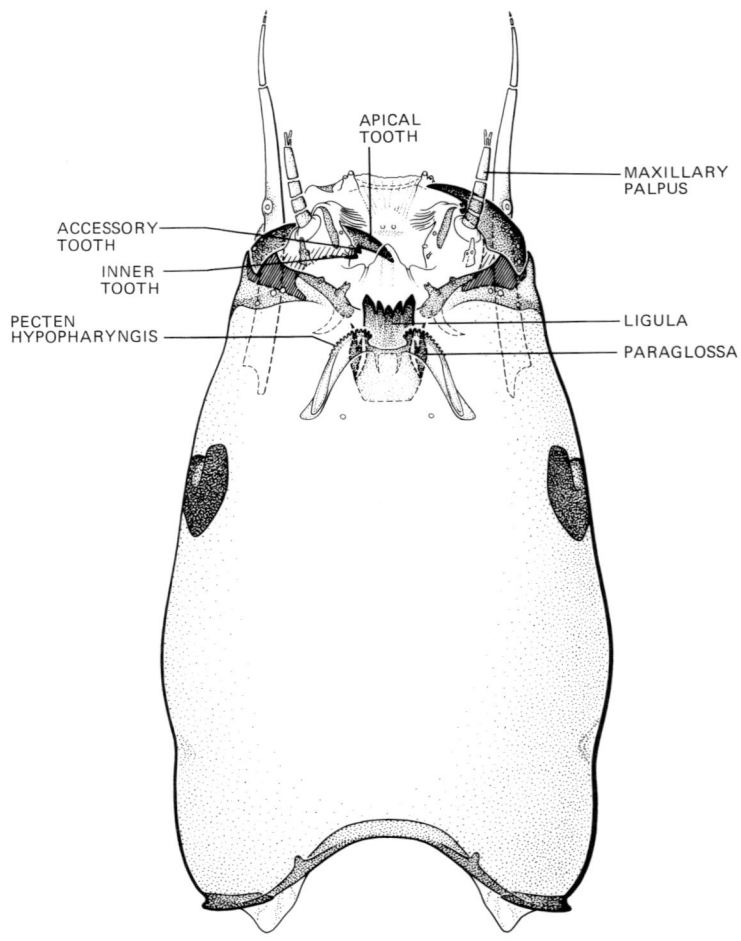

Fig. 5. *Ablabesmyia* — head capsule, ventral view.

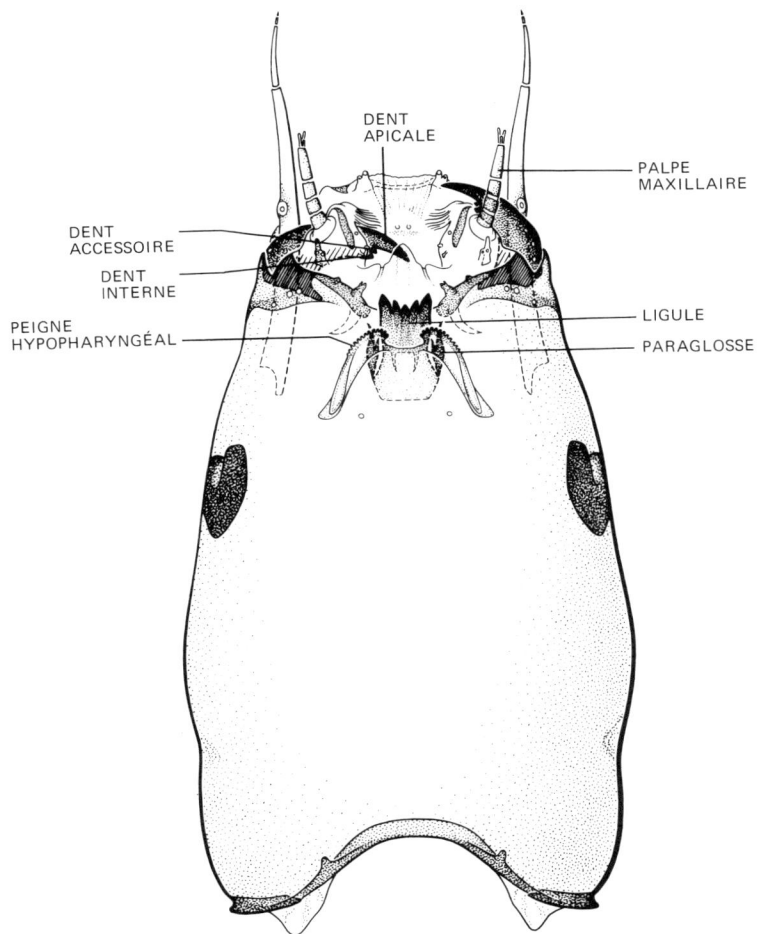

Fig. 5. *Ablabesmyia* – capsule céphalique, vue ventrale.

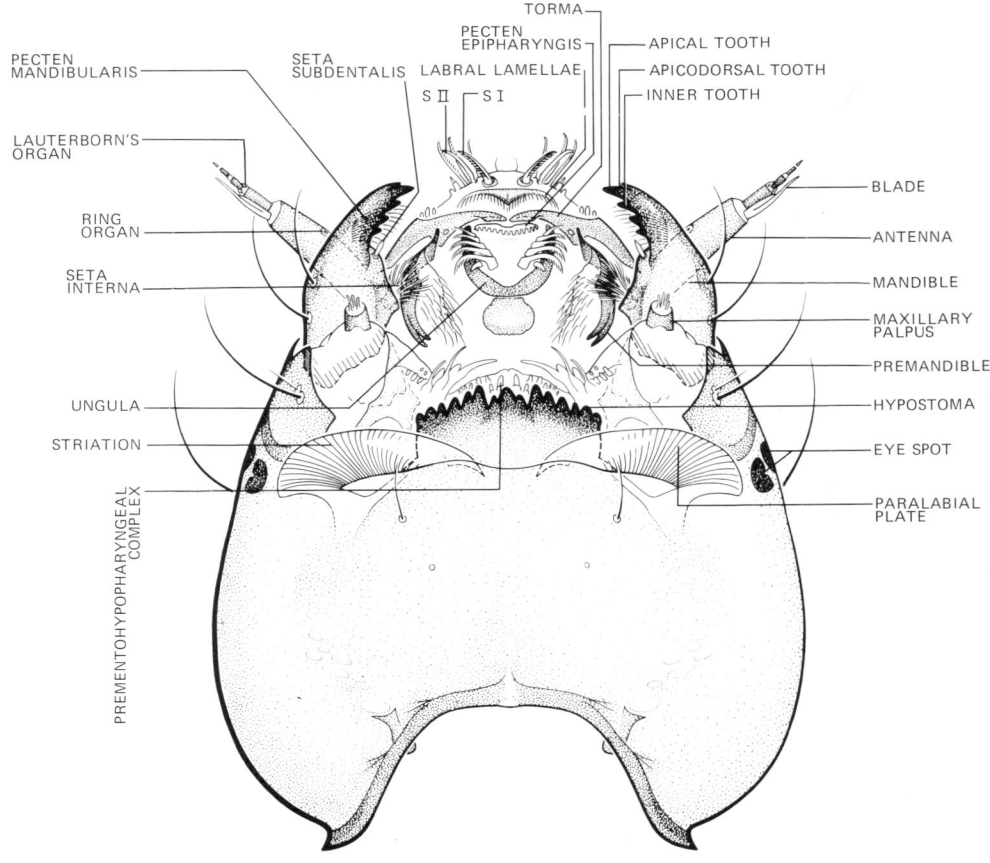

Fig. 6. *Chironomus* – head capsule, ventral view.

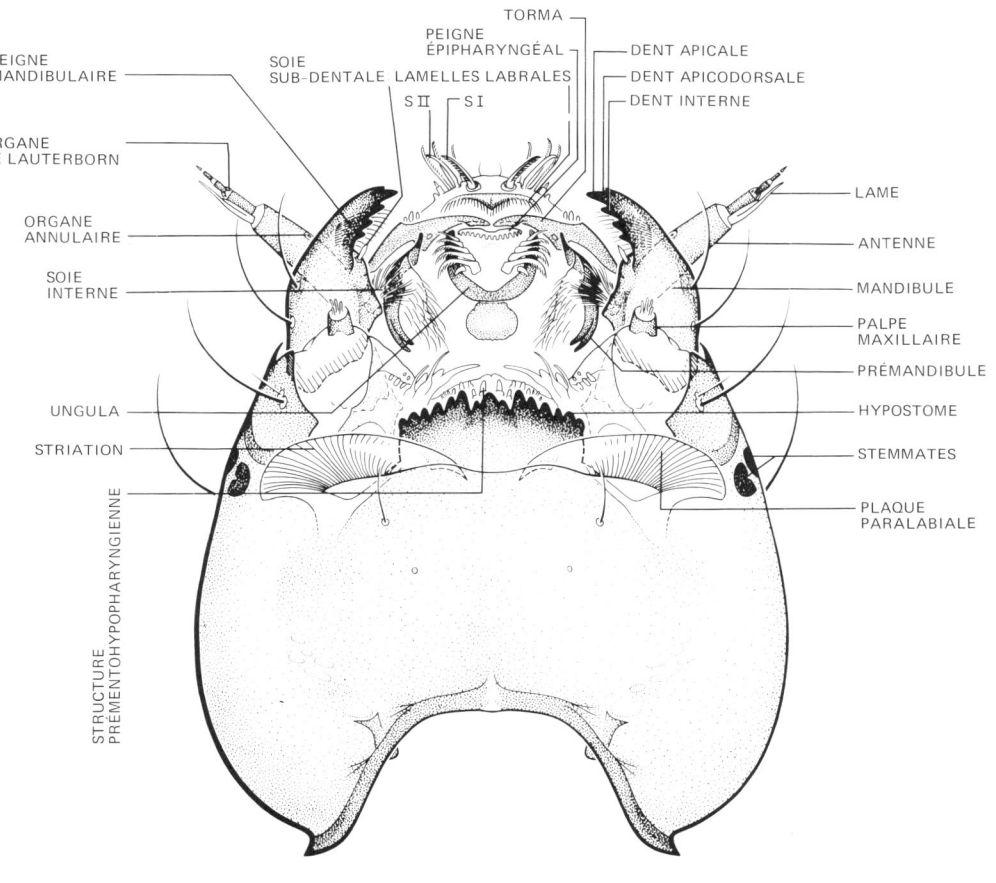

Fig. 6. *Chironomus* – capsule céphalique, vue ventrale.

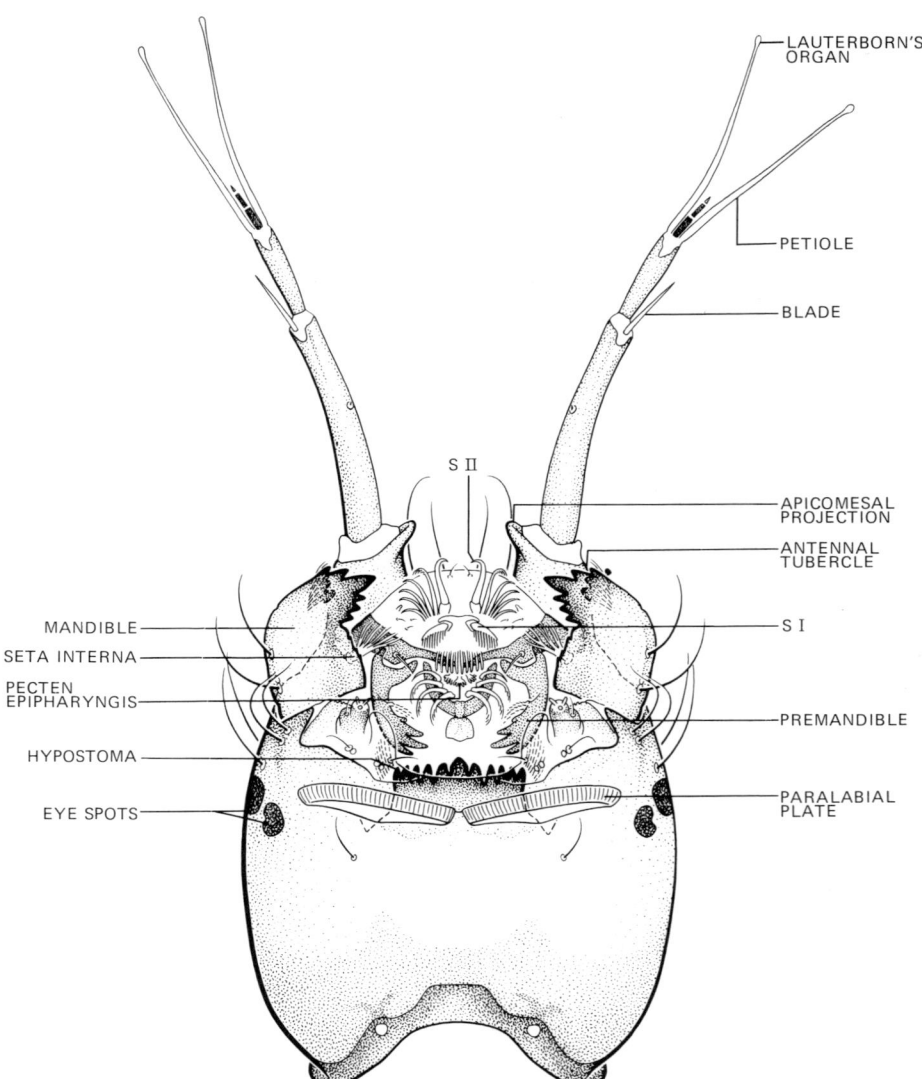

Fig. 7. *Micropsectra* – head capsule, ventral view.

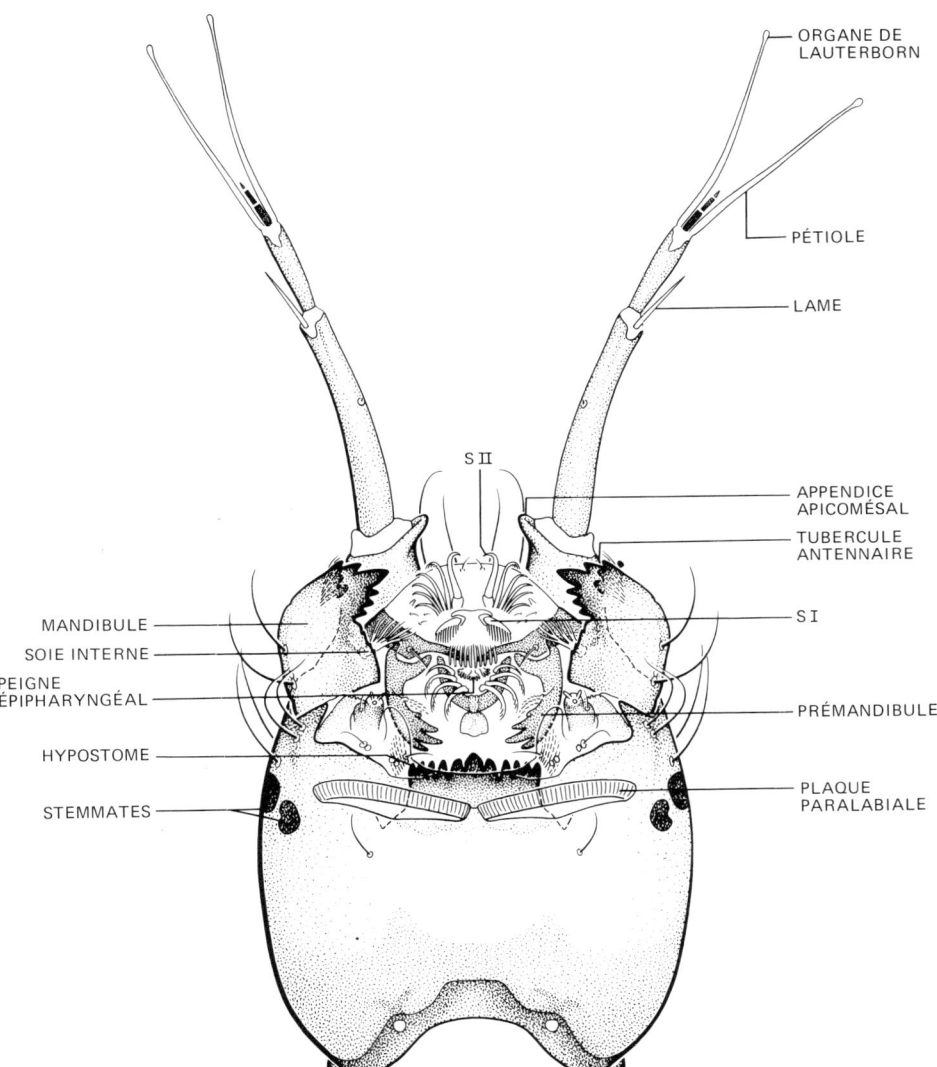

Fig. 7. *Micropsectra* – capsule céphalique, vue ventrale.

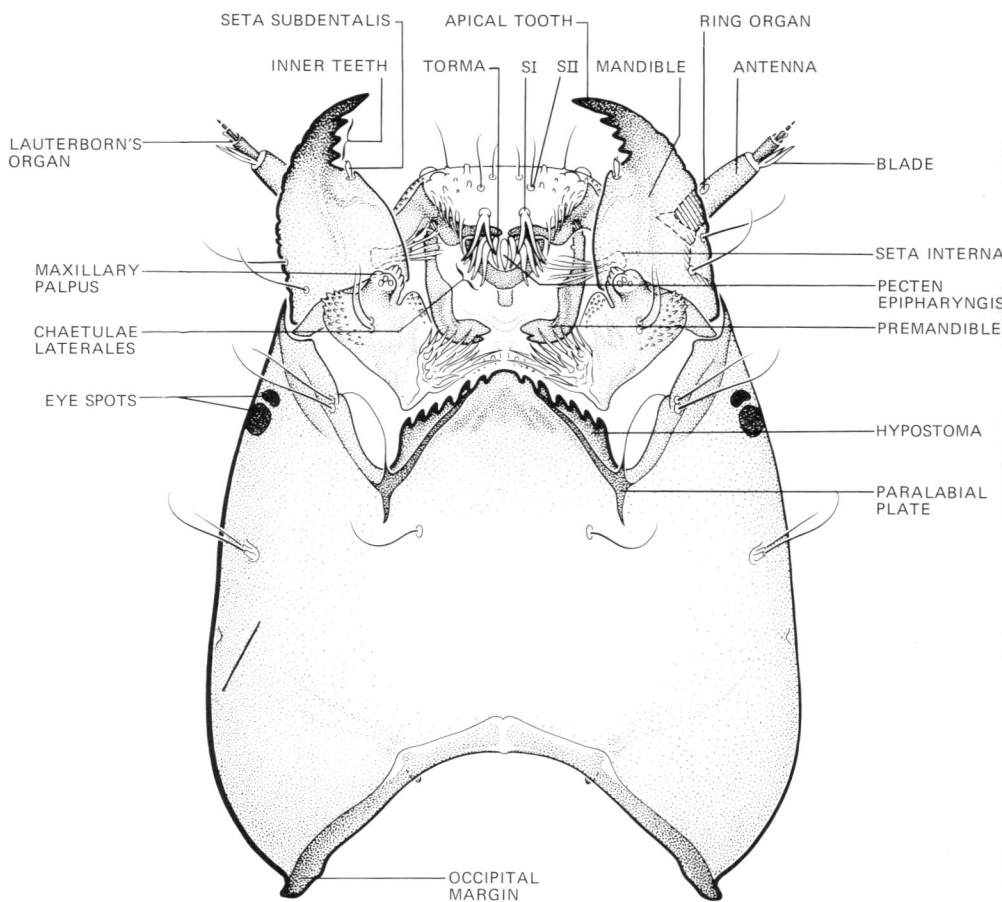

Fig. 8. *Cricotopus* – head capsule, ventral view.

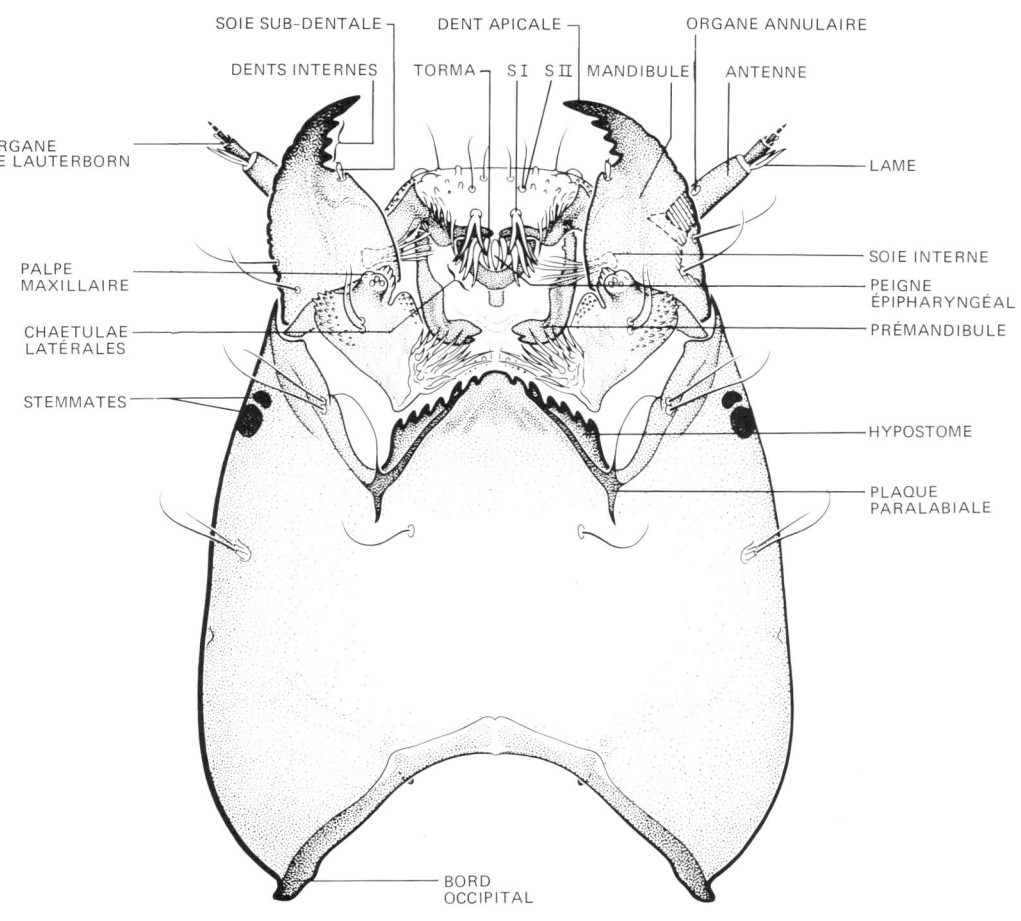

Fig. 8.  *Cricotopus* – capsule céphalique, vue ventrale.

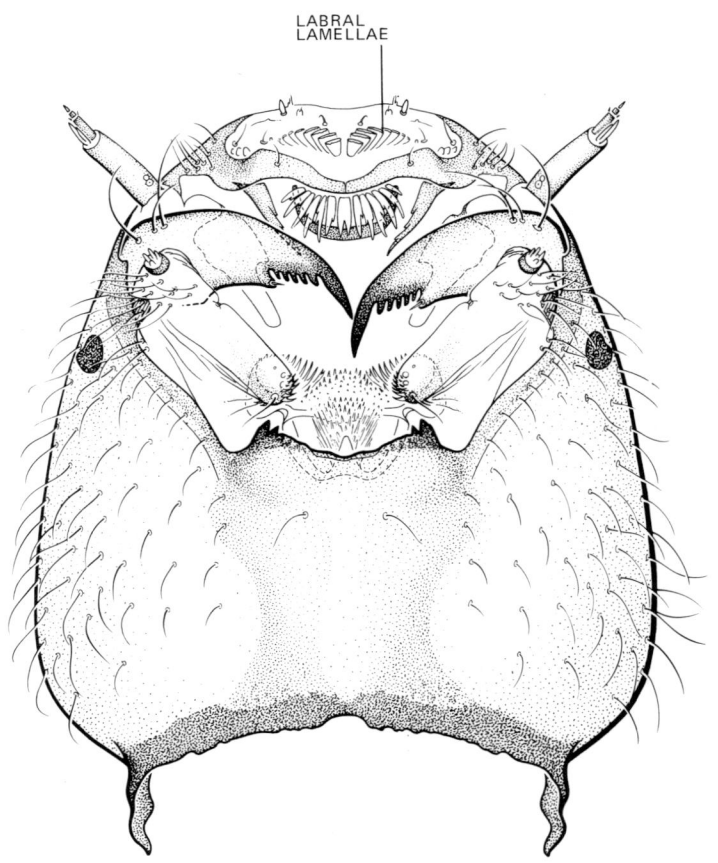

Fig. 9. *Protanypus* – head capsule, ventral view.

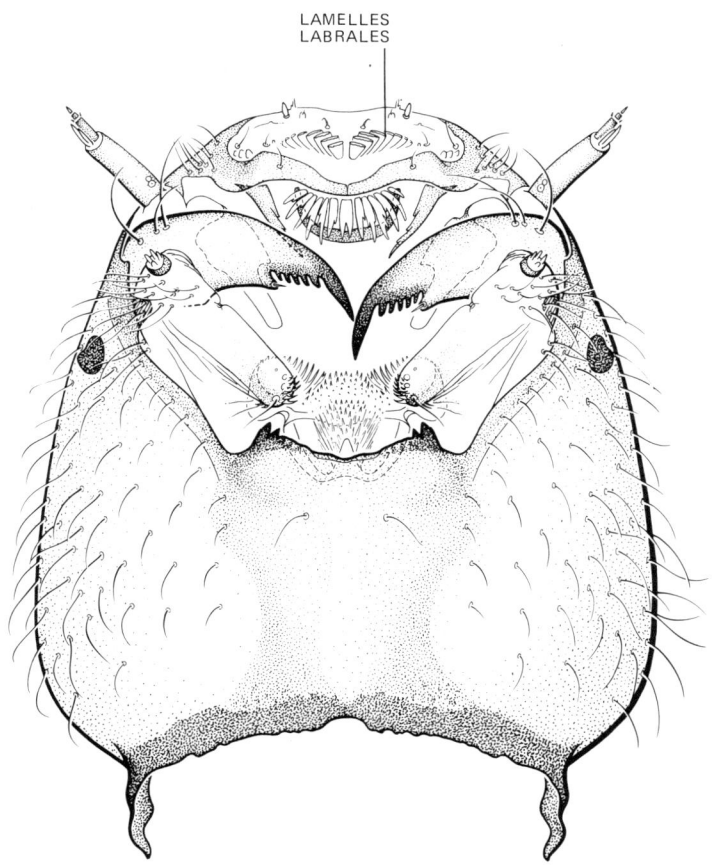

Fig. 9. *Protanypus* – capsule céphalique, vue ventrale.

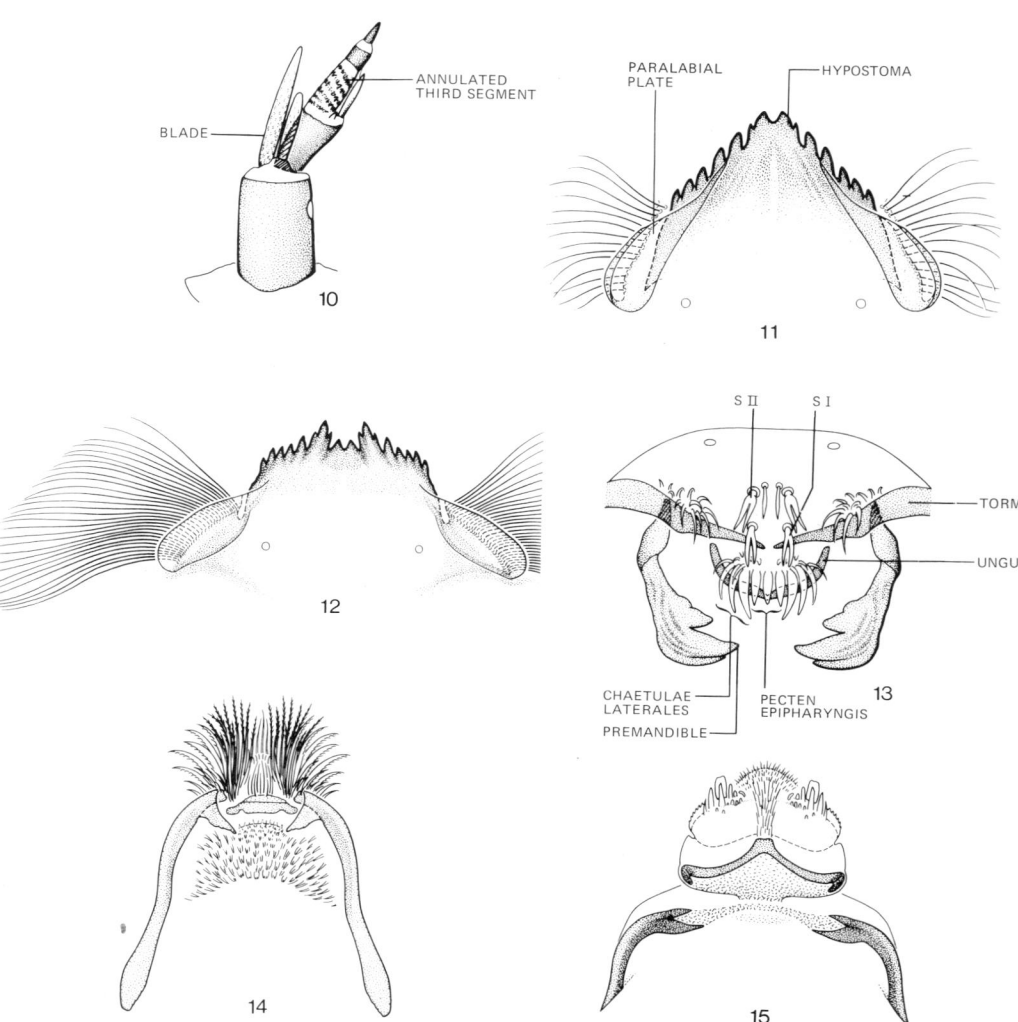

Figs. 10–15. 10, *Diamesa* – antenna; 11, *Rheocricotopus* – hypostoma and paralabial plates; 12, *Prodiamesa* – hypostoma and paralabial plates; 13, *Pseudosmittia* – labrum and premandibles; 14, *Paraclunio* – prementohypopharyngeal complex; 15, *Pseudosmittia* – prementohypopharyngeal complex.

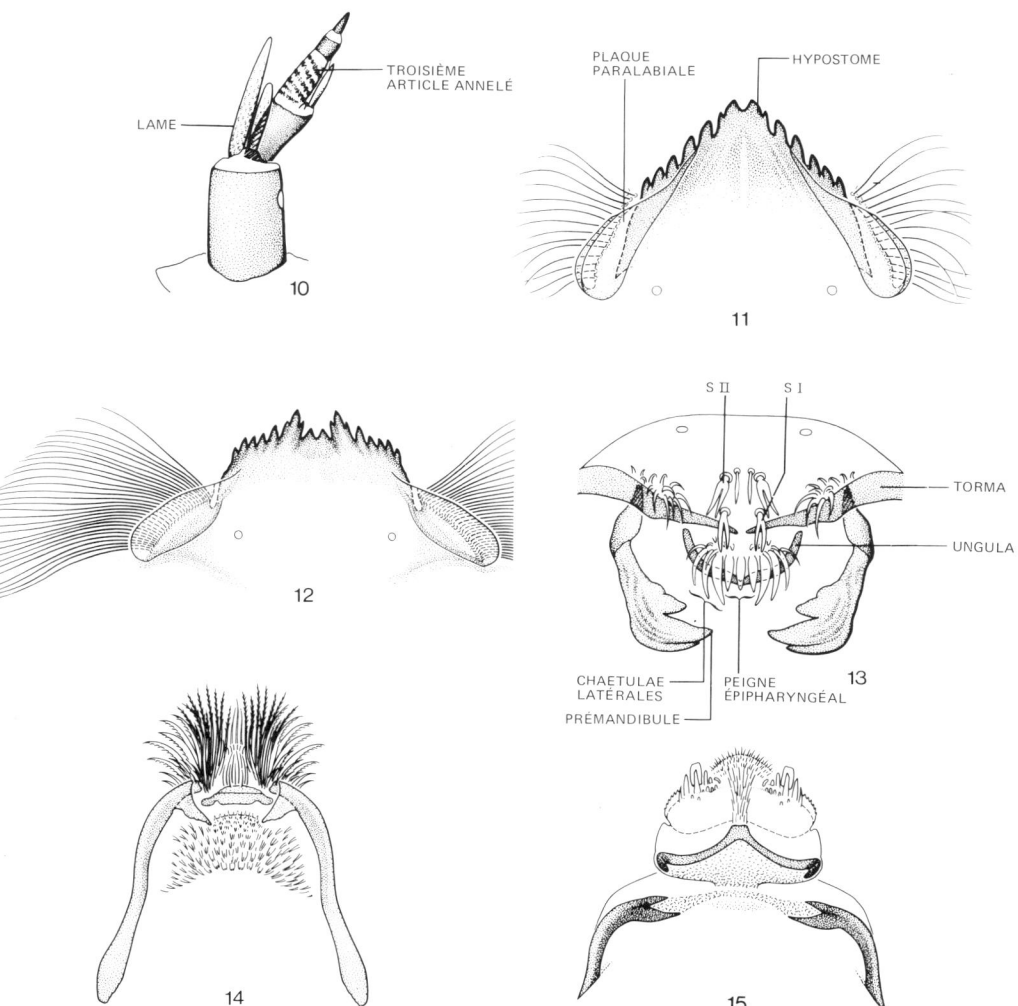

Fig. 10–15. 10, *Diamesa* – antenne; 11, *Rheocricotopus* – hypostome et plaques paralabiales; 12, *Prodiamesa* – hypostome et plaques paralabiales; 13, *Pseudosmittia* – labre et prémandibules; 14, *Paraclunio* – structure prémentohypopharyngienne; 15, *Pseudosmittia* – structure prémentohypopharyngienne.

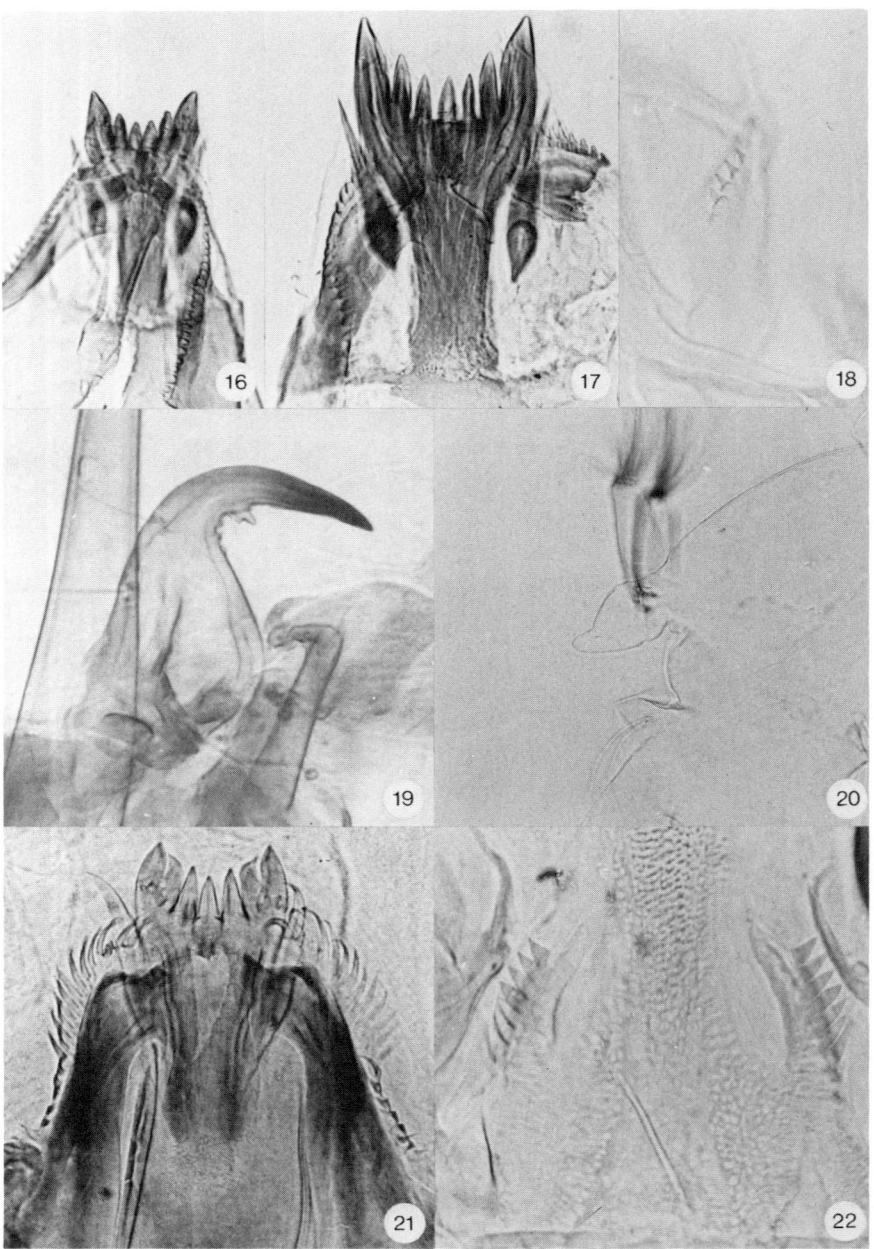

Figs. 16–22. 16, 17, *Clinotanypus* – ligula; 18, *Clinotanypus* – paralabial teeth; 19, *Clinotanypus* – mandible; 20, *Clinotanypus* – conical projection between bases of procerci, lateral view; 21, *Coelotanypus* – ligula; 22, *Coelotanypus* – paralabial plate.

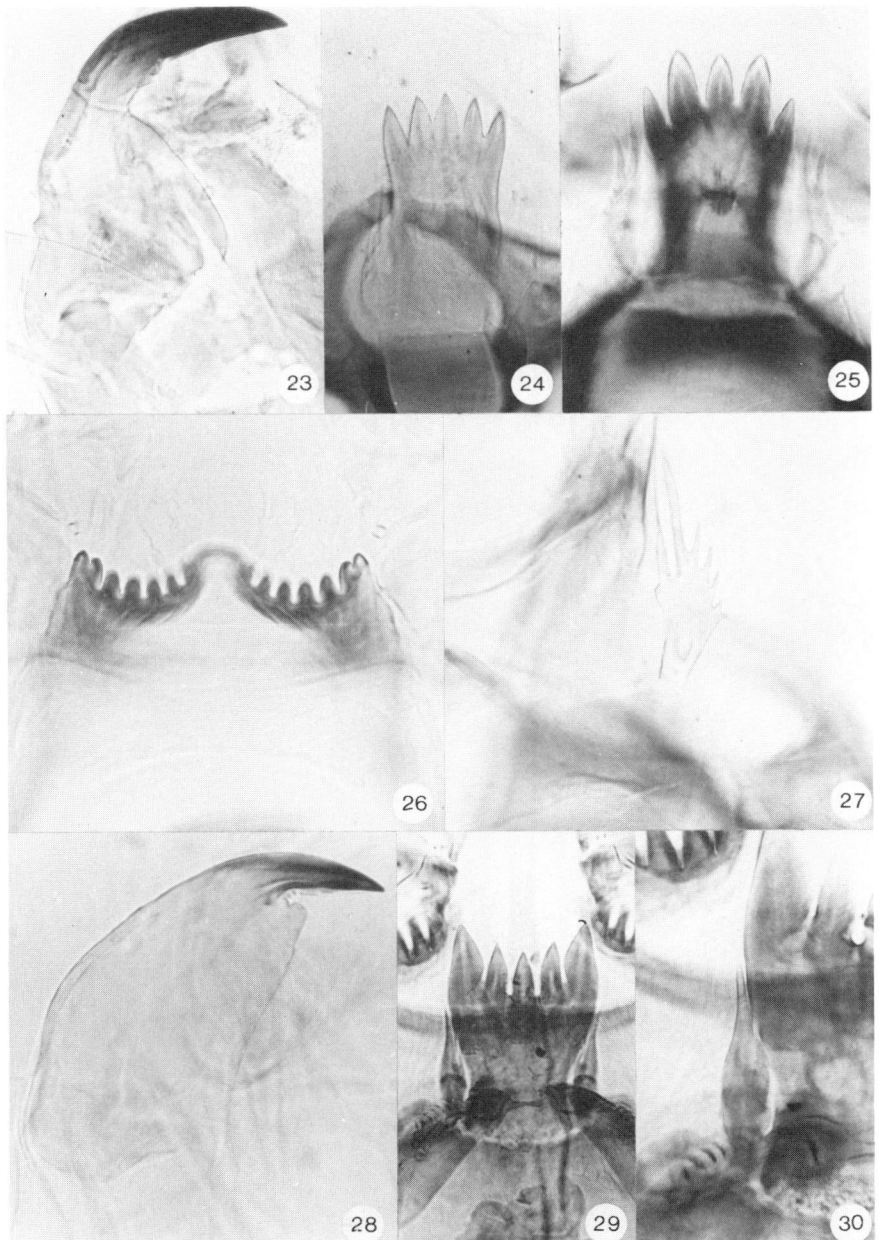

Figs. 23–30. 23, *Coelotanypus* – mandible; 24, *Tanypus (Tanypus)* – ligula; 25, *Tanypus (Apelopia)* – ligula; 26, *Tanypus* – paralabial plates; 27, *Tanypus* – paraglossa; 28, *Tanypus* – mandible; 29, *Apsectrotanypus* – ligula; 30, *Apsectrotanypus* – paraglossa.

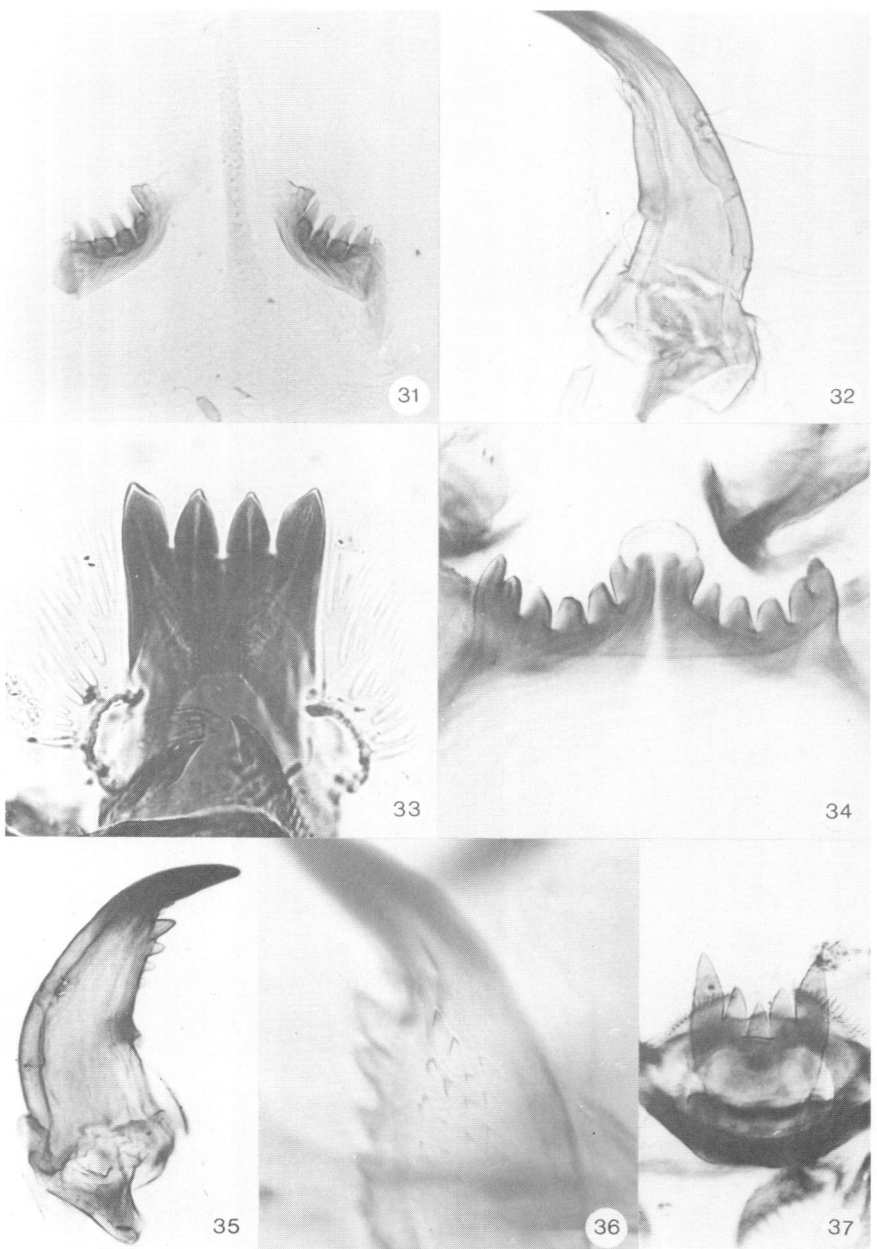

Figs. 31–37. 31, *Apsectrotanypus* – paralabial plates; 32, *Apsectrotanypus* – mandible; 33, *Derotanypus* – ligula and paraglossa; 34, *Derotanypus* – paralabial plates; 35, *Derotanypus* – mandible; 36, *Derotanypus* – mandible, small spines on dorsal surface; 37, *Macropelopia* – ligula.

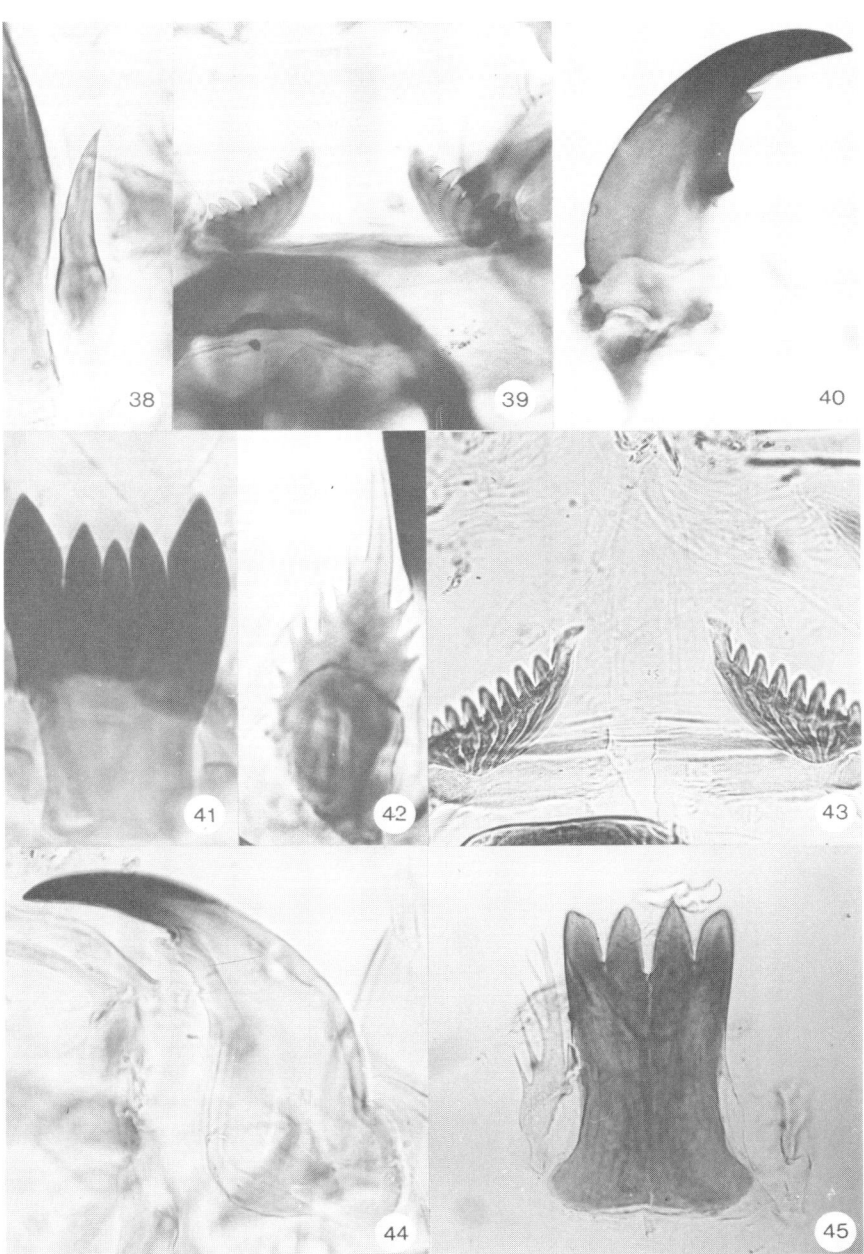

Figs. 38–45. 38, *Macropelopia* – paraglossa; 39, *Macropelopia* – paralabial plates; 40, *Macropelopia* – mandible; 41, *Procladius* – ligula; 42, *Procladius* – paraglossa; 43, *Procladius* – paralabial plates; 44, *Procladius* – mandible; 45, *Psectrotanypus* – ligula and paraglossa.

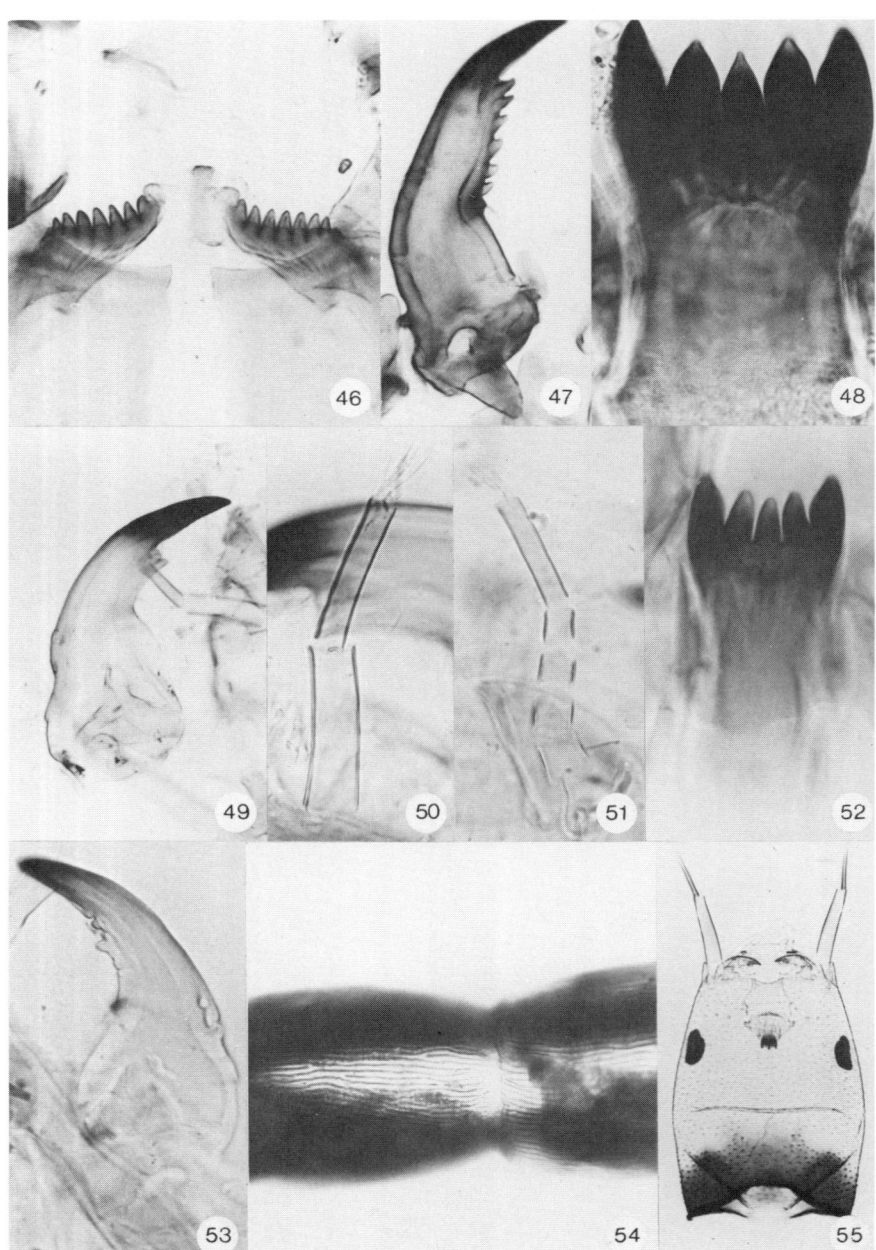

Figs. 46–55. 46, *Psectrotanypus* – paralabial plates; 47, *Psectrotanypus* – mandible; 48, *Abiabesmyia* – ligula; 49, *Ablabesmyia* – mandible; 50, 51, *Ablabesmyia* – maxillary palpus; 52, *Guttipelopia* – ligula; 53, *Guttipelopia* – mandible; 54, *Guttipelopia* – abdominal segments 6 and 7; 55, *Labrundinia* – head capsule, ventral view.

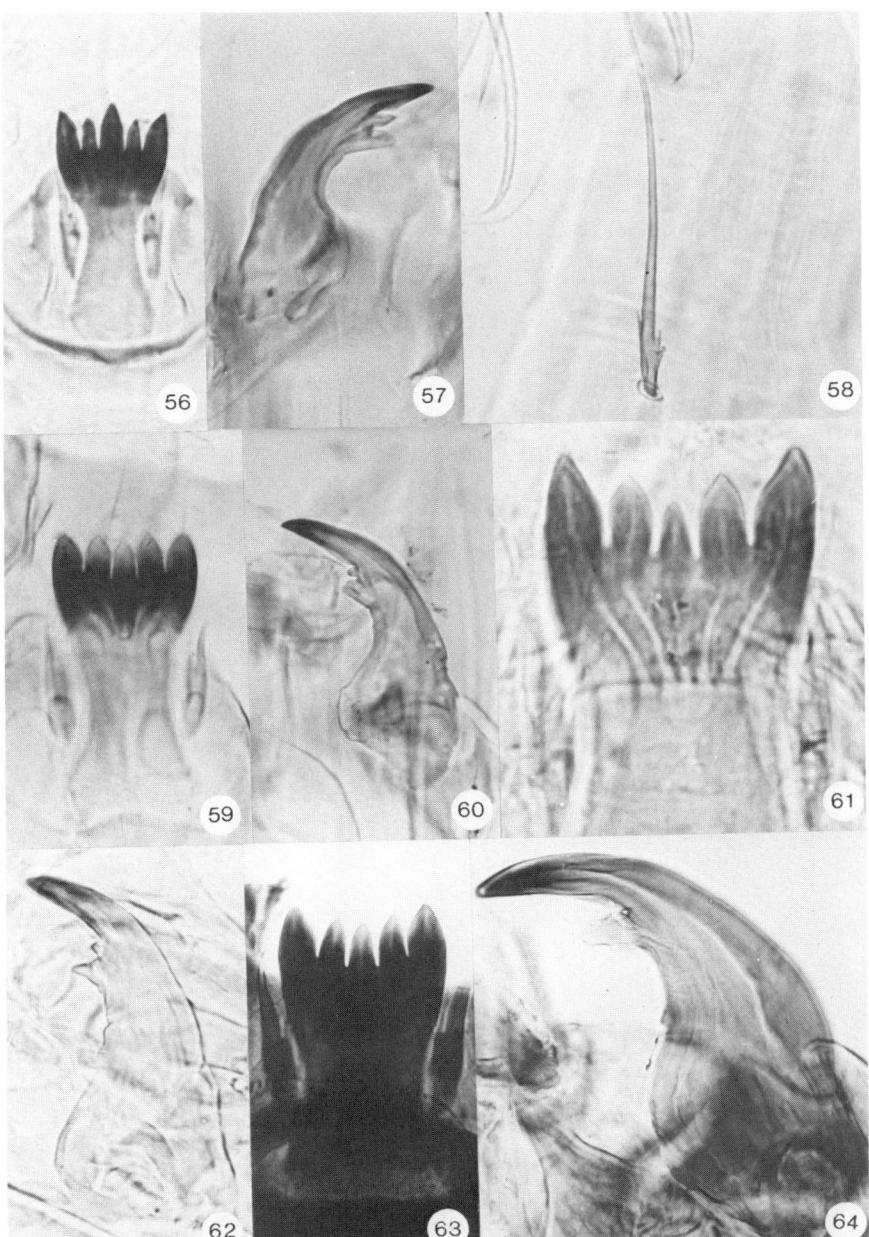

Figs. 56–64. 56, *Labrundinia* – ligula; 57, *Labrundinia* – mandible; 58, *Labrundinia* – seta on posterior parapod; 59, *Larsia* – ligula; 60, *Larsia* – mandible; 61, *Monopelopia* – ligula; 62, *Monopelopia* – mandible; 63, *Natarsia* – ligula; 64, *Natarsia* – mandible.

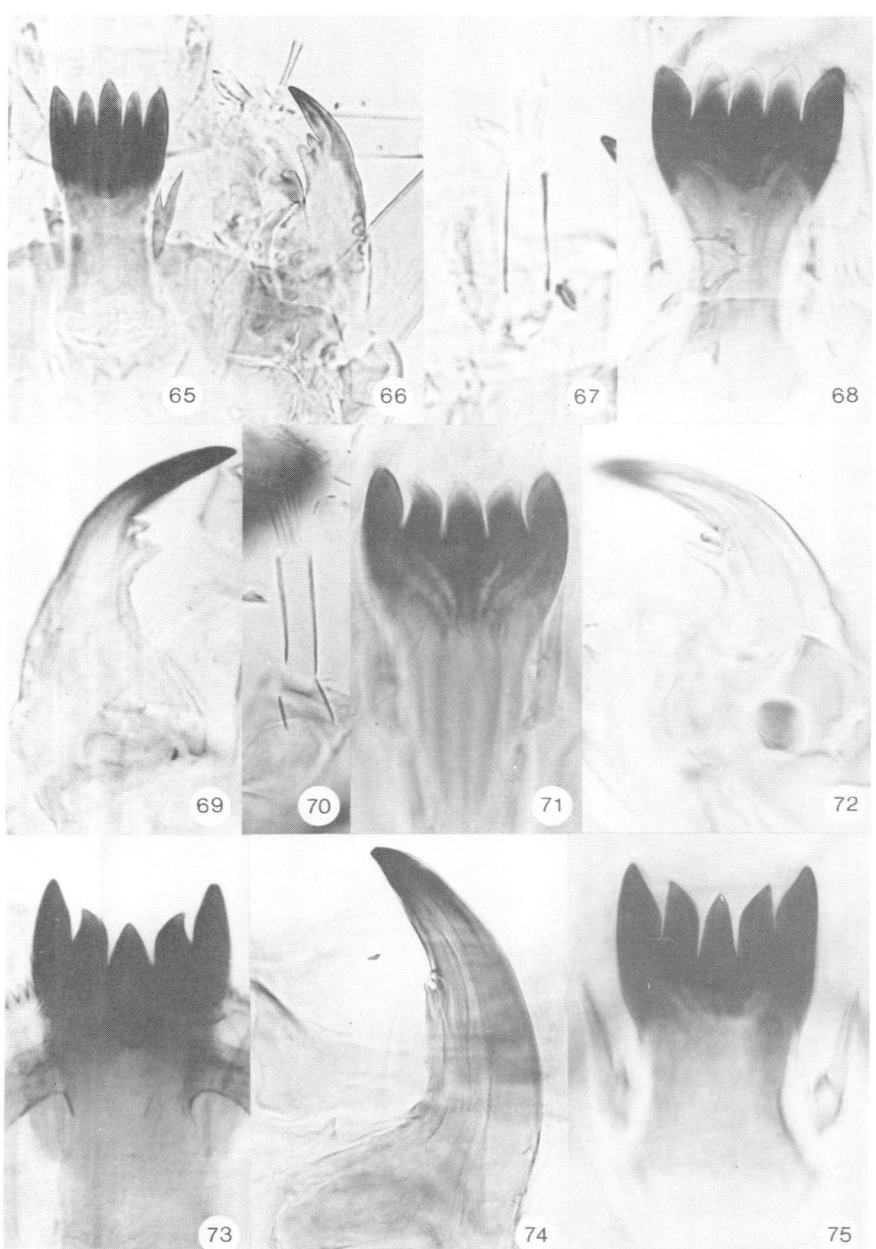

Figs. 65–75. 65, *Nilotanypus* – ligula and paraglossa; 66, *Nilotanypus* – mandible; 67, *Nilotanypus* – maxillary palpus; 68, *Paramerina* – ligula; 69, *Paramerina* – mandible; 70, *Paramerina* – maxillary palpus; 71, *Pentaneura* – ligula; 72, *Pentaneura* – mandible; 73, *Thienemannimyia* – ligula; 74, *Thienemannimyia* – mandible; 75, *Trissopelopia* – ligula.

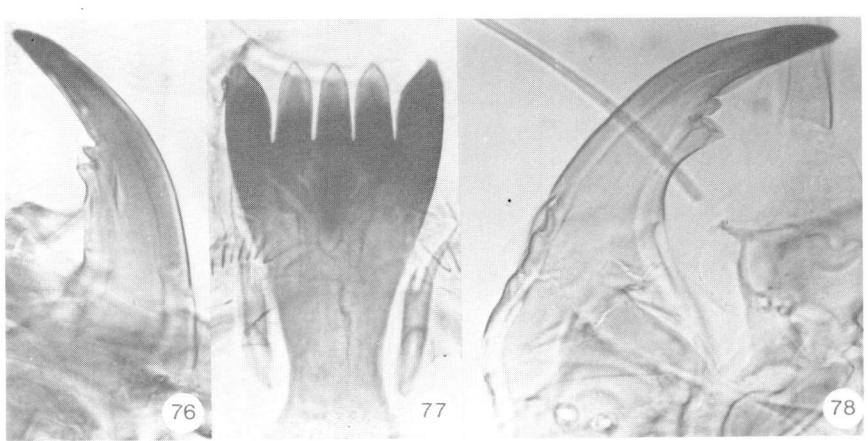

Figs. 76–78. 76, *Trissopelopia* – mandible; 77, *Zavrelimyia* – ligula and paraglossa; 78, *Zavrelimyia* – mandible.

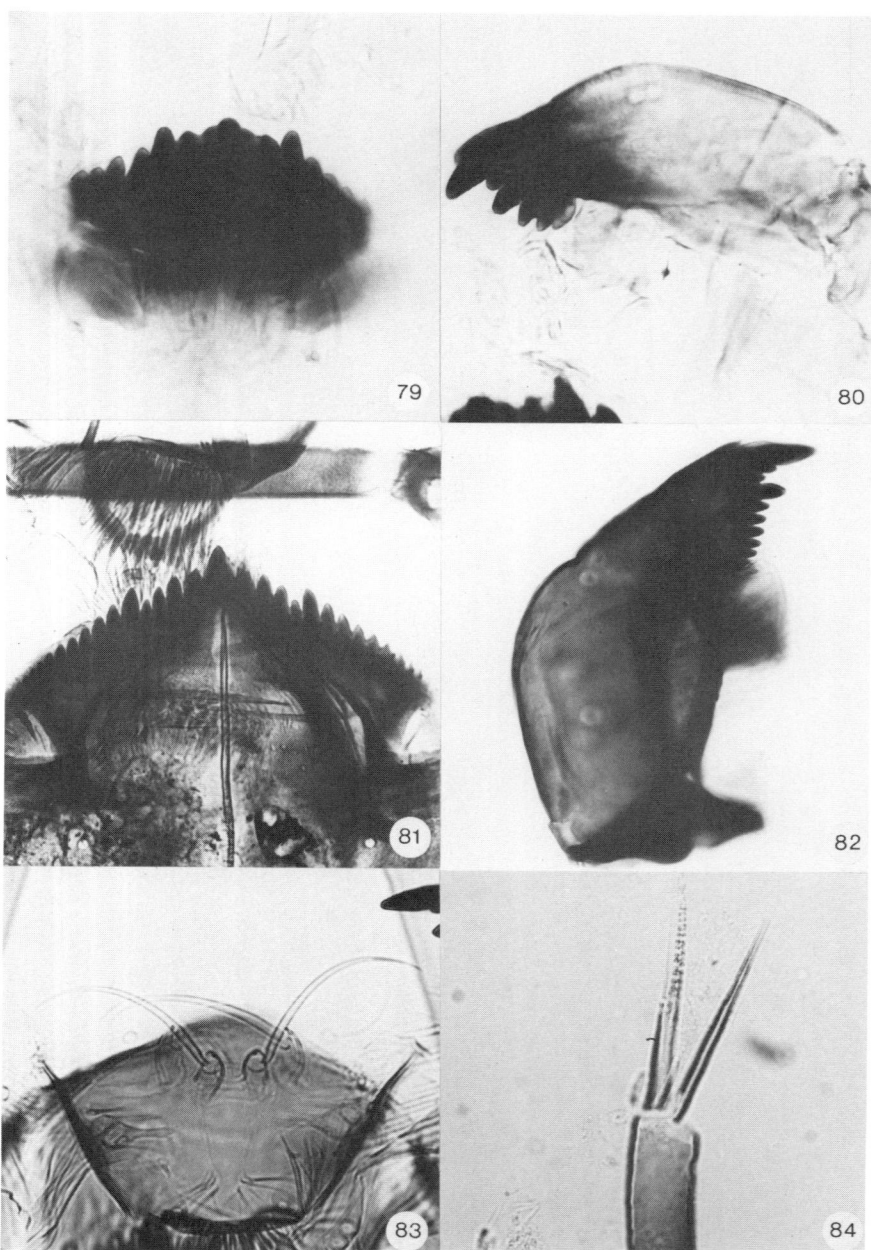

Figs. 79–84. 79, *Parochlus* – hypostoma; 80, *Parochlus* – mandible; 81, *Lasiodiamesa* – hypostoma; 82, *Lasiodiamesa* – mandible; 83, *Lasiodiamesa* – SI; 84, *Lasiodiamesa* – antenna.

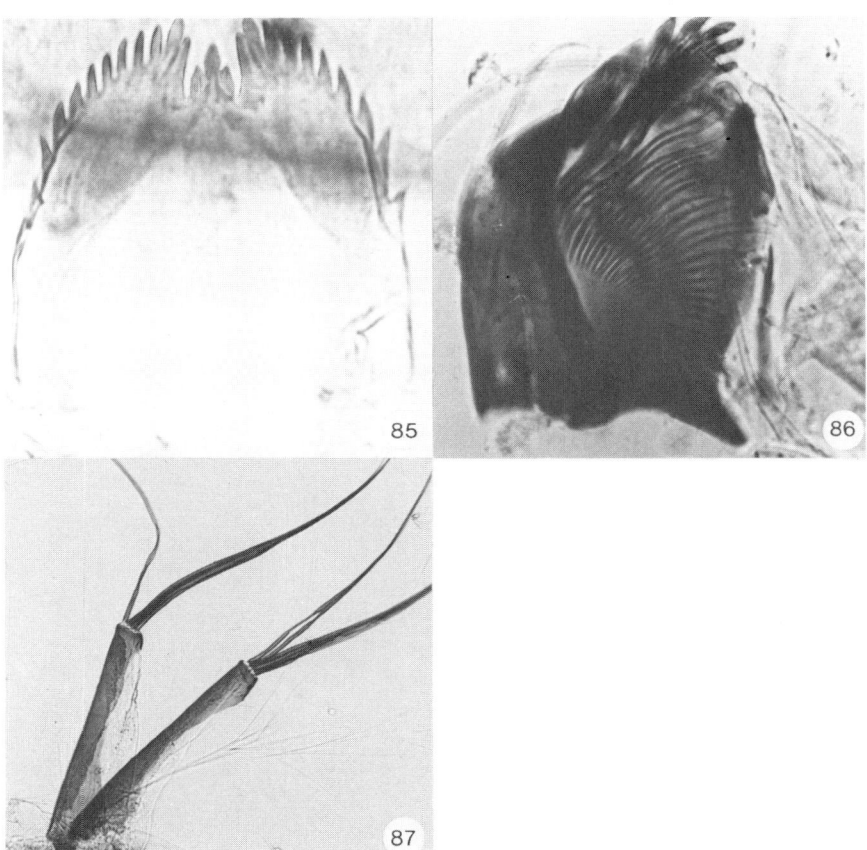

Figs. 85–87. 85, *Trichotanypus* – hypostoma; 86, *Trichotanypus* – mandible; 87, *Trichotanypus* – procercus.

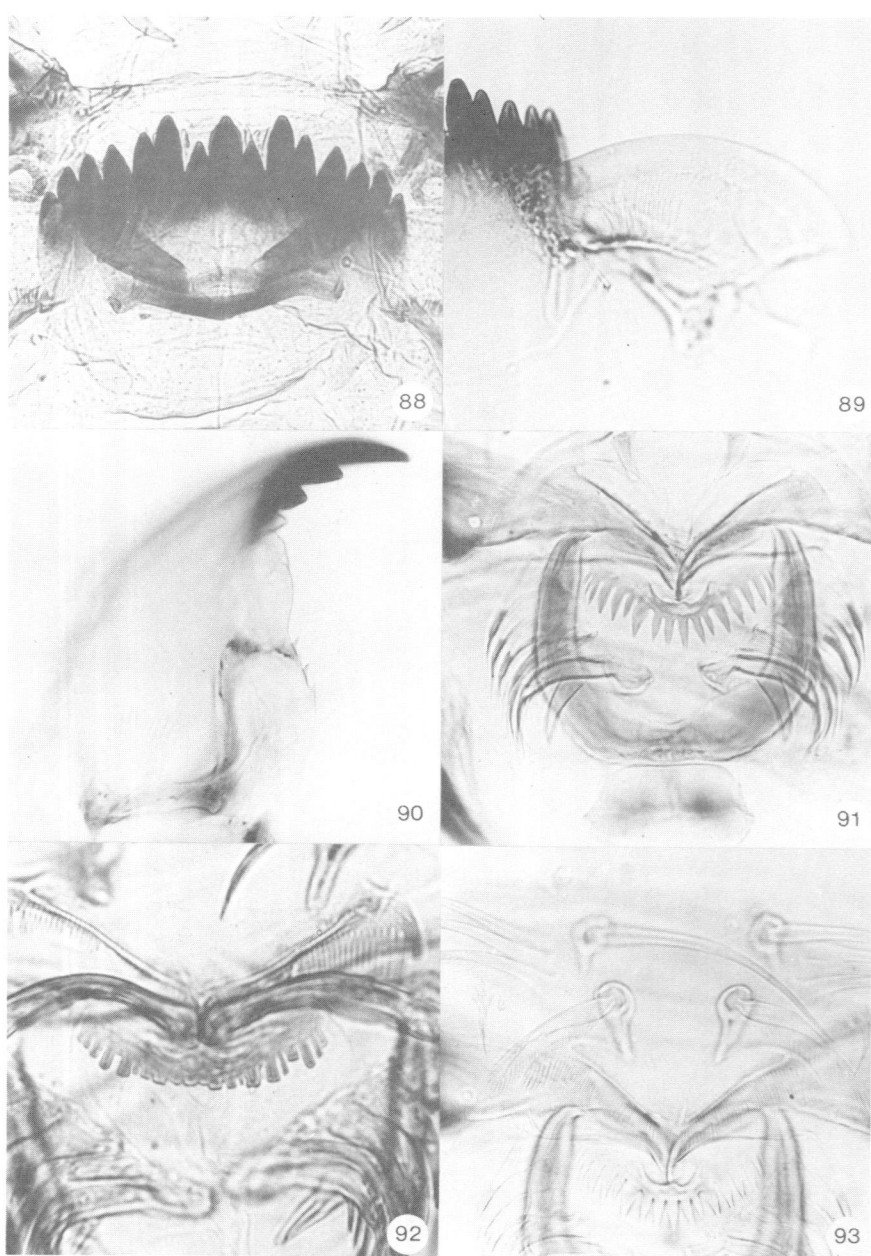

Figs. 88–93. 88, *Chironomus* – hypostoma; 89, *Chironomus* – paralabial plate; 90, *Chironomus* – mandible; 91, *Chironomus; plumosus* group – pecten epipharyngis; 92, *Chironomus* (*Chaetolabis*) – pecten epipharyngis; 93, *Chironomus* – SI and SII.

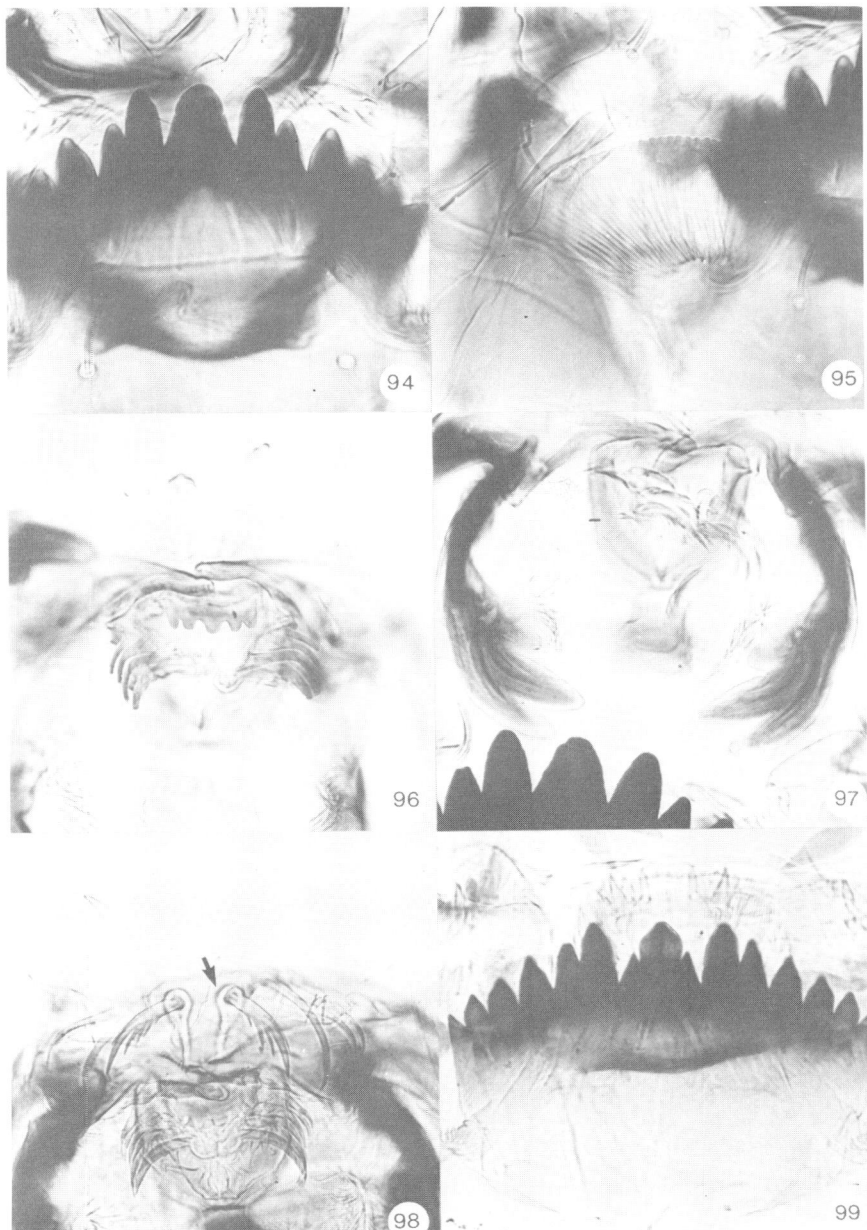

Figs. 94–99. 94, *Dicrotendipes* – hypostoma; 95, *Dicrotendipes* – paralabial plate; 96, *Dicrotendipes* – pecten epipharyngis; 97, *Dicrotendipes* – premandibles; 98, *Dicrotendipes* – SI; 99, *Einfeldia* – hypostoma.

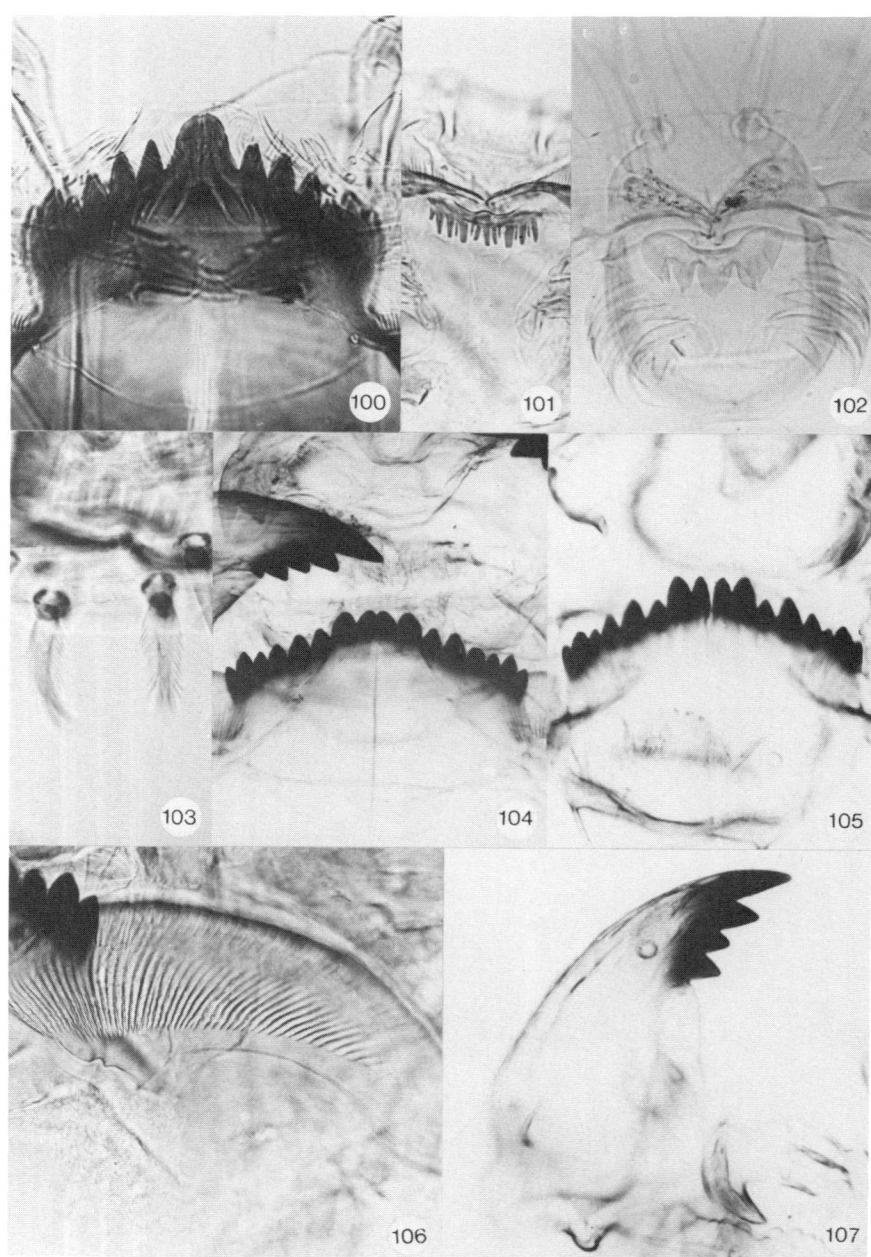

Figs. 100–107. 100, *Einfeldia* – hypostoma; 101, 102, *Einfeldia* – pecten epipharyngis; 103, *Einfeldia* – SI; 104, 105, *Endochironomus* – hypostoma; 106, *Endochironomus* – paralabial plate; 107, *Endochironomus* – mandible.

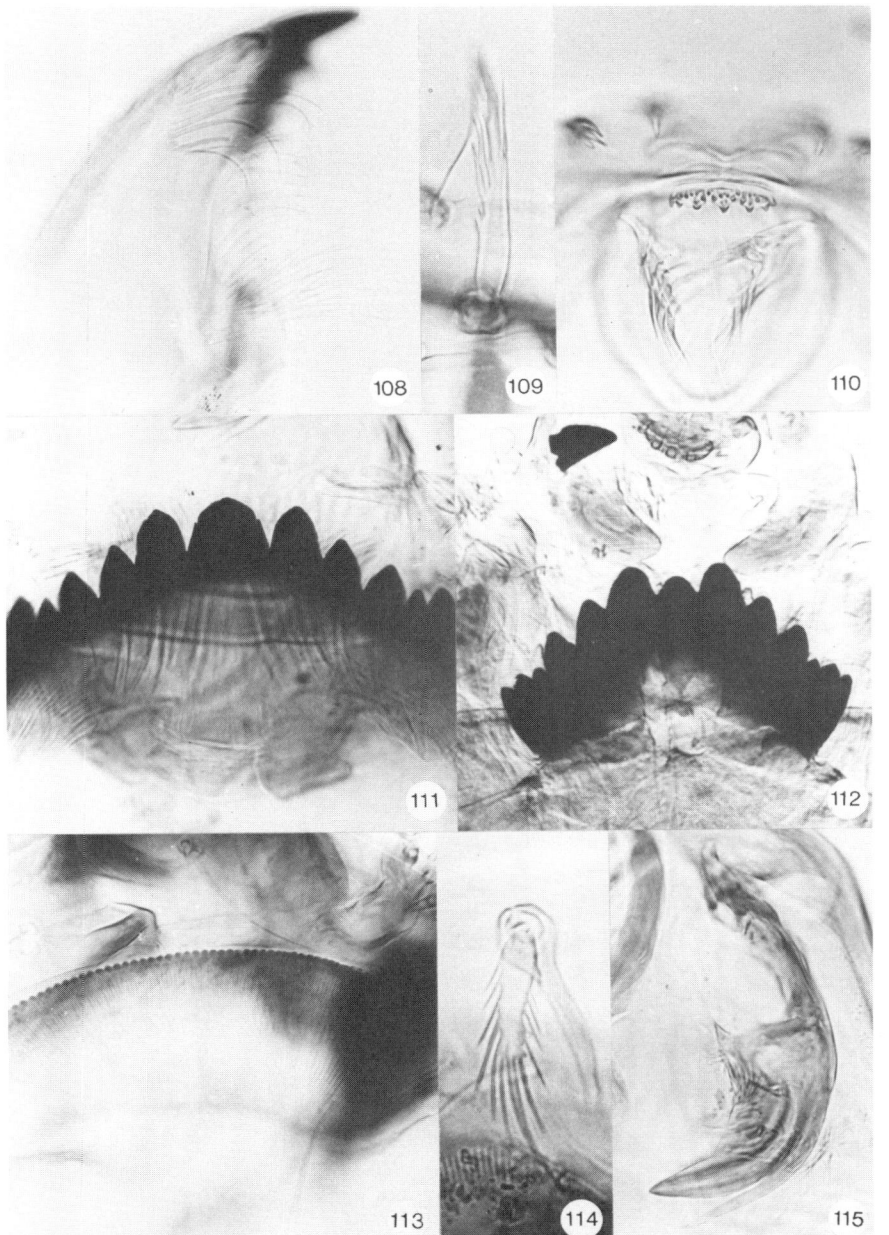

Figs. 108–115. 108, *Endochironomus* – pecten mandibularis; 109, *Endochironomus* – SII; 110, *Endochironomus* – pecten epipharyngis; 111, *Glyptotendipes* (*Phytotendipes*) – hypostoma; 112, *Glyptotendipes* (*Glyptotendipes*) – hypostoma; 113, *Glyptotendipes* (*Phytotendipes*) – paralabial plate; 114, *Glyptotendipes* (*Phytotendipes*) – SI; 115, *Glyptotendipes* (*Phytotendipes*) – premandible.

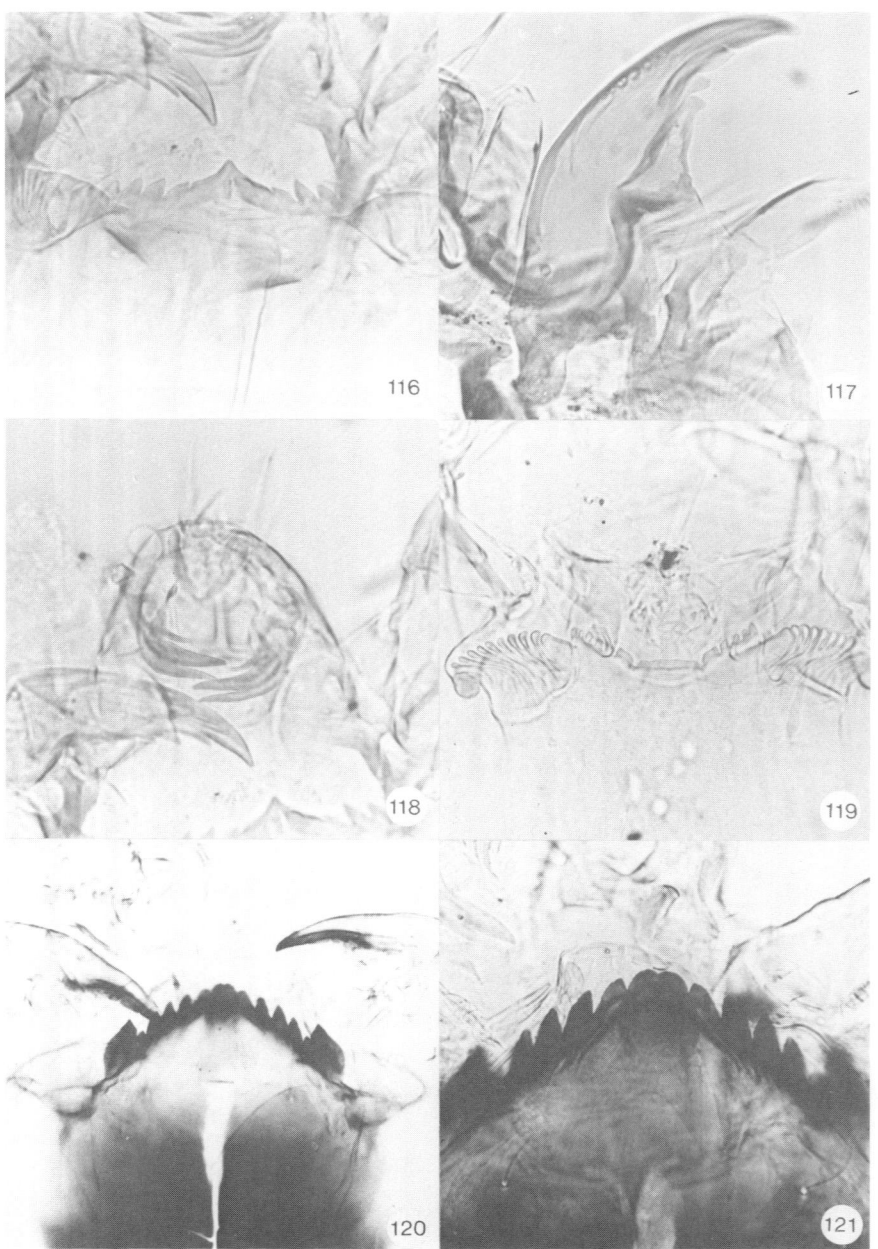

Figs. 116–121. 116, *Acalcarella* – hypostoma; 117, *Acalcarella* – mandible; 118, *Acalcarella* – premandibles; 119, *Chernovskiia* – hypostoma; 120, 121, *Cladopelma* – hypostoma.

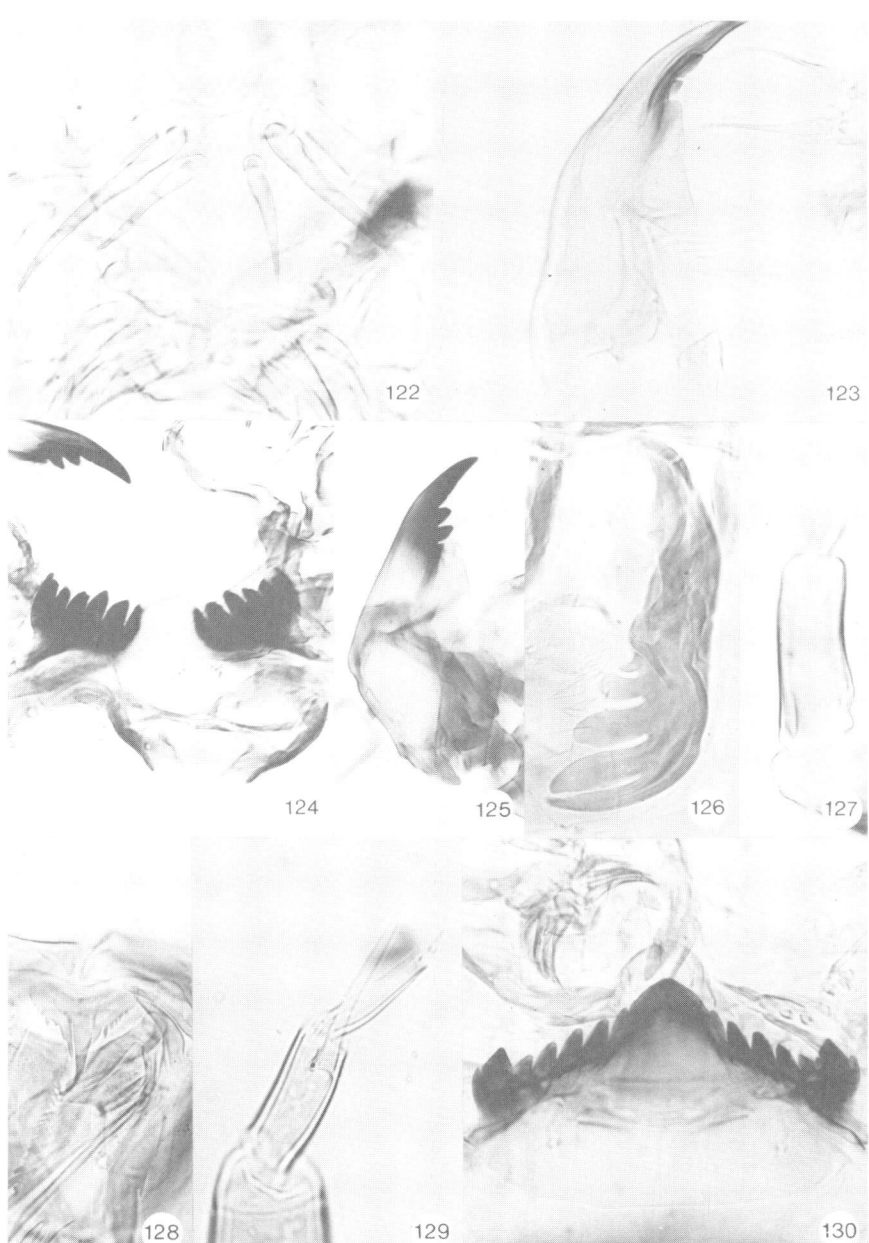

Figs. 122–130. 122, *Cladopelma* – SI and SII; 123, *Cladopelma* – mandible; 124, *Cryptochironomus* – hypostoma; 125, *Cryptochironomus* – mandible; 126, *Cryptochironomus* – premandible; 127, *Cryptochironomus* – maxillary palpus; 128, *Cryptochironomus* – pecten epipharyngis; 129, *Cryptochironomus* – antennae; 130, *Cryptotendipes* – hypostoma.

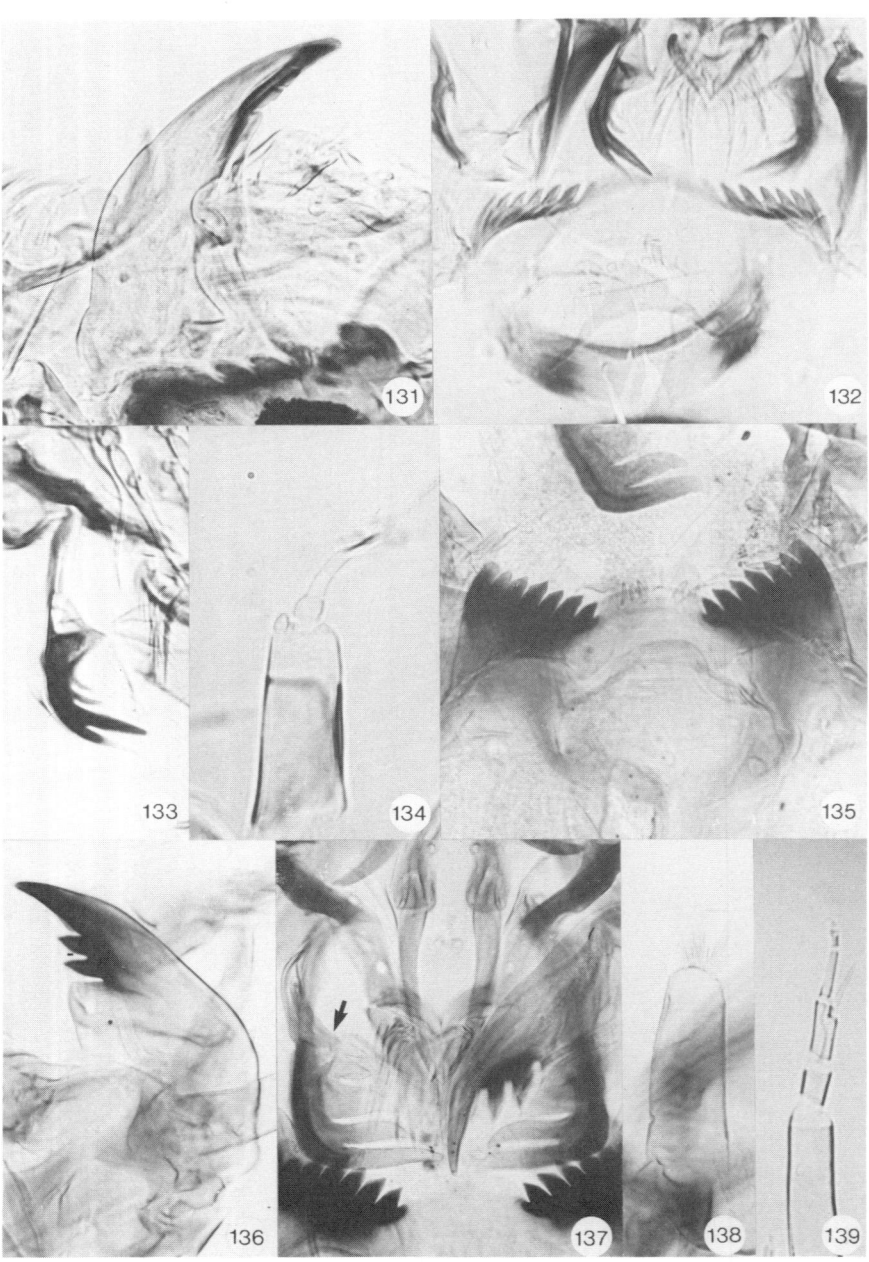

Figs. 131–139. 131, *Cryptotendipes* – mandible; 132, *Cyphomella* – hypostoma; 133, *Cyphomella* – premandible; 134, *Cyphomella* antenna; 135, *Demicryptochironomus* – hypostoma; 136, *Demicryptochironomus* – mandible; 137, *Demicryptochironomus* – premandibles; 138, *Demicryptochironomus* – maxillary palpus; 139, *Demicryptochironomus* – antenna.

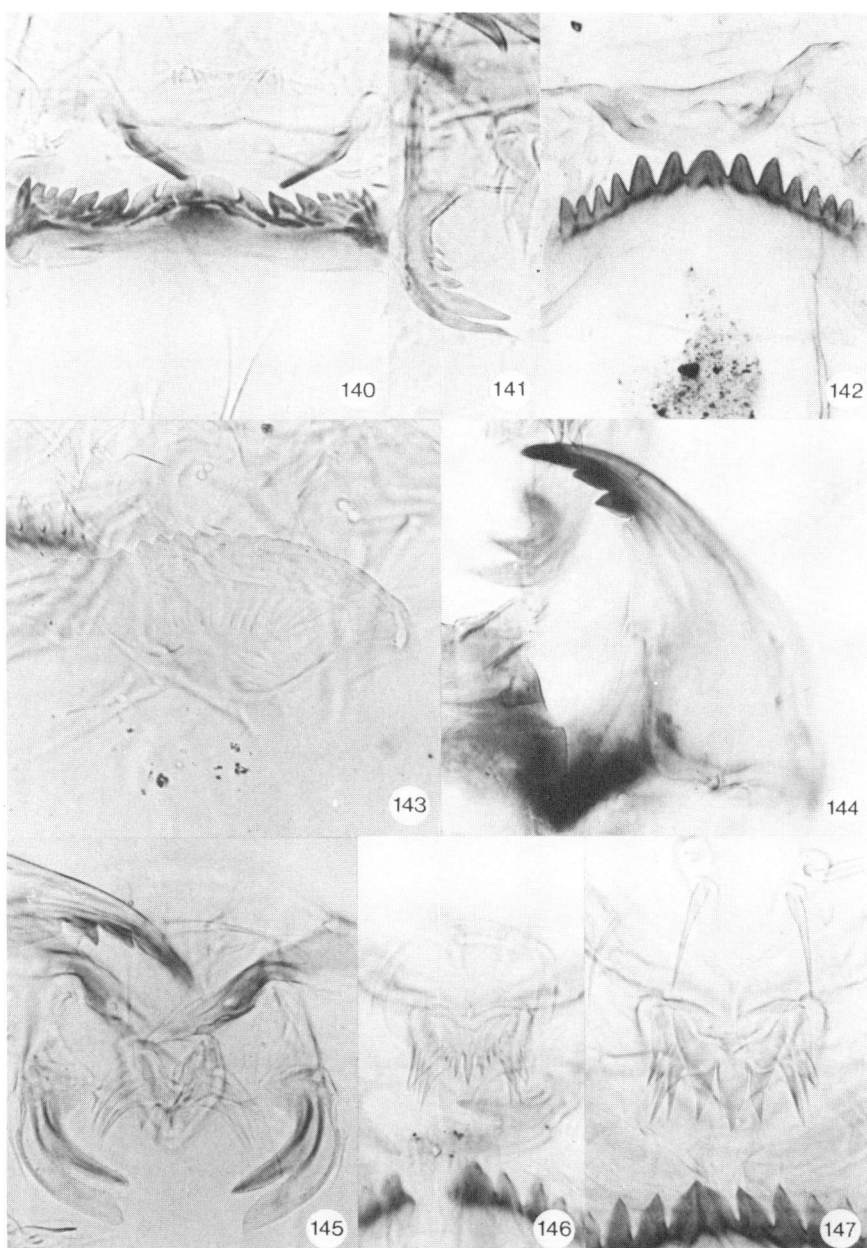

Figs. 140–147. 140, *Harnischia* – hypostoma; 141, *Harnischia* – premandible; 142, *Parachironomus* – hypostoma; 143, *Parachironomus* – paralabial plate; 144, *Parachironomus* – mandible; 145, *Parachironomus* – premandibles; 146, *Parachironomus* – pecten epipharyngis; 147, *Parachironomus* – pecten epipharyngis and SI.

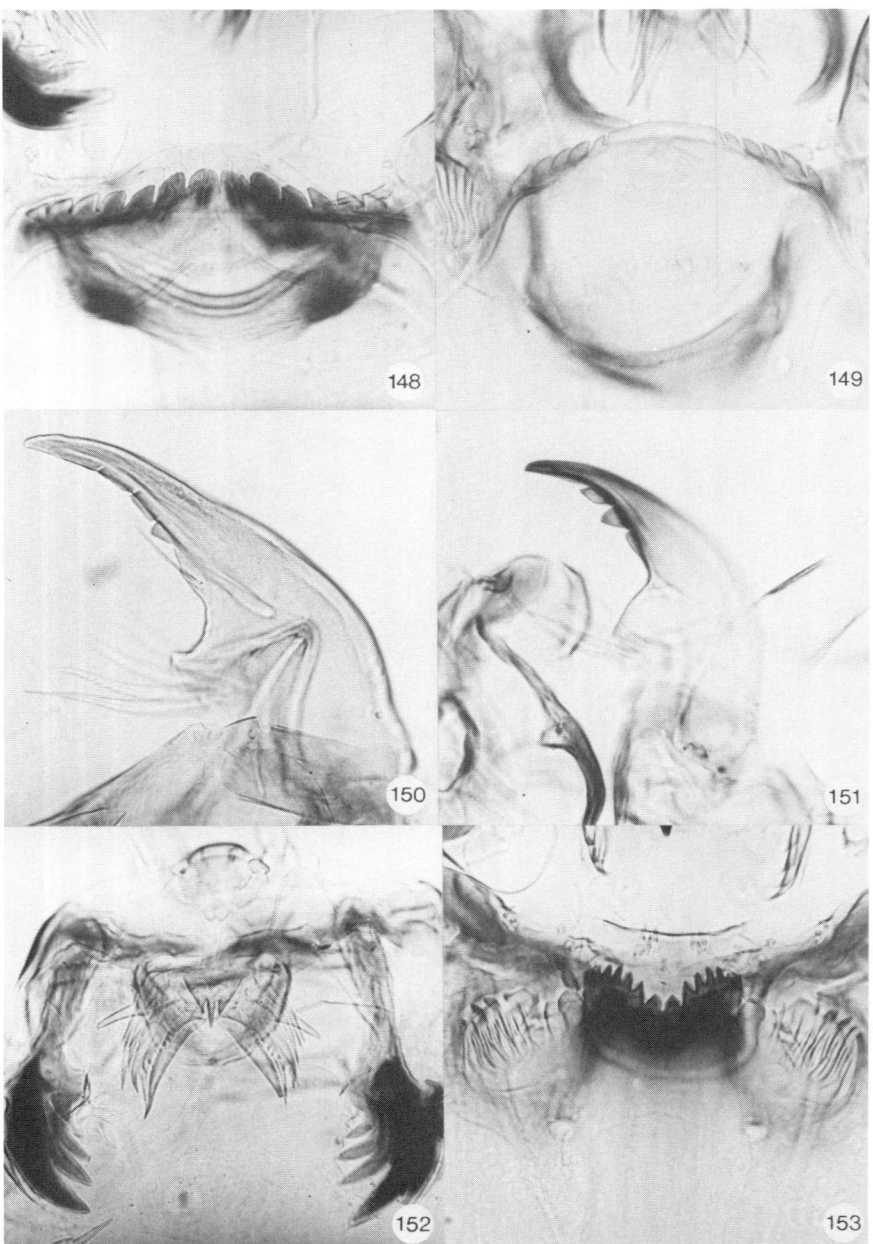

Figs. 148–153. 148, 149, *Paracladopelma* – hypostoma; 150, 151, *Paracladopelma* – mandible; 152, *Paracladopelma* – pecten epipharyngis and premandibles; 153, *Robackia* – hypostoma.

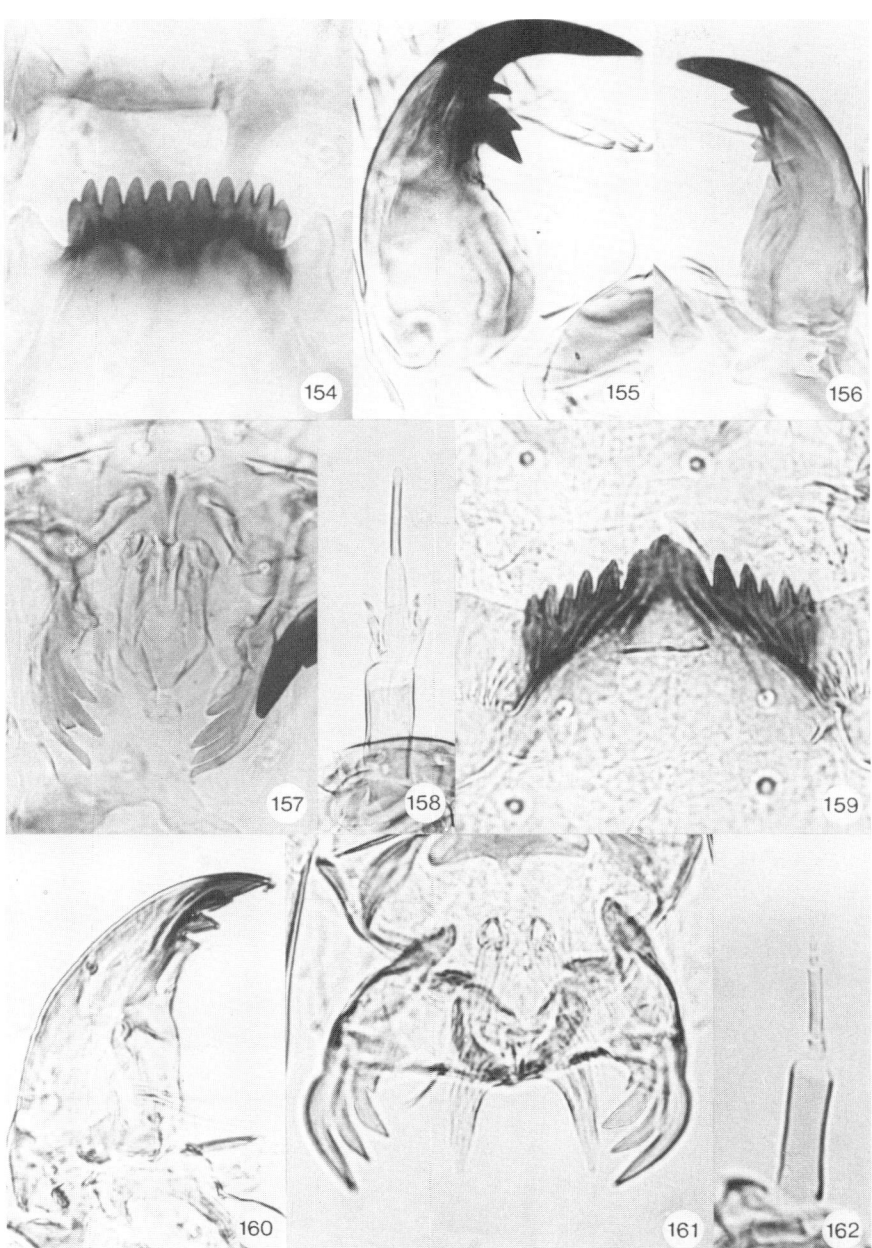

Figs. 154–162. 154, *Robackia* – hypostoma; 155, 156, *Robackia* – mandible; 157, *Robackia* – premandibles; 158, *Robackia* – maxillary palpus; 159, *Saetheria* – hypostoma; 160, *Saetheria* – mandible; 161, *Saetheria* – premandibles; 162, *Saetheria* – antenna.

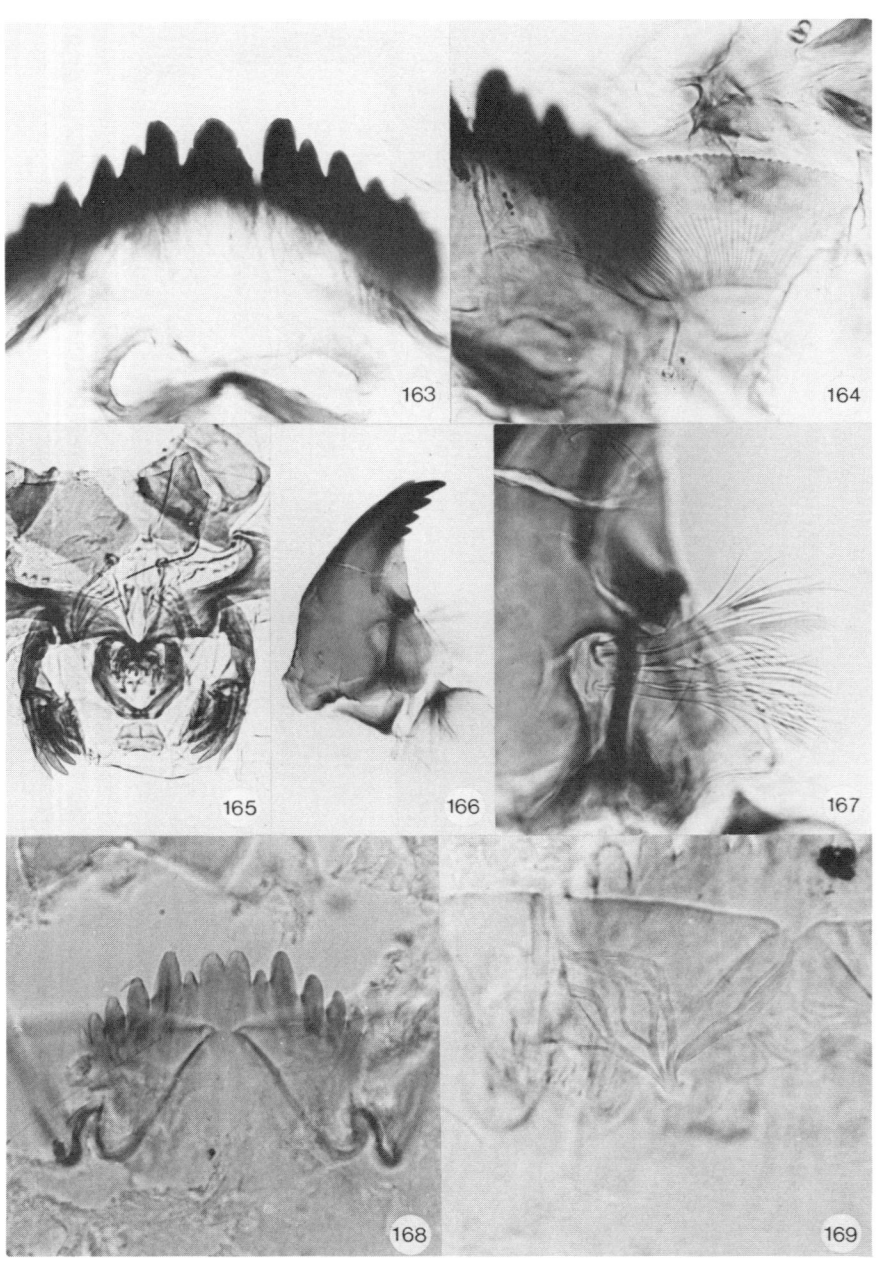

Figs. 163–169. 163, *Kiefferulus* – hypostoma; 164, *Kiefferulus* – paralabial plate; 165, *Kiefferulus* – premandibles; 166, *Kiefferulus* – mandible; 167, *Kiefferulus* – seta interna; 168, *Lauterborniella* – hypostoma and apices of paralabial plates; 169, *Lauterborniella* – plumose seta posterior to hypostoma.

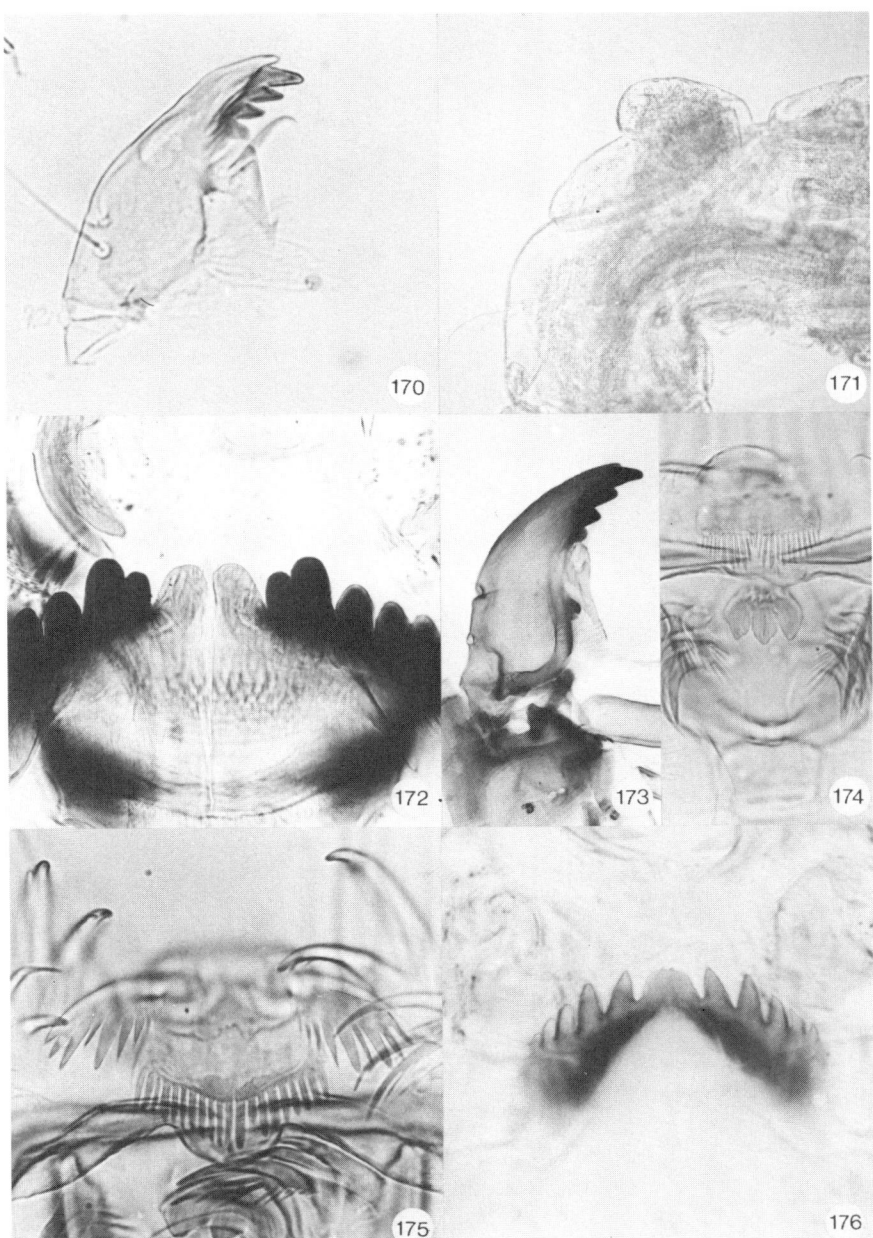

Figs. 170–176. 170, *Lauterborniella* – mandible; 171, *Lauterborniella* – abdominal segment 8; dorsomedial conical projection; 172, *Microtendipes* – hypostoma; 173, *Microtendipes* – mandible; 174, *Microtendipes* – pecten epipharyngis; 175, *Microtendipes* – SI and labral lamellae; 176, *Nilothauma* – hypostoma.

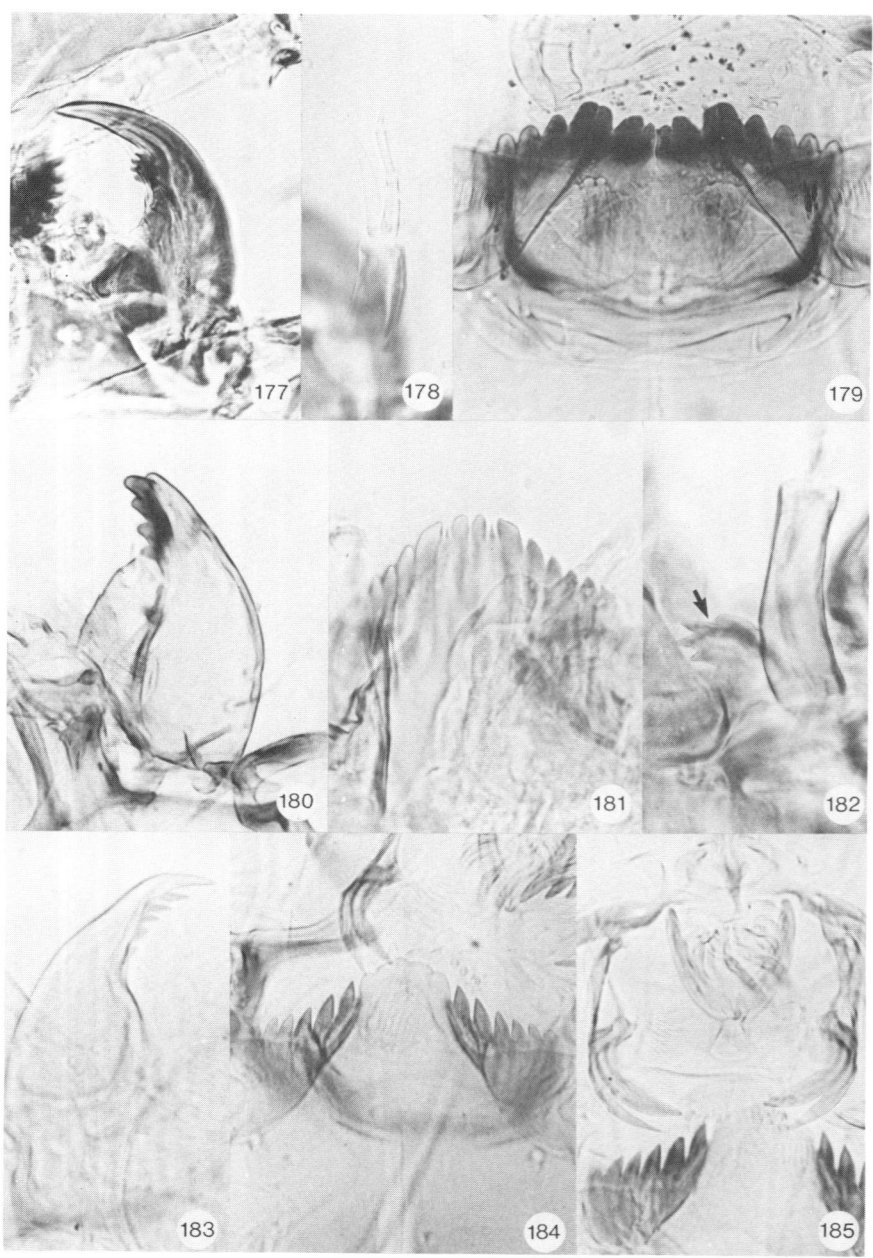

Figs. 177–185. 177, *Nilothauma* – mandible; 178, *Nilothauma* – antenna; 179, *Omisus* – hypostoma; 180, *Omisus* – mandible; 181, *Pagastiella* – hypostoma; 182, *Pagastiella* – antenna and curved projection; 183, *Pagastiella* – mandible; 184, *Paralauterborniella* – hypostoma; 185, *Paralauterborniella* – premandibles.

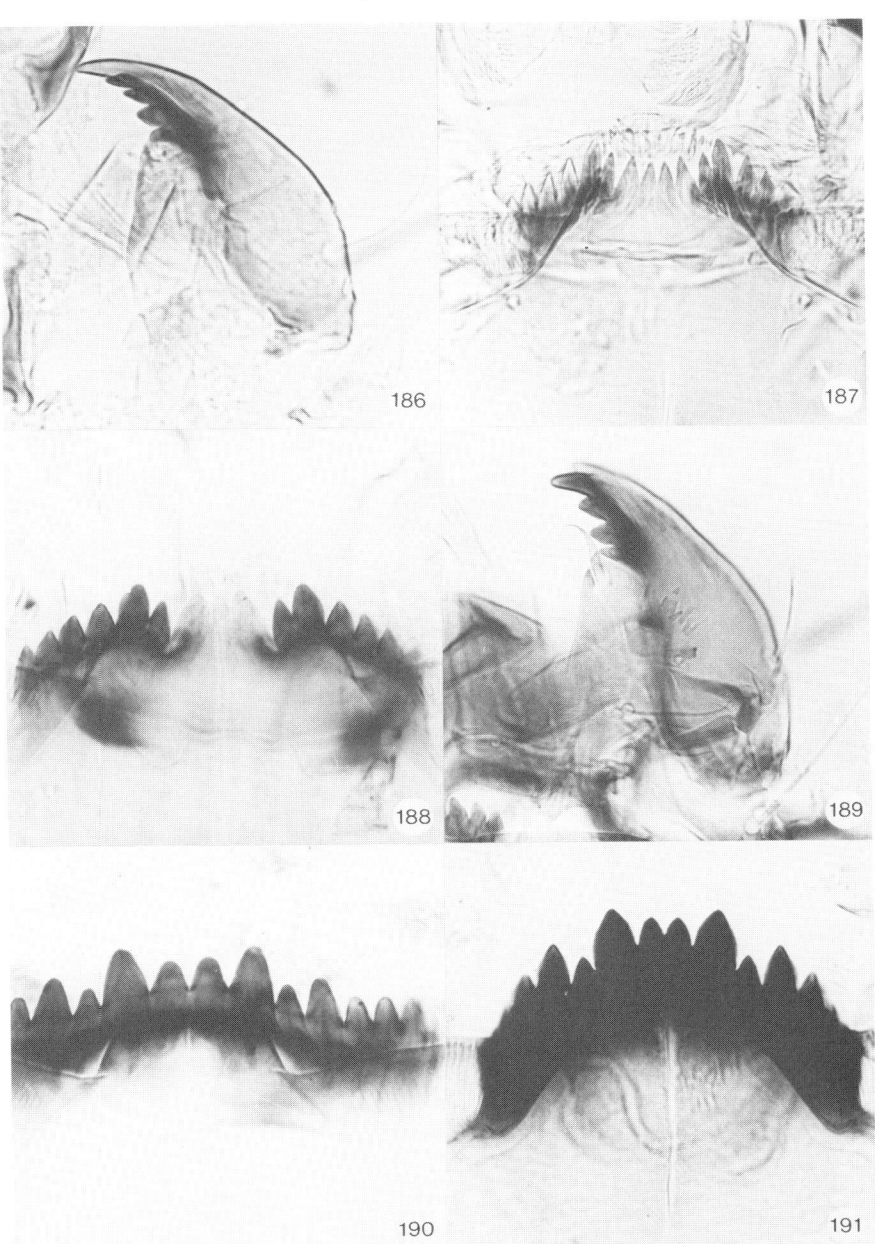

Figs. 186–191. 186, *Paralauterborniella* – mandible; 187, 188, *Paratendipes* – hypostoma; 189, *Paratendipes* – mandible; 190, 191, *Phaenopsectra* – hypostoma.

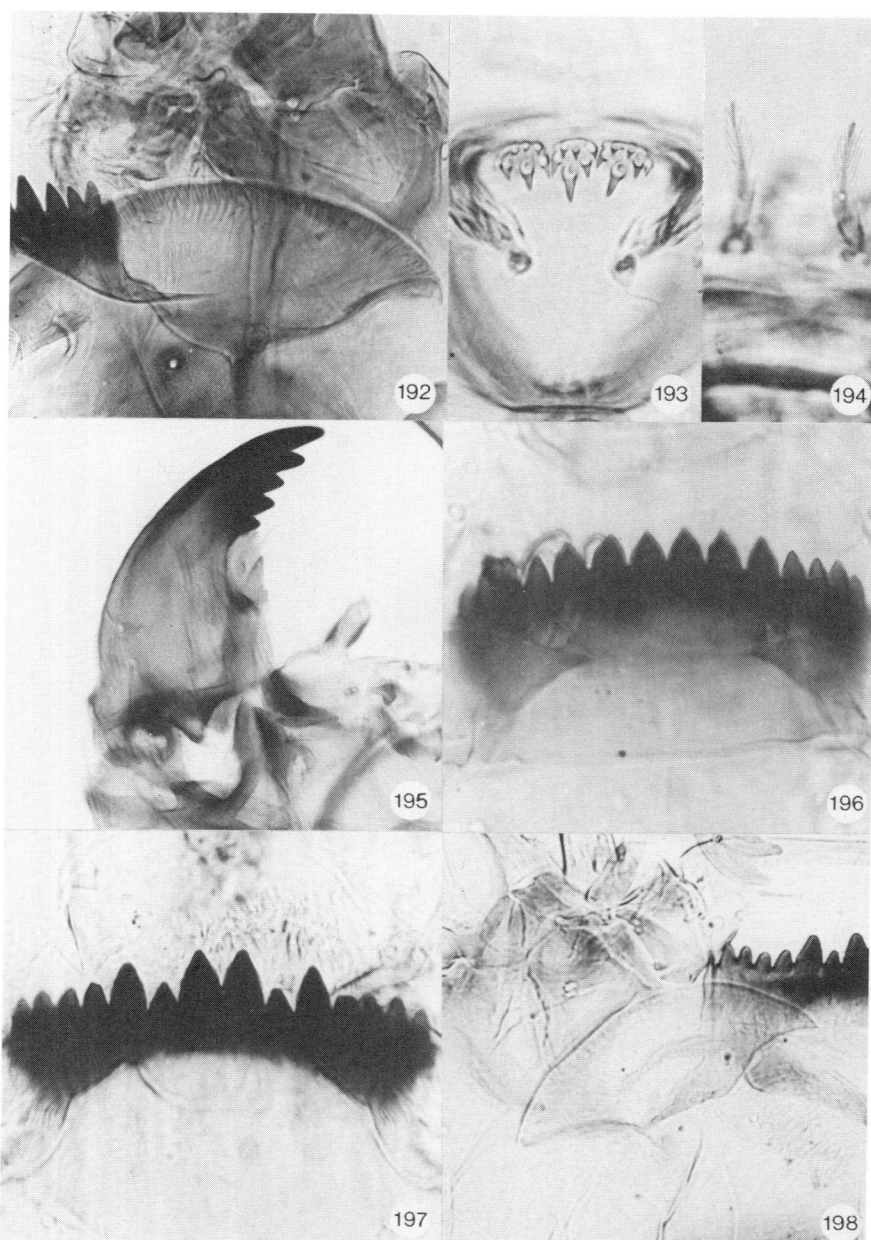

Figs. 192–198. 192, *Phaenopsectra* – paralabial plate; 193, *Phaenopsectra* – pecten epipharyngis; 194, *Phaenopsectra* – SII; 195, *Phaenopsectra* – mandible; 196, 197, *Polypedilum (Pentapedilum)* – hypostoma; 198, *Polypedilum (Polypedilum)* – paralabial plate.

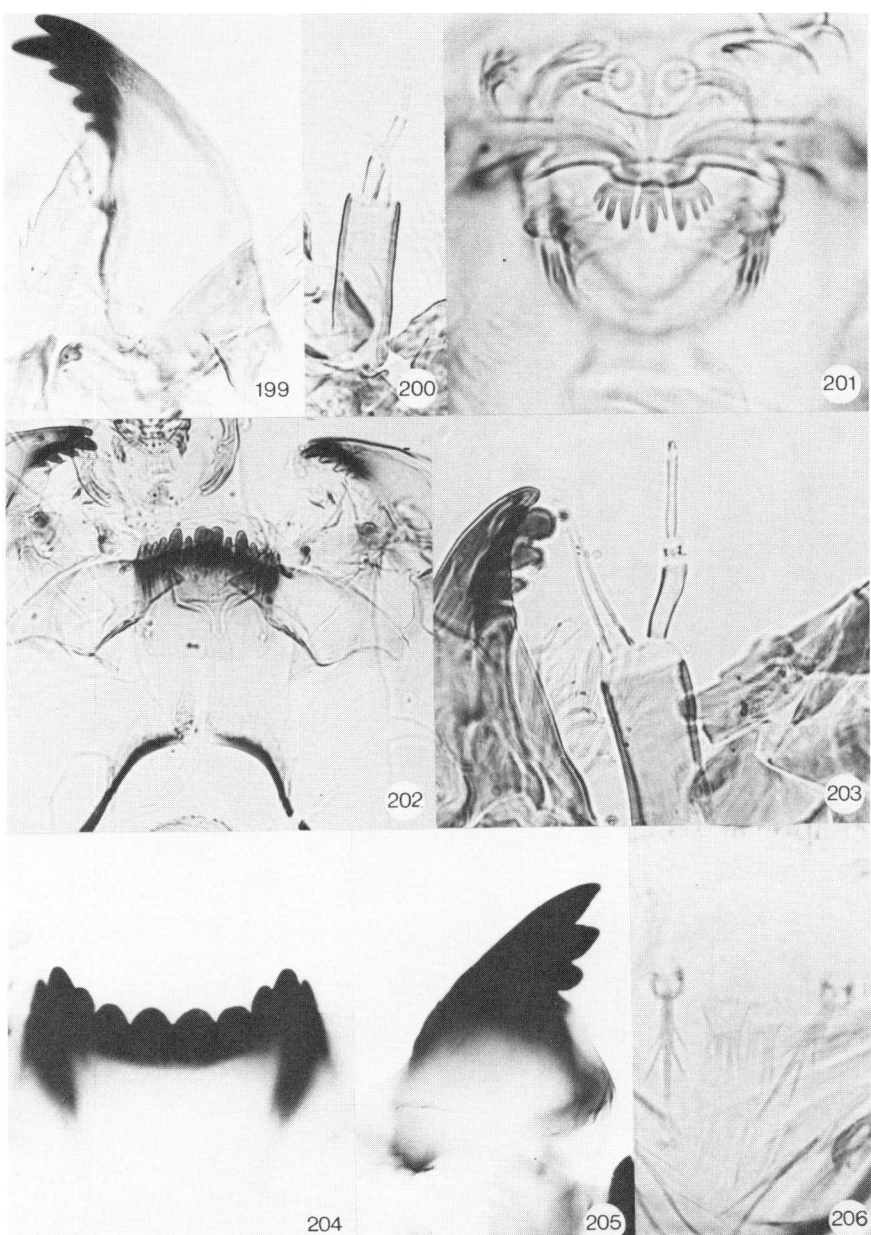

Figs. 199–206. 199, *Polypedilum (Polypedilum)* – mandible; 200, *Polypedilum (Polypedilum)* – antenna; 201, *Polypedilum (Tripodura)* – pecten epipharyngis; 202, *Polypedilum (Tripodura)* – paralabial plates; 203, *Polypedilum (Tripodura)* – antenna; 204, *Stenochironomus* – hypostoma; 205, *Stenochironomus* – mandible; 206, *Stenochironomus* – SI.

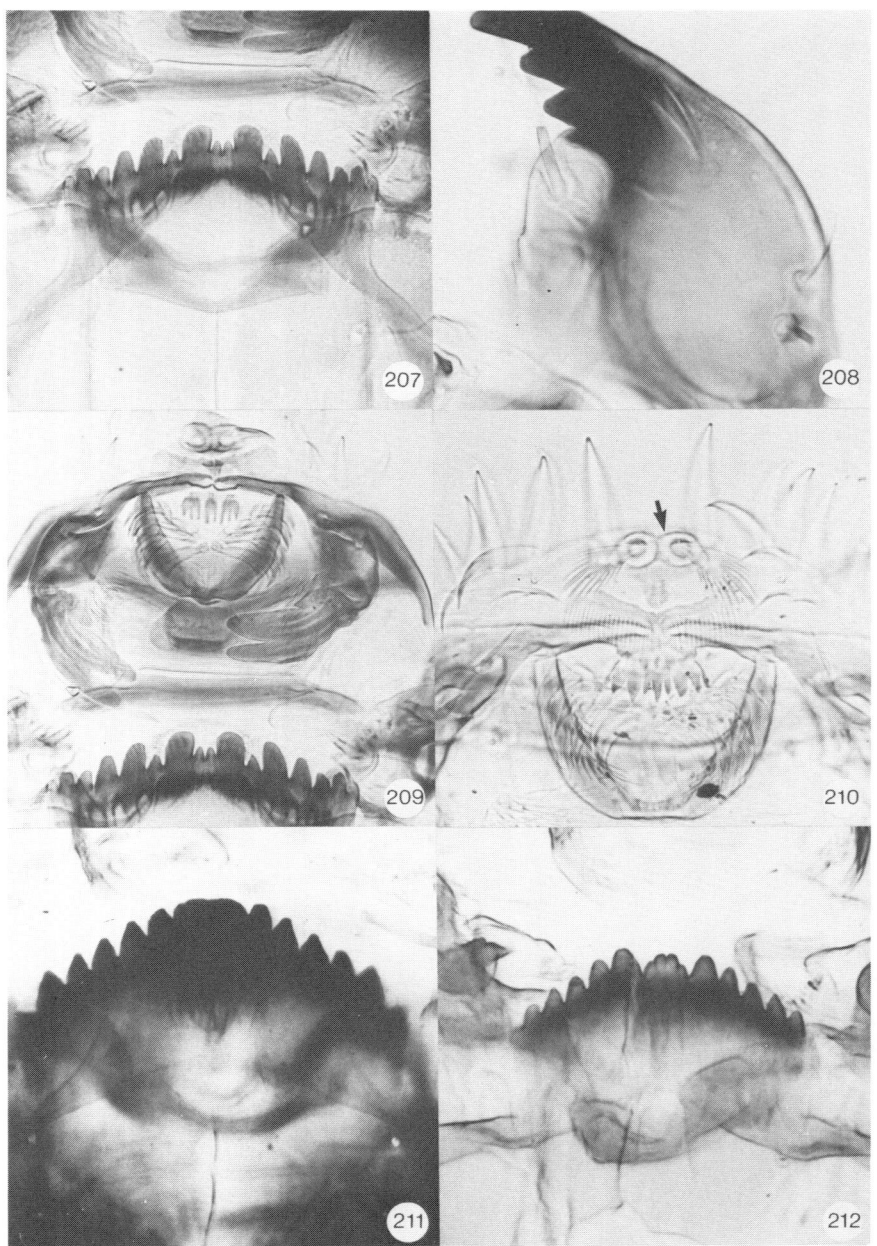

Figs. 207–212. 207, *Stictochironomus* – hypostoma; 208, *Stictochironomus* – mandible; 209, *Stictochironomus* – premandibles and pecten epipharyngis; 210, *Stictochironomus* – SI; 211, *Xenochironomus* (*Axarus*) – hypostoma; 212, *Xenochironomus* (*Axarus*) – hypostoma.

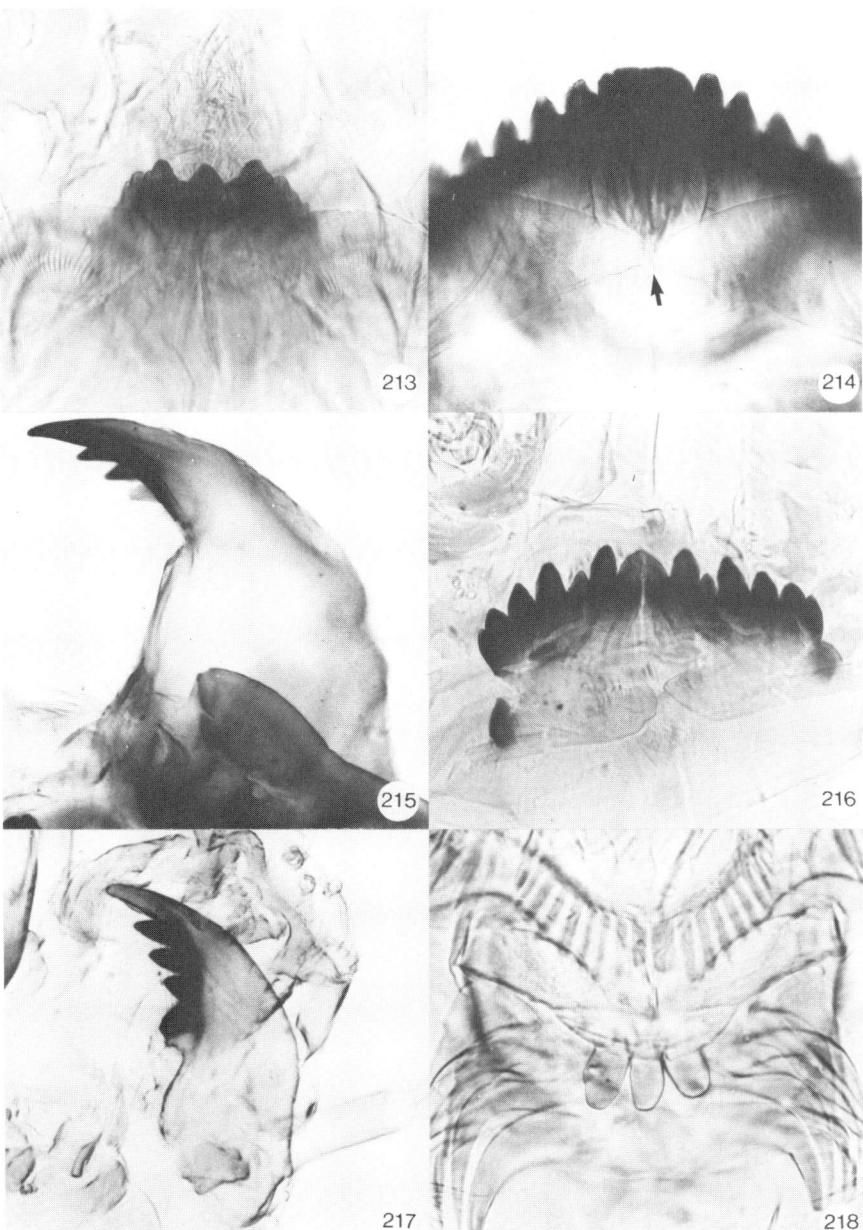

Figs. 213–218. 213, *Xenochironomus* (*Xenochironomus*) – hypostoma; 214, *Xenochironomus* (*Axarus*) – apices of paralabial plates; 215, *Xenochironomus* (*Axarus*) – mandible; 216, *Pseudochironomus* – hypostoma and paralabial plates; 217, *Pseudochironomus* – mandible; 218, *Pseudochironomus* – pecten epipharyngis.

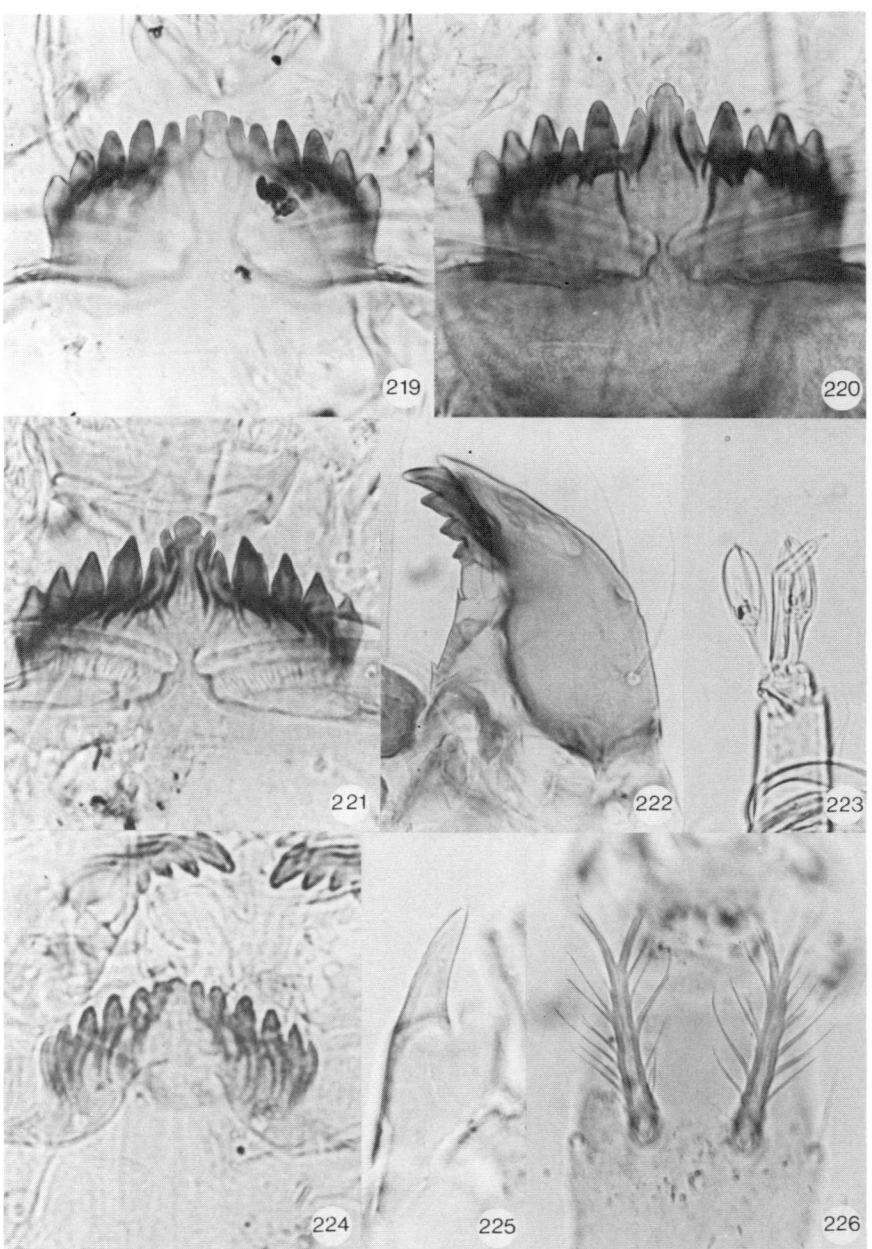

Figs. 219–226. 219–221, *Cladotanytarsus* – hypostoma; 222, *Cladotanytarsus* – mandible; 223, *Cladotanytarsus* – Lauterborn's organs; 224, *Constempellina* – hypostoma; 225, *Constempellina* – apicomesal projection on antennal tubercle; 226, *Constempellina* – plumose setae on frontoclypeal apotome.

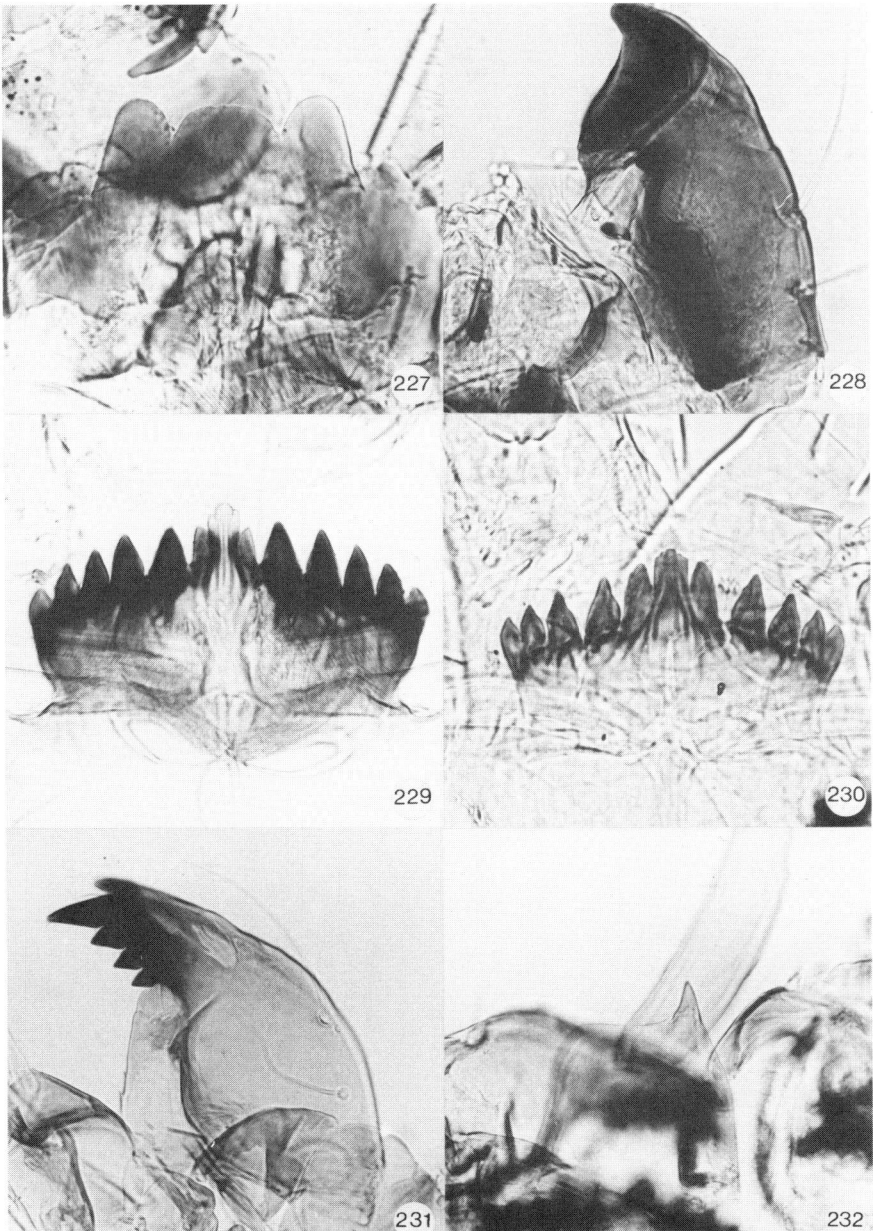

Figs. 227–232. 227, *Corynocera* – hypostoma; 228, *Corynocera* – mandible; 229, 230, *Micropsectra* – hypostoma; 231, *Micropsectra* – mandible; 232, *Micropsectra* – apicomesal projection on antennal tubercle.

205

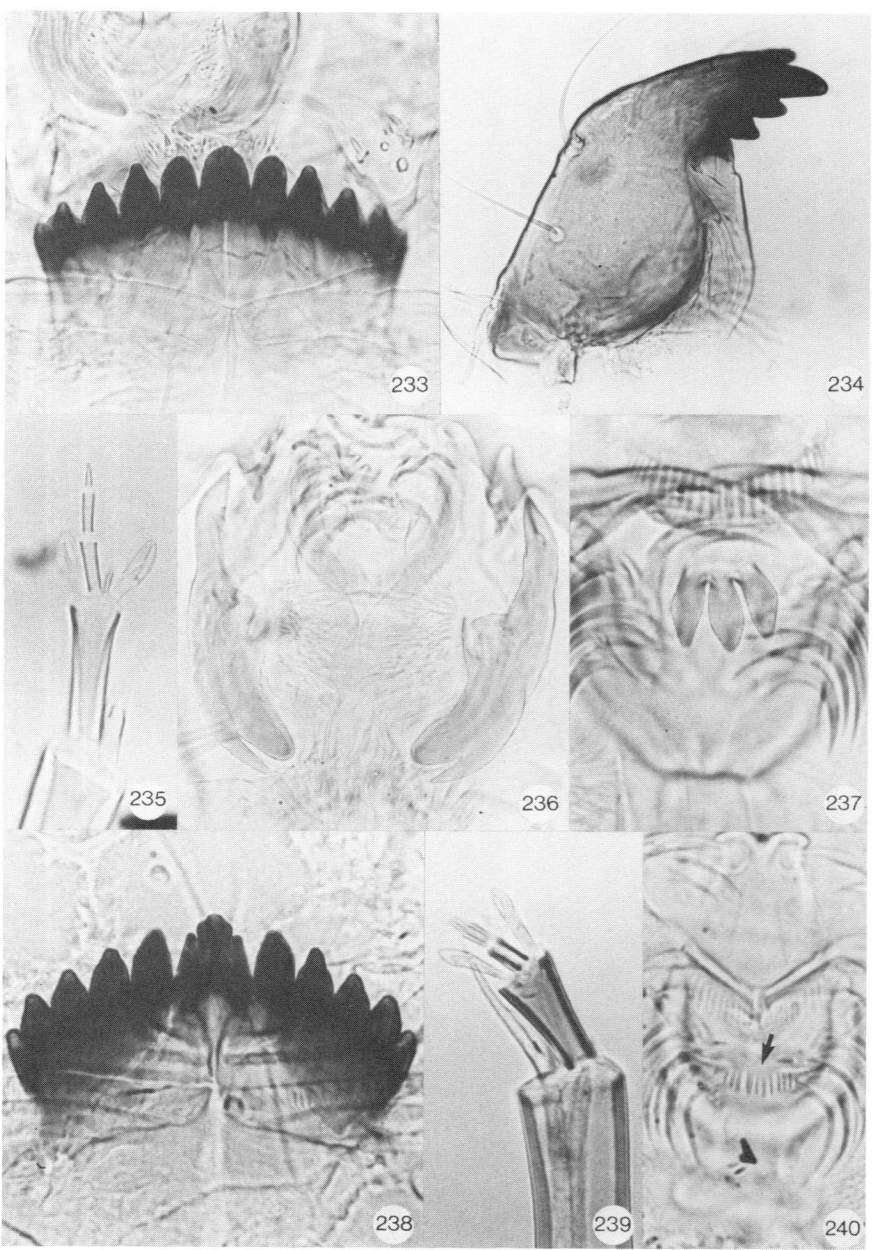

Figs. 233–240. 233, *Paratanytarsus* – hypostoma; 234, *Paratanytarsus* – mandible; 235, *Paratanytarsus* – Lauterborn's organs; 236, *Paratanytarsus* – premandibles; 237, *Paratanytarsus* – pecten epipharyngis; 238, *Rheotanytarsus* – hypostoma; 239, *Rheotanytarsus* – Lauterborn's organs; 240, *Rheotanytarsus* – pecten epipharyngis.

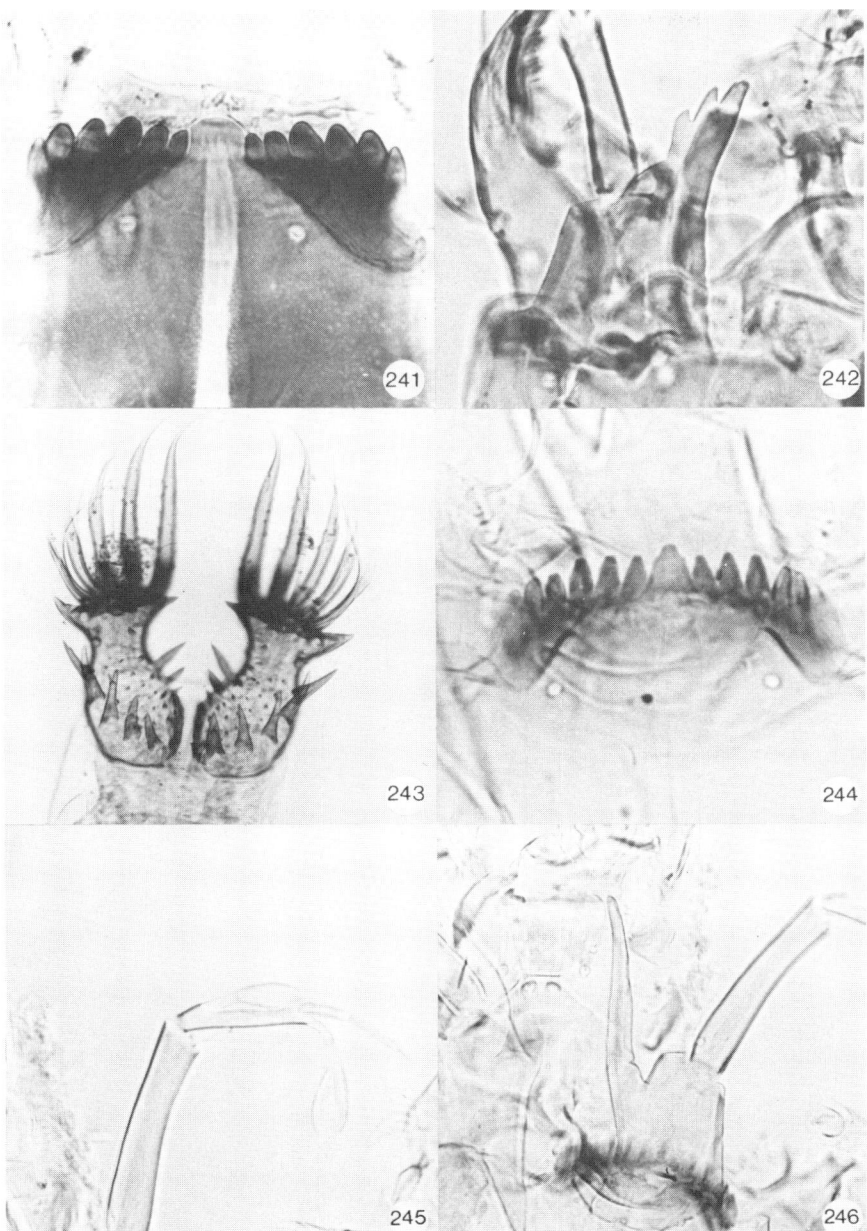

Figs. 241–246. 241, *Stempellina* – hypostoma; 242, *Stempellina* – apicomesal projection on antennal tubercle; 243, *Stempellina* – procercus; 244, *Stempellinella* – hypostoma and apices of paralabial plates; 245, *Stempellinella* – Lauterborn's organs; 246, *Stempellinella* – apicomesal projection on antennal tubercle.

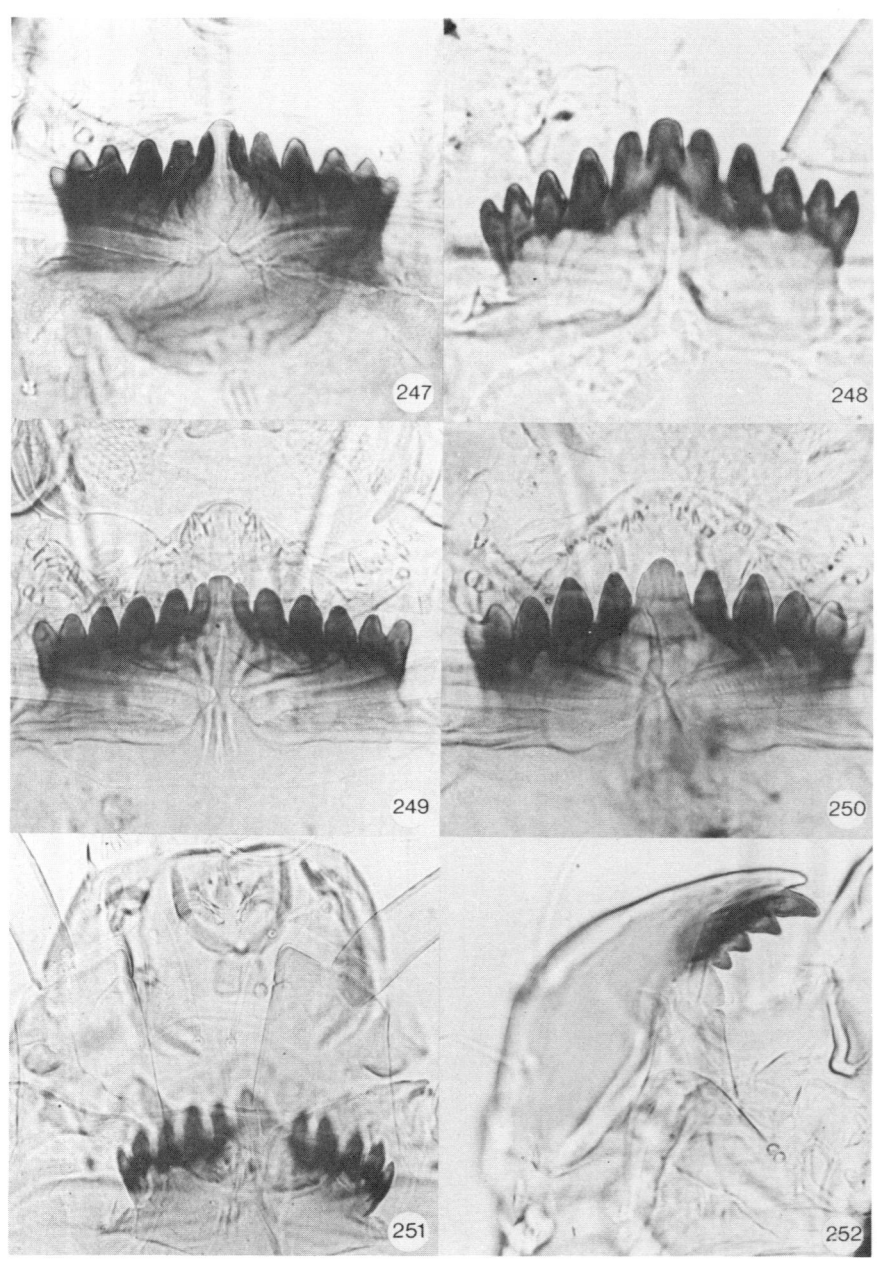

Figs. 247–252. 247–250, *Tanytarsus* – hypostoma; 251, *Tanytarsus* – antennal tubercle; 252, *Tanytarsus* – mandible.

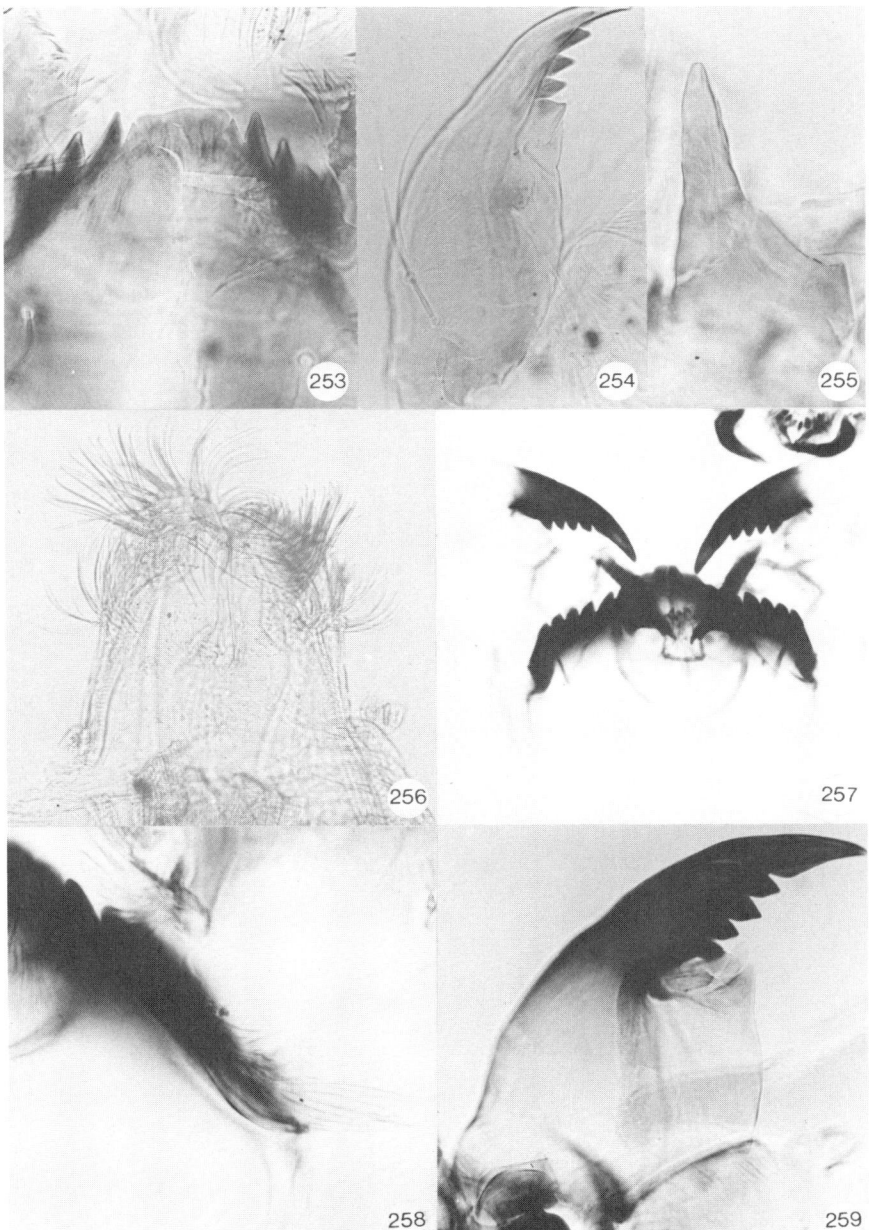

Figs. 253–259. 253, *Abiskomyia* – hypostoma; 254, *Abiskomyia* – mandible; 255, *Abiskomyia* – apicomesal projection on antennal tubercle; 256, *Abiskomyia* – apical and preapical claws on anterior parapod; 257, *Acricotopus* – hypostoma; 258, *Acricotopus* – paralabial plate with setae; 259, *Acricotopus* – mandible.

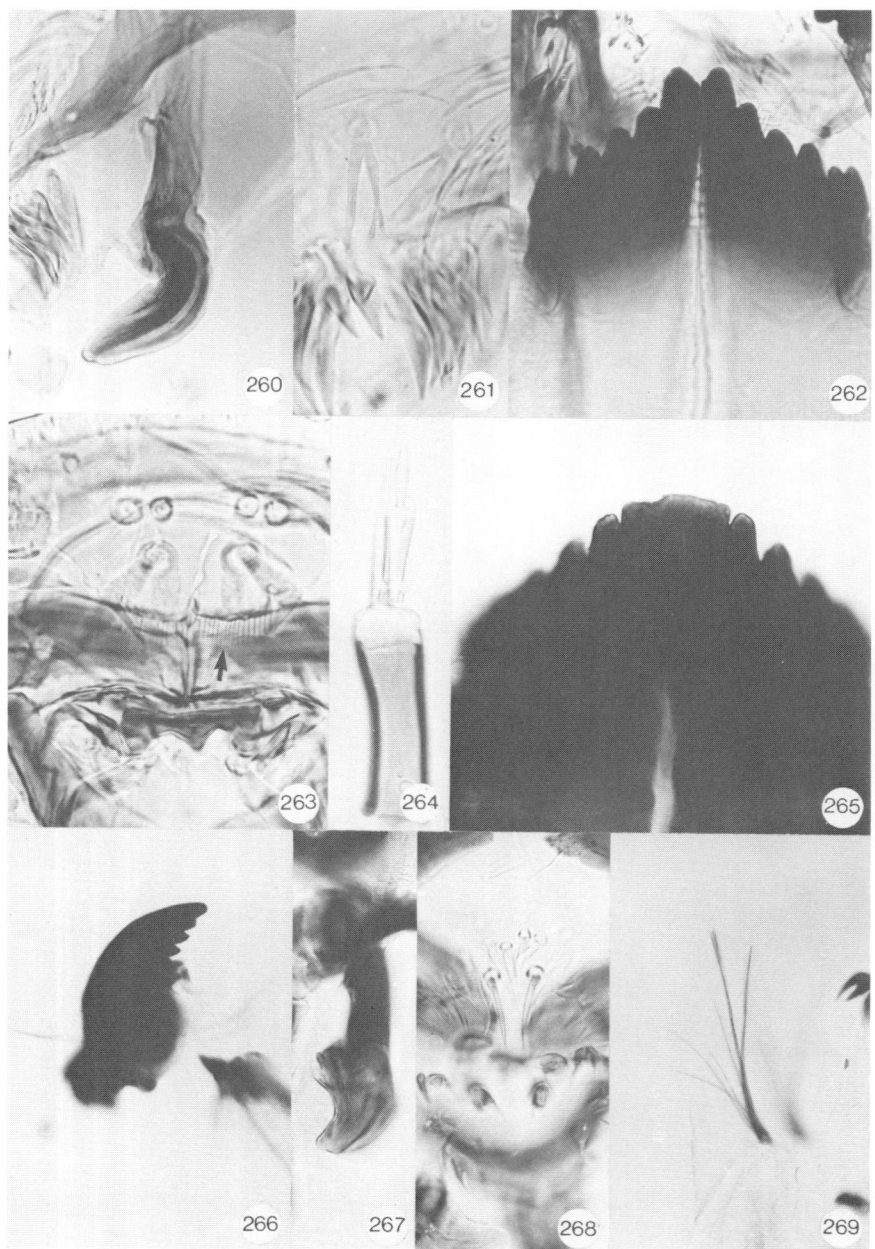

Figs. 260–269. 260, *Acricotopus* – premandible; 261, *Acricotopus* – SI; 262, *Brillia* – hypostoma; 263, *Brillia* – labral lamellae; 264, *Brillia* – antenna; 265, *Cardiocladius* – hypostoma; 266, *Cardiocladius* – mandible; 267, *Cardiocladius* – premandible; 268, *Cardiocladius* – SI; 269, *Cardiocladius* – posterior parapod and anal setae of procerci.

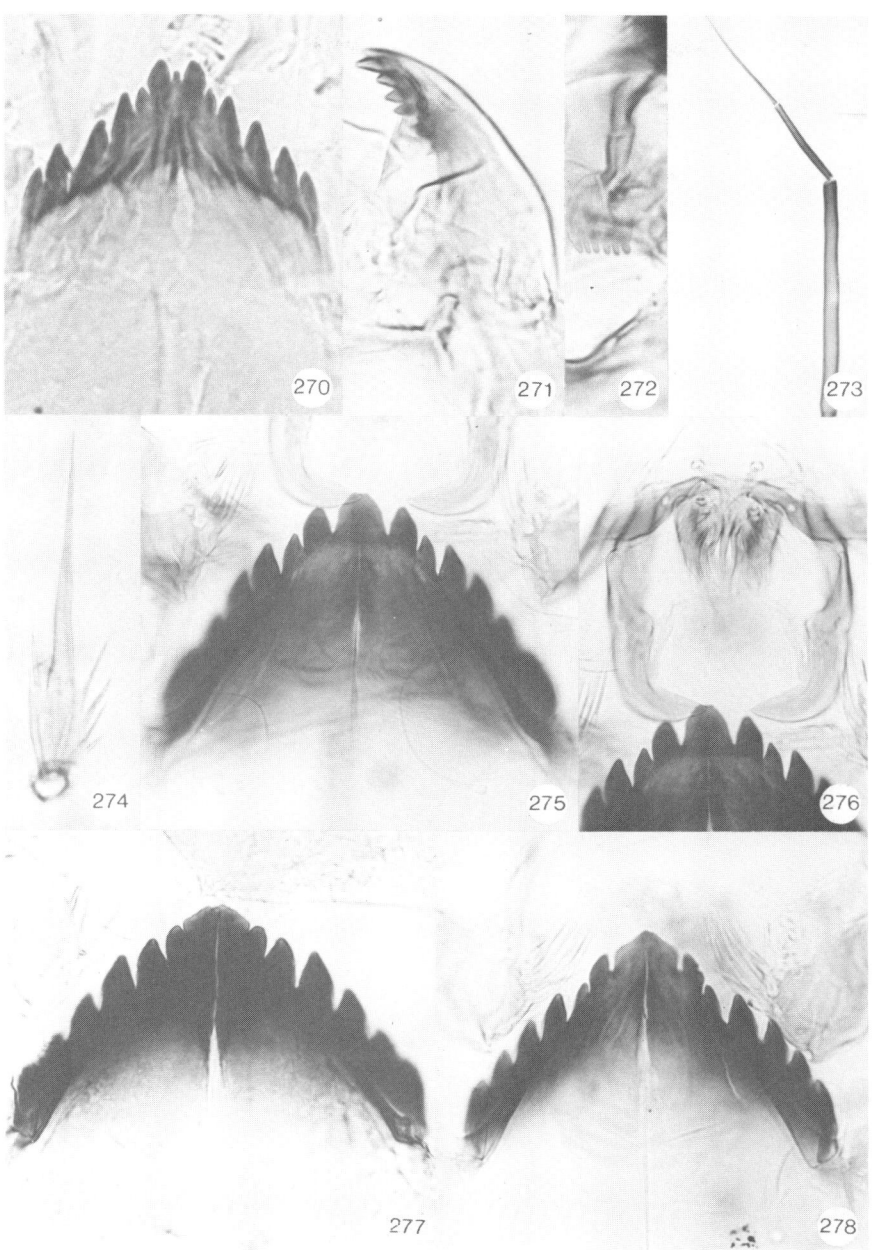

Figs. 270–278. 270, *Corynoneura* – hypostoma; 271, *Corynoneura* – mandible; 272, *Corynoneura* – premandible; 273, *Corynoneura* – antennae; 274, *Corynoneura* – plumose seta on posterior parapod; 275, *Cricotopus (Cricotopus)*; *tremulus* group – hypostoma; 276, *Cricotopus (Cricotopus)*; *tremulus* group – premandible; 277, *Cricotopus (Cricotopus)*; *trifascia* group – hypostoma; 278, *Cricotopus (Cricotopus)*; *cylindraceus* group – hypostoma.

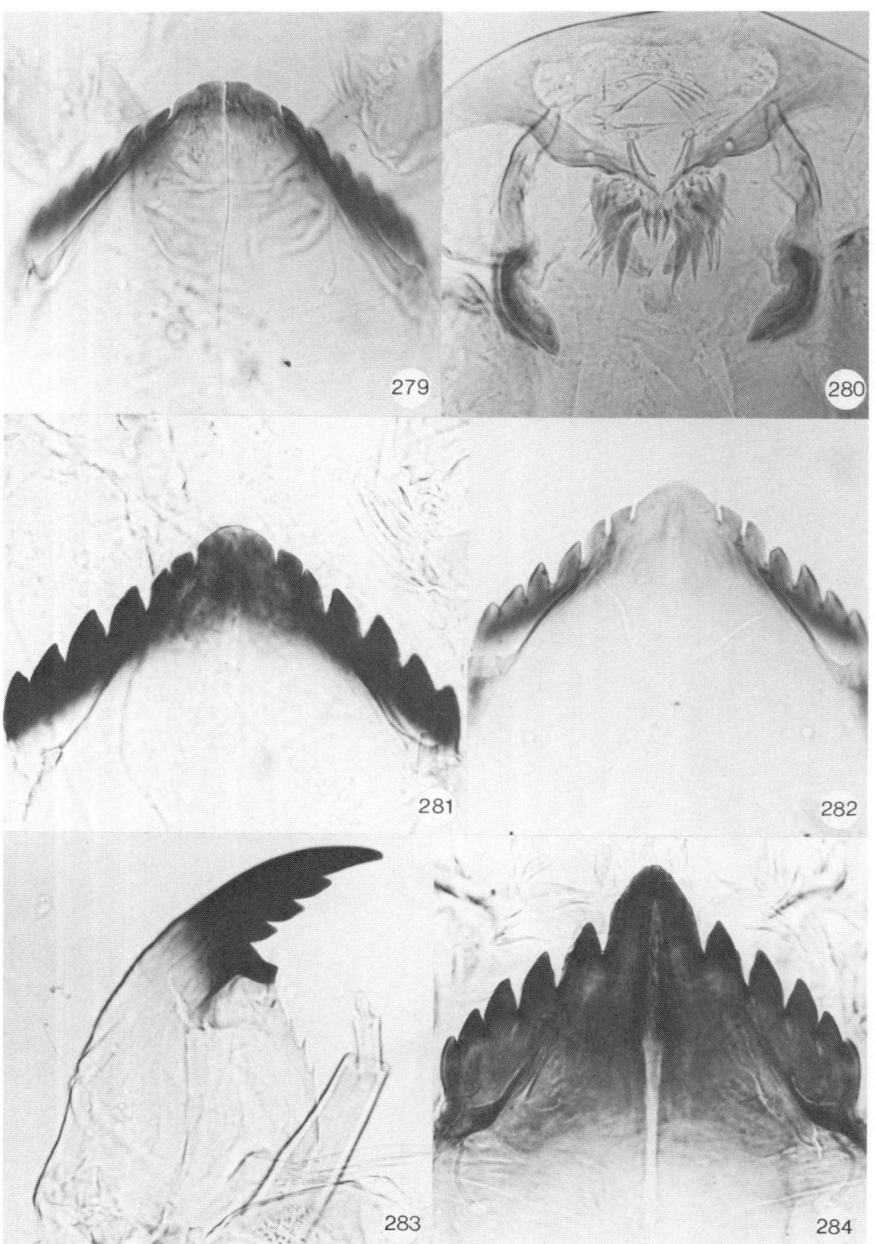

Figs. 279–284. 279, *Cricotopus (Cricotopus)*; *festivellus* group – hypostoma; 280, *Cricotopus* or *Orthocladius* – palatum and premandible; 281, 282, *Cricotopus (Cricotopus)*; *bicinctus* group – hypostoma; 283, *Cricotopus (Cricotopus)*; *bicinctus* group – mandible; 284, *Cricotopus (Isocladius)*; *sylvestris* group – hypostoma.

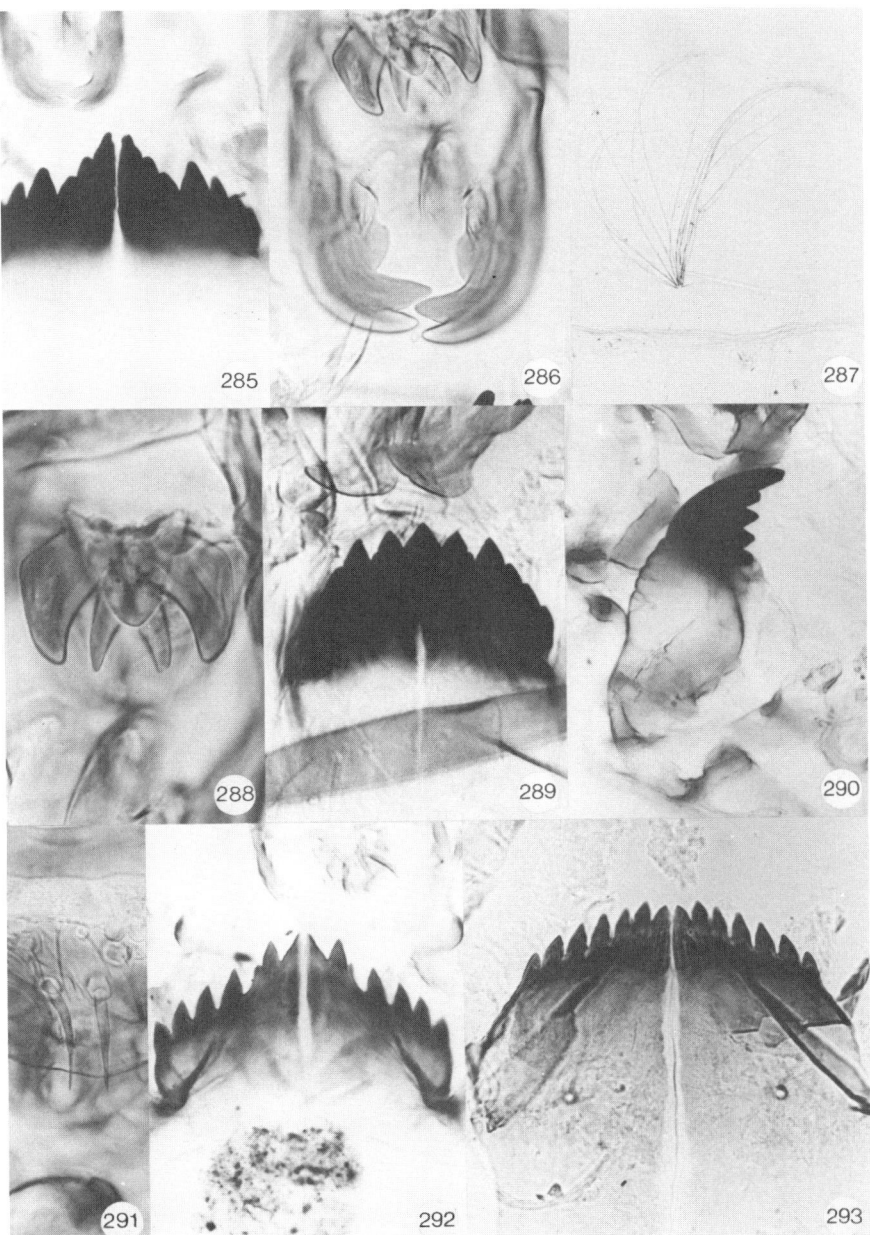

Figs. 285–293. 285, *Cricotopus (Isocladius)*; *sylvestris* group – hypostoma; 286, *Cricotopus (Isocladius)*; *sylvestris* group – premandibles; 287, *Cricotopus (Isocladius)*; *sylvestris* group – plumose abdominal setae; 288, *Cricotopus (Isocladius)*; *sylvestris* group – pecten epipharyngis; 289, *Cricotopus (Isocladius)*; *obnixus* group – hypostoma; 290, *Cricotopus (Isocladius)*; *obnixus* group – mandible; 291, *Cricotopus (Isocladius)*; *obnixus* group – SI; 292, *Cricotopus (Isocladius)*; *laricomalis* group – hypostoma; 293, *Diplocladius* – hypostoma.

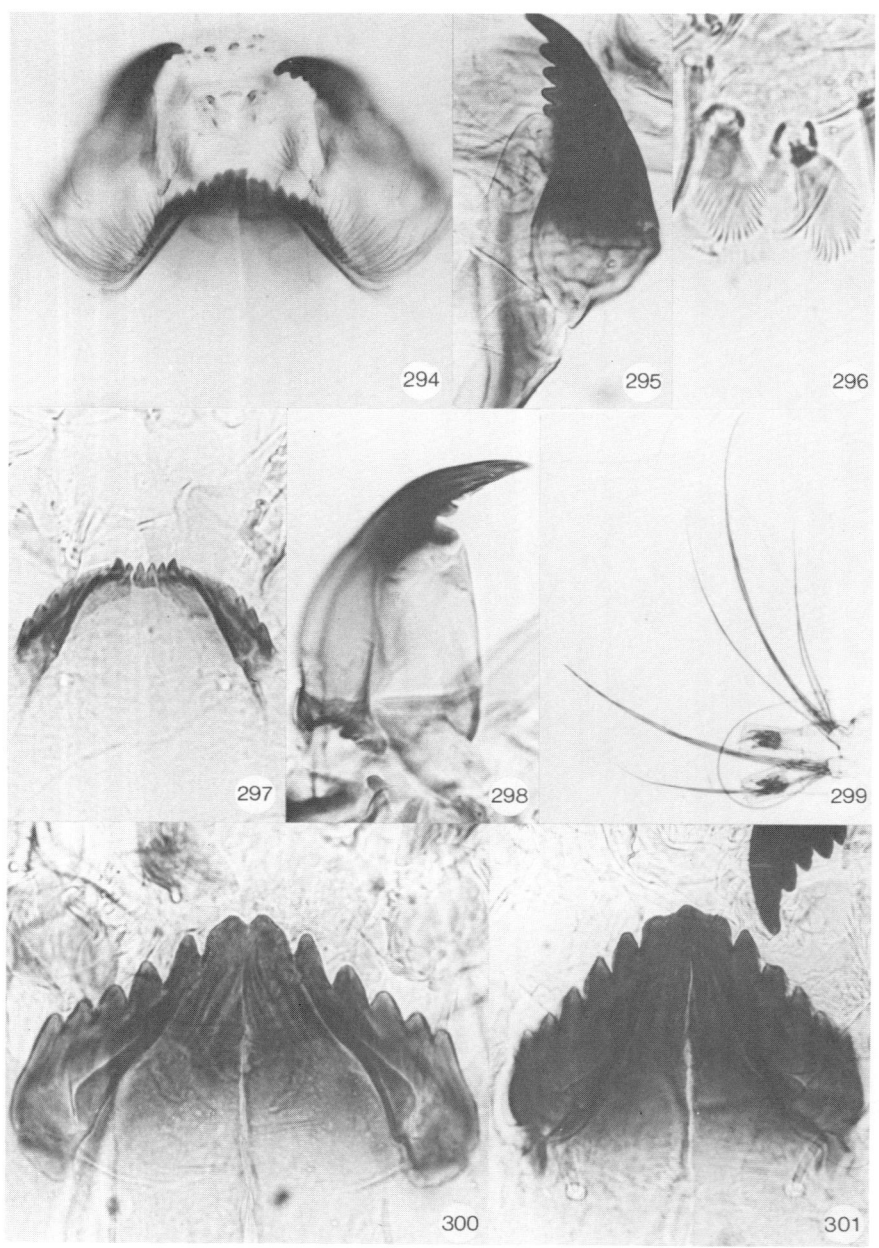

Figs. 294–301. 294, *Diplocladius* – paralabial plate with setae; 295, *Diplocladius* – mandible; 296, *Diplocladius* – SI; 297, *Epoicocladius* – hypostoma; 298, *Epoicocladius* – mandible; 299, *Epoicocladius* – procerci with anal setae; 300, 301, *Eukiefferiella* – hypostoma.

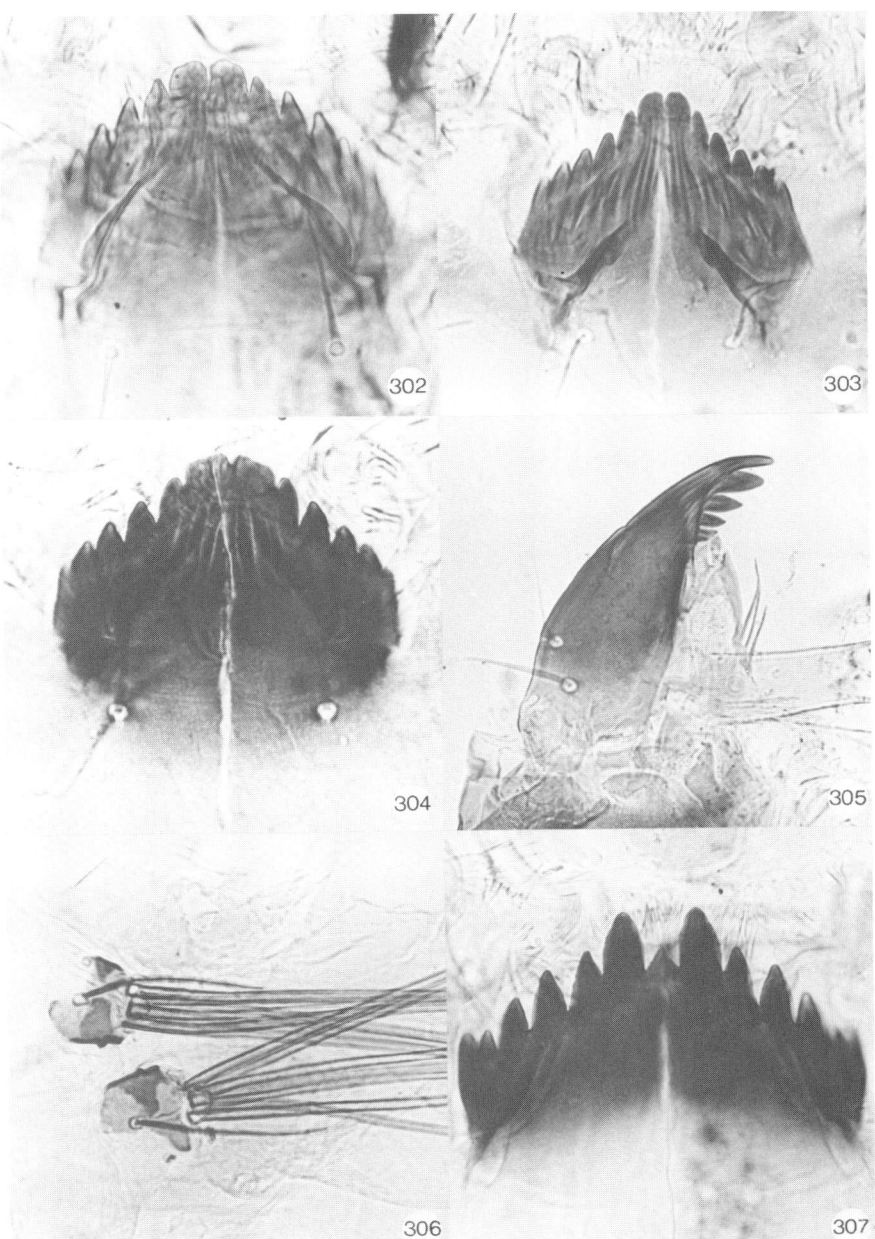

Figs. 302–307. 302–304, *Eukiefferiella* – hypostoma; 305, *Eukiefferiella* – mandible; 306, *Eukiefferiella* – porcerci; 307, *Euryhapsis* – hypostoma.

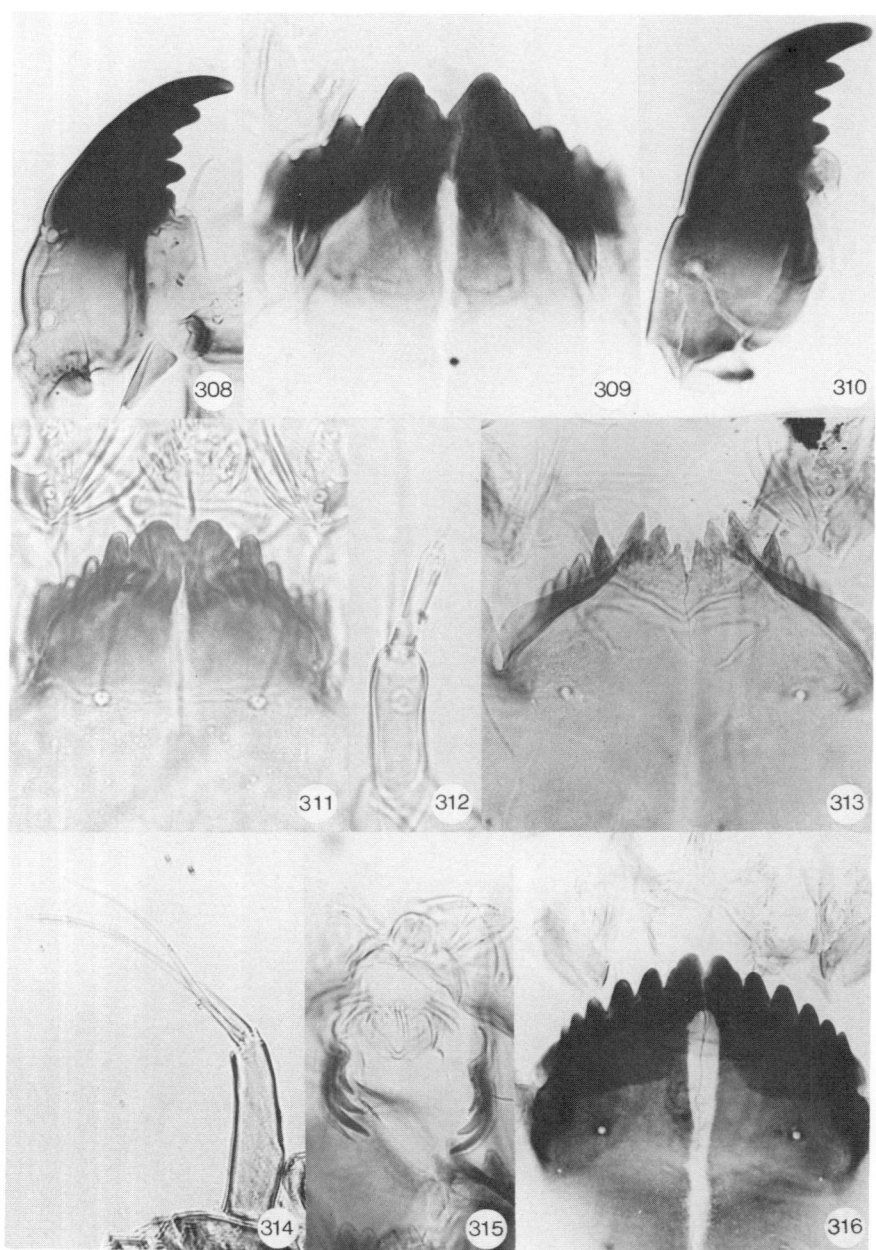

Figs. 308–316. 308, *Euryhapsis* – mandible; 309, *Gymnometriocnemus* – hypostoma; 310, *Gymnometriocnemus* – mandible; 311, *Heleniella* – hypostoma; 312, *Heleniella* – antenna; 313, *Heterotanytarsus* – hypostoma; 314, *Heterotanytarsus* – antenna; 315, *Heterotanytarsus* – premandibles; 316, *Heterotrissocladius* – hypostoma.

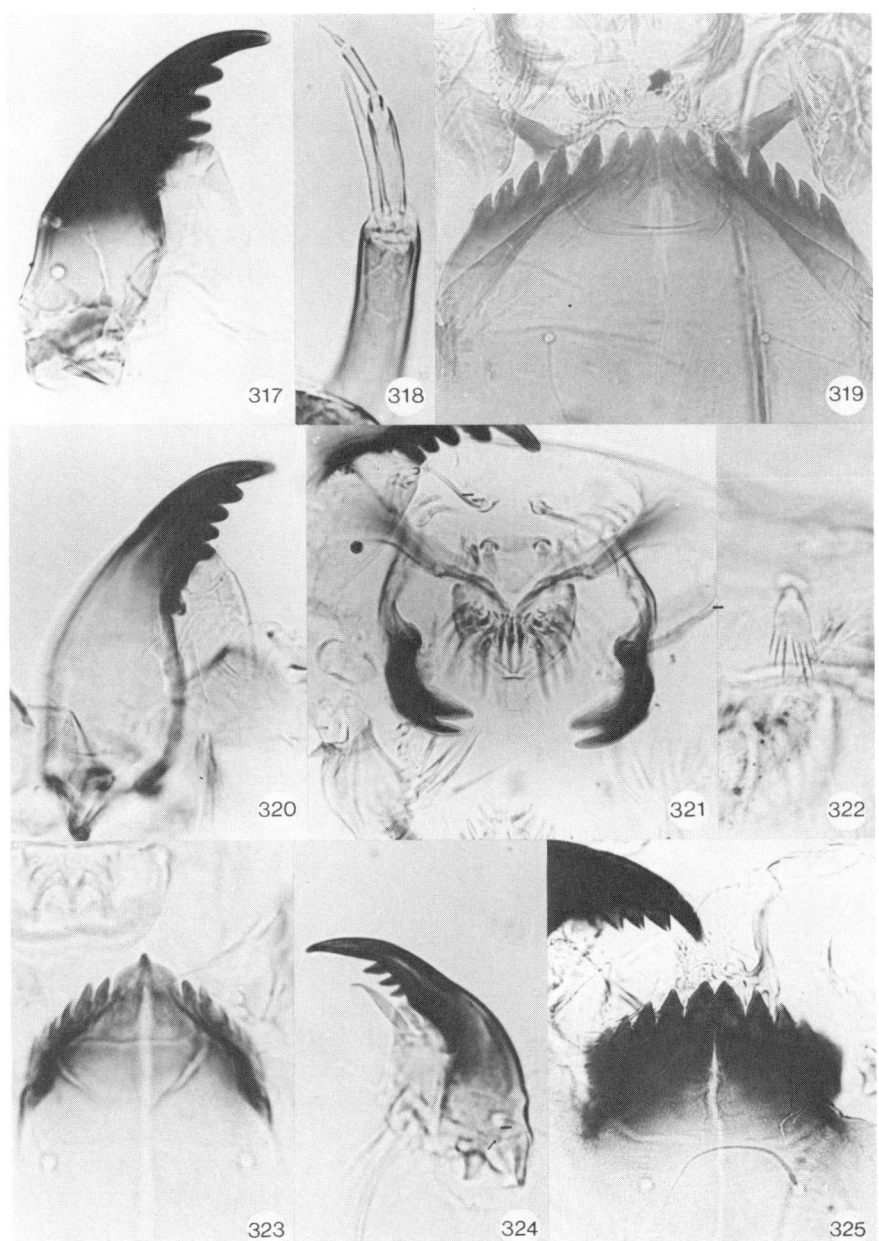

Figs. 317–325. 317, *Heterotrissocladius* – mandible; 318, *Heterotrissocladius* – antenna; 319, *Hydrobaenus* – hypostoma; 320, *Hydrobaenus* – mandible; 321, *Hydrobaenus* – premandibles; 322, *Hydrobaenus* – SI; 323, *Krenosmittia* – hypostoma; 324, *Krenosmittia* – mandible; 325, *Limnophyes* – hypostoma.

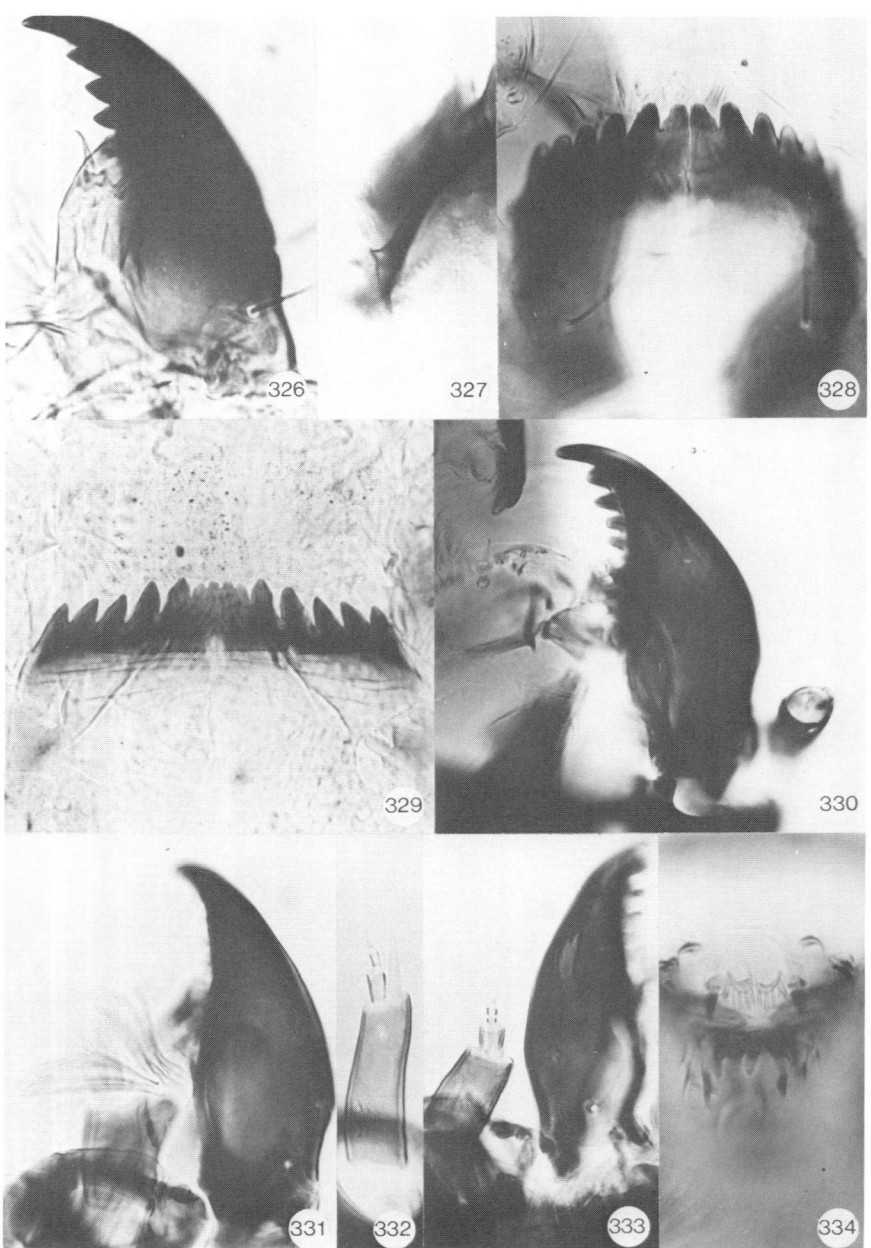

Figs. 326–334. 326, *Limnophyes* – mandible; 327, *Limnophyes* – distolateral corner of paralabial plate; 328, 329, *Metriocnemus* – hypostoma; 330, *Metriocnemus* – mandible; 331, *Metriocnemus* – seta interna; 332, 333, *Metriocnemus* – antenna; 334, *Metriocnemus* – labral lamellae.

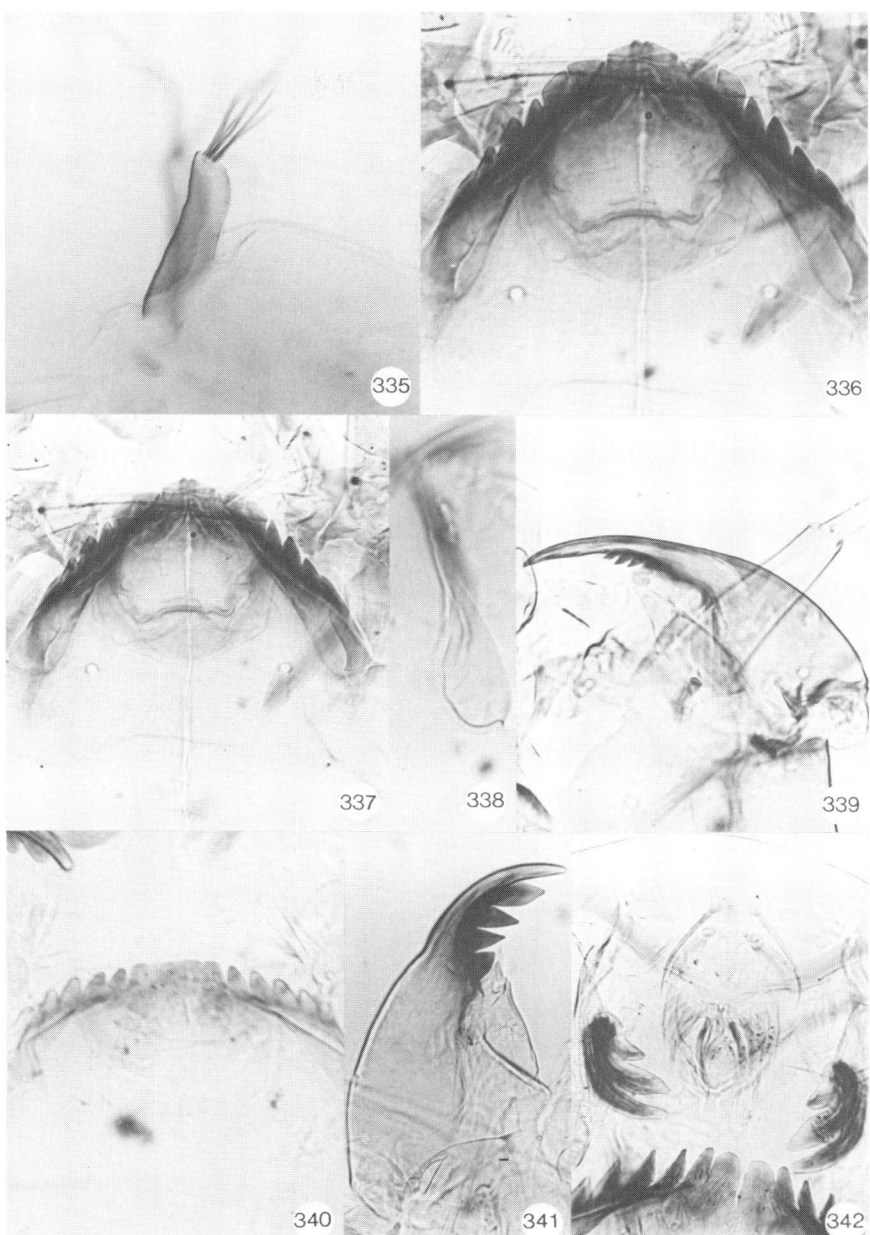

Figs. 335–342. 335, *Metriocnemus* – procercus; 336, *Nanocladius* (*Plecopteracoluthus*) – hypostoma; 337, *Nanocladius* (*Plecopteracoluthus*) – hypostoma and caudolateral apex of paralabial plate; 338, *Nanocladius* (*Nanocladius*) – caudolateral apex of paralabial plate; 339, *Nanocladius* (*Nanocladius*) – mandible; 340, *Oliveridia* – hypostoma; 341, *Oliveridia* – mandible; 342, *Oliveridia* – premandibles.

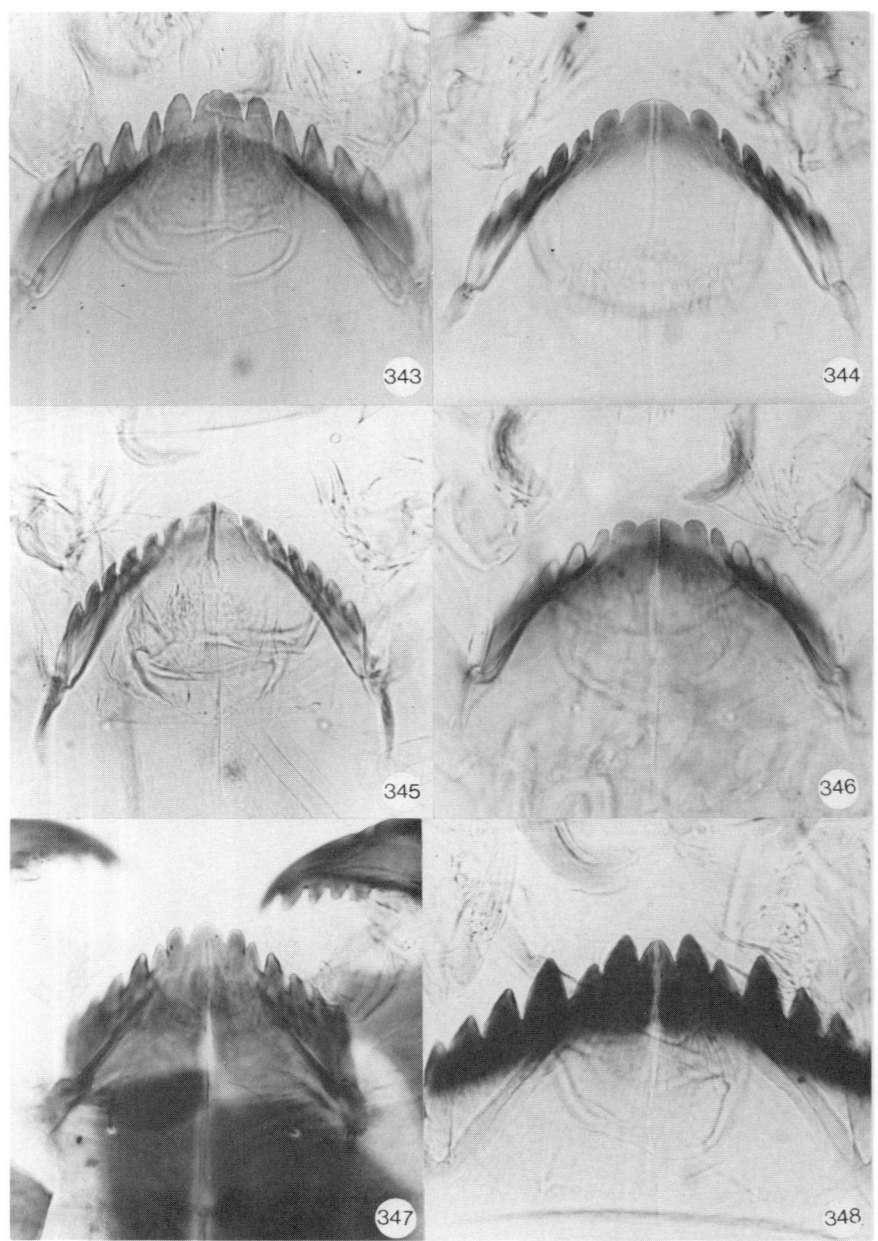

Figs. 343–348. 343–346, *Orthocladius* (*Orthocladius*) – hypostoma; 347, *Orthocladius* (*Eudactylocladius*) – hypostoma; 348, *Orthocladius* (*Pogonocladius*) – hypostoma.

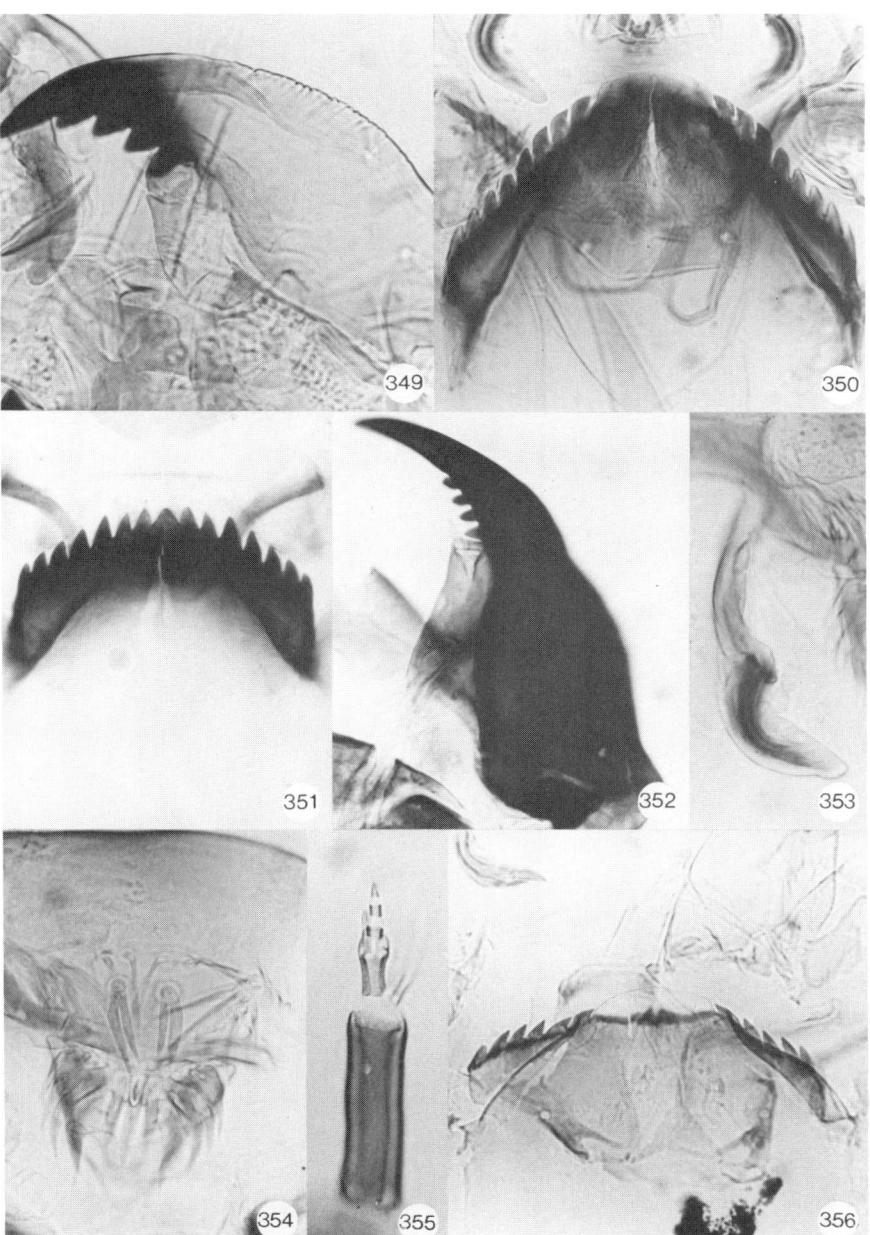

Figs. 349–356. 349, *Orthocladius (Pogonocladius)* – mandible; 350, 351, *Orthocladius (Euorthocladius)* – hypostoma; 352, *Orthocladius (Euorthocladius)* – mandible; 353, *Orthocladius (Euorthocladius)* – premandible; 354, *Orthocladius (Euorthocladius)* – SI; 355, *Orthocladius (Euorthocladius)* – antenna; 356, *Paracladius* – hypostoma and paralabial plate.

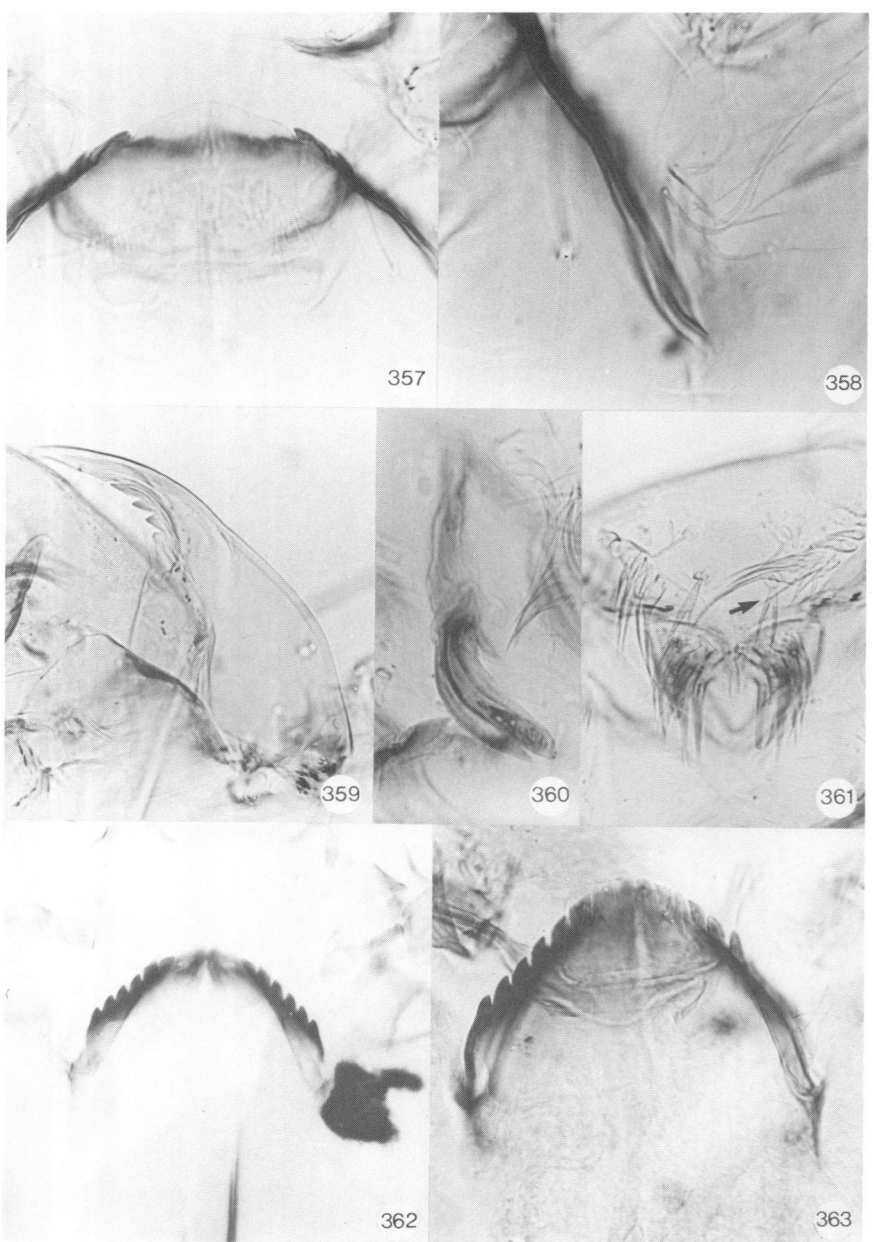

Figs. 357–363. 357, *Paracladius* – median tooth of hypostoma; 358, *Paracladius* – paralabial plate with setae; 359, *Paracladius* – mandible; 360, *Paracladius* – premandible; 361, *Paracladius* – SI; 362, *Paracricotopus* – hypostoma; 363, *Paracricotopus* – hypostoma and paralabial plate with seta.

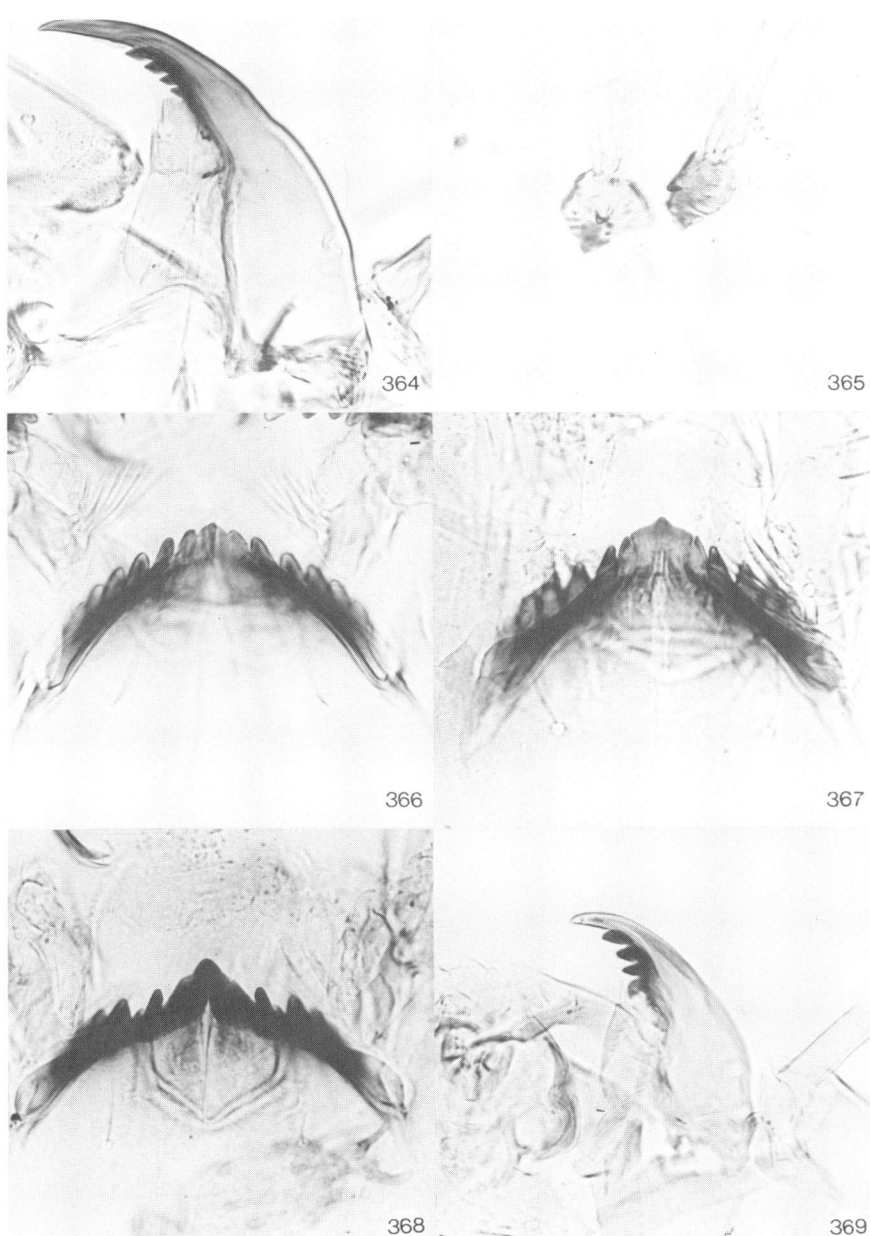

Figs. 364–369. 364, *Paracricotopus* – mandible; 365, *Paracricotopus* – procerci with spurs; 366–368, *Parakiefferiella* – hypostoma; 369, *Parakiefferiella* – mandible.

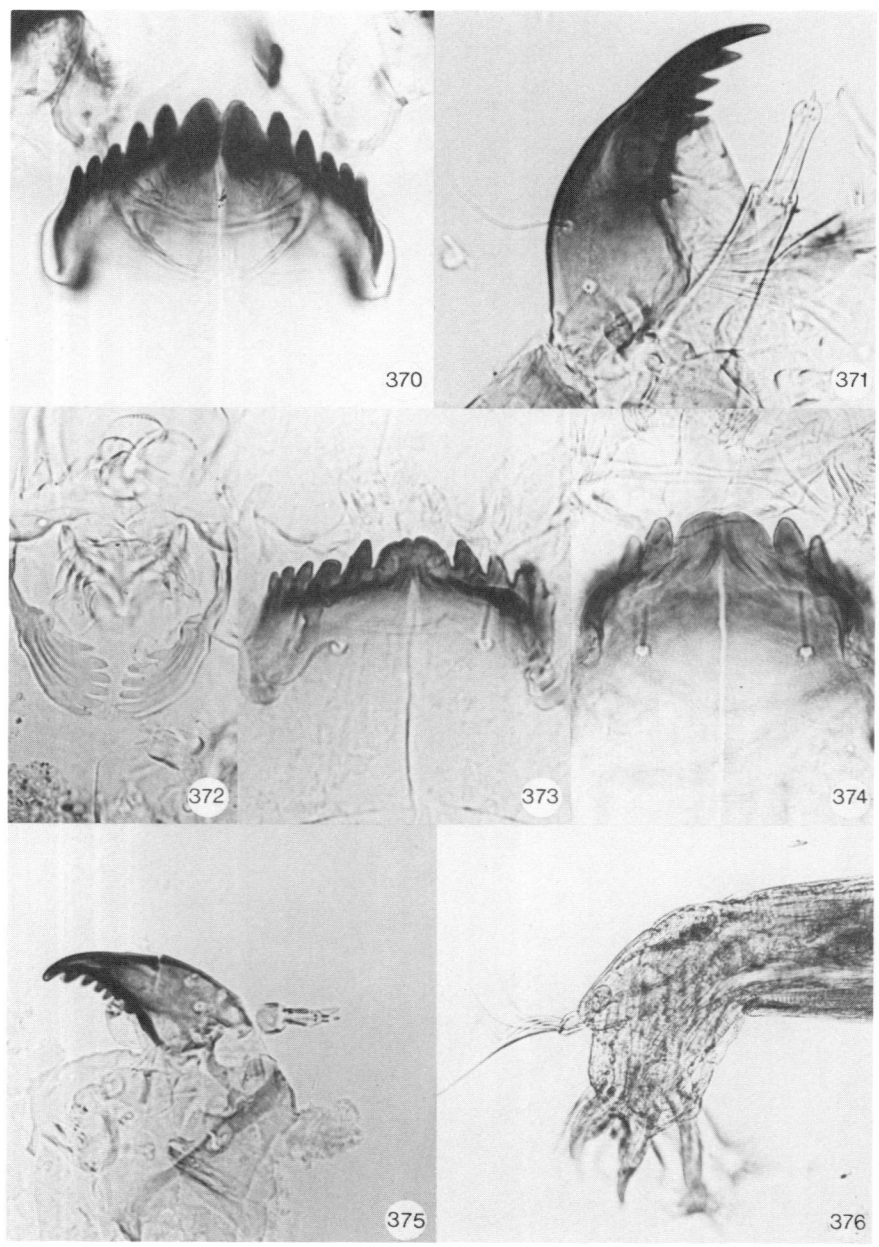

Figs. 370–376. 370, *Parametriocnemus* – hypostoma and paralabial plates; 371, *Parametriocnemus* – mandible and antenna; 372, *Parametriocnemus* – premandibles; 373, *Paraphaenocladius* – hypostoma; 374, *Paraphaenocladius* – median tooth of hypostoma; 375, *Paraphaenocladius* – antenna and mandible; 376, *Paraphaenocladius* – anal end.

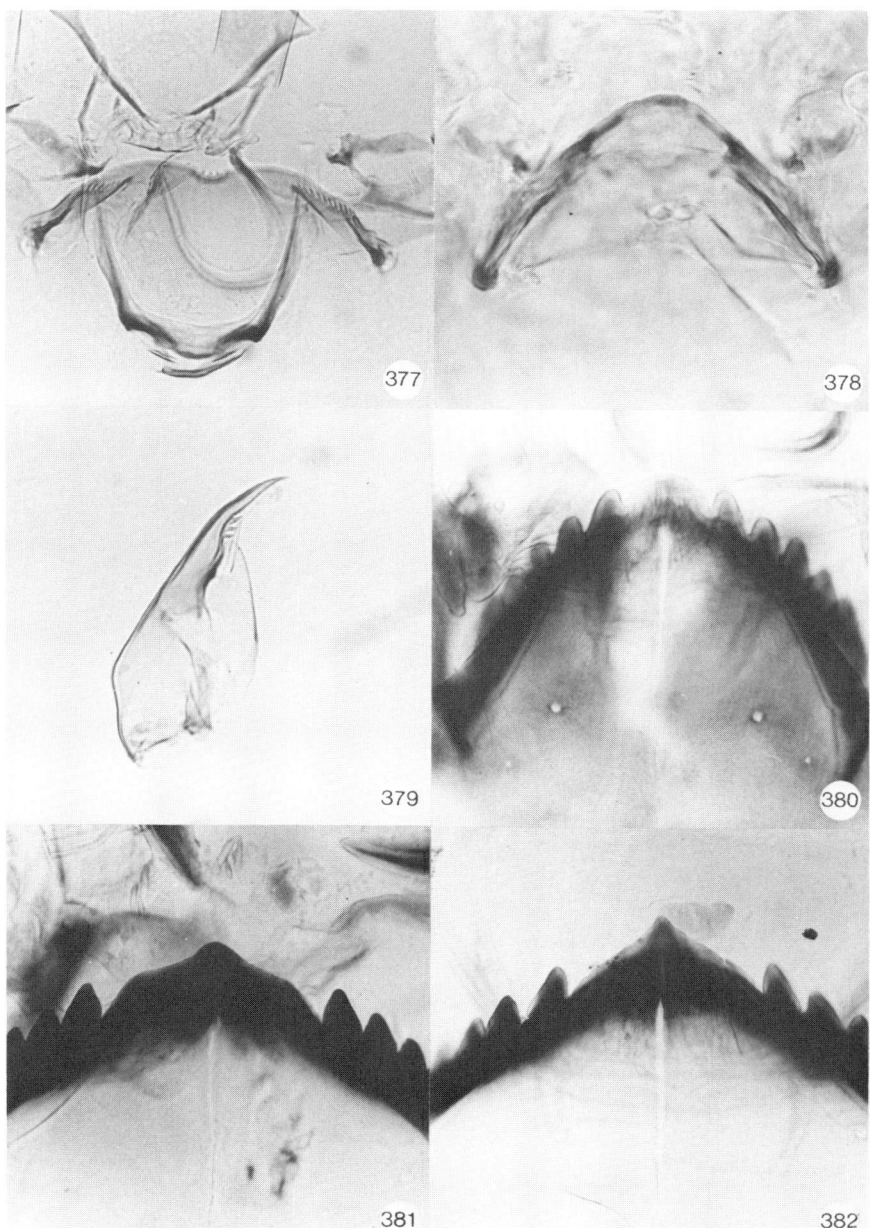

Figs. 377–382. 377, *Phycoidella* – hypostoma and paralabial plates; 378, *Phycoidella* – median area of hypostoma; 379, *Phycoidella* – mandible; 380–382, *Psectrocladius* – hypostoma.

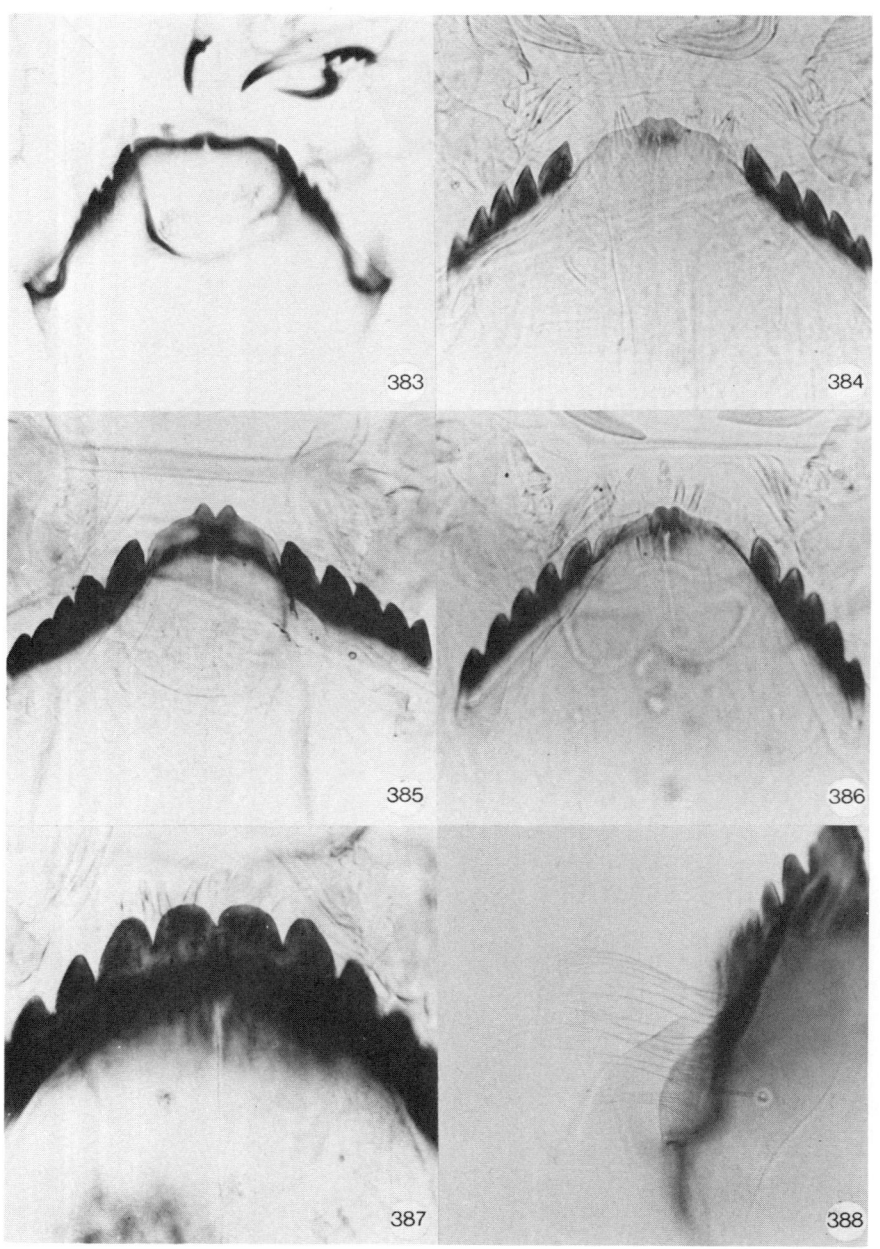

Figs. 383–388. 383–387, *Psectrocladius* – hypostoma; 388, *Psectrocladius* – paralabial plate with setae.

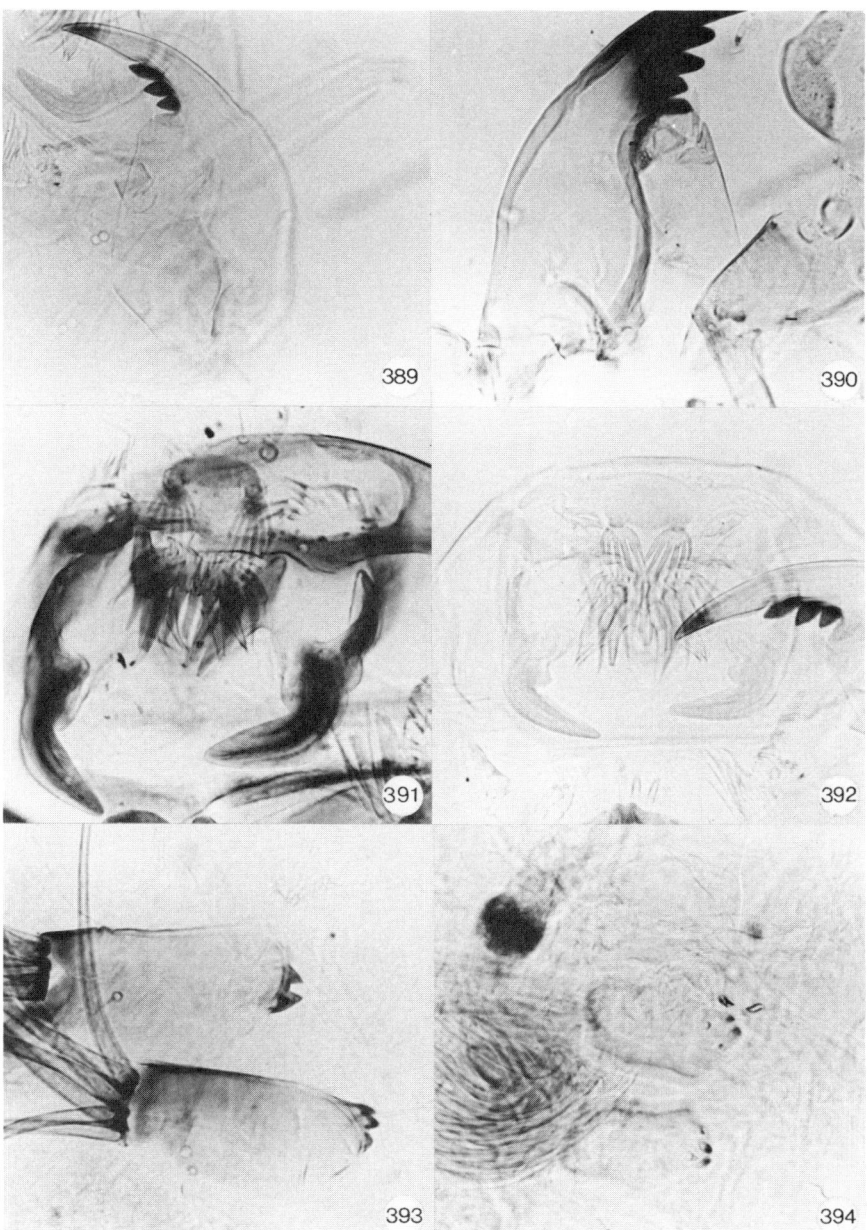

Figs. 389–394. 389, 390, *Psectrocladius* – mandible; 391, *Psectrocladius* – premandibles; 392, *Psectrocladius* – SI; 393, 394, *Psectrocladius* – procerci with spurs.

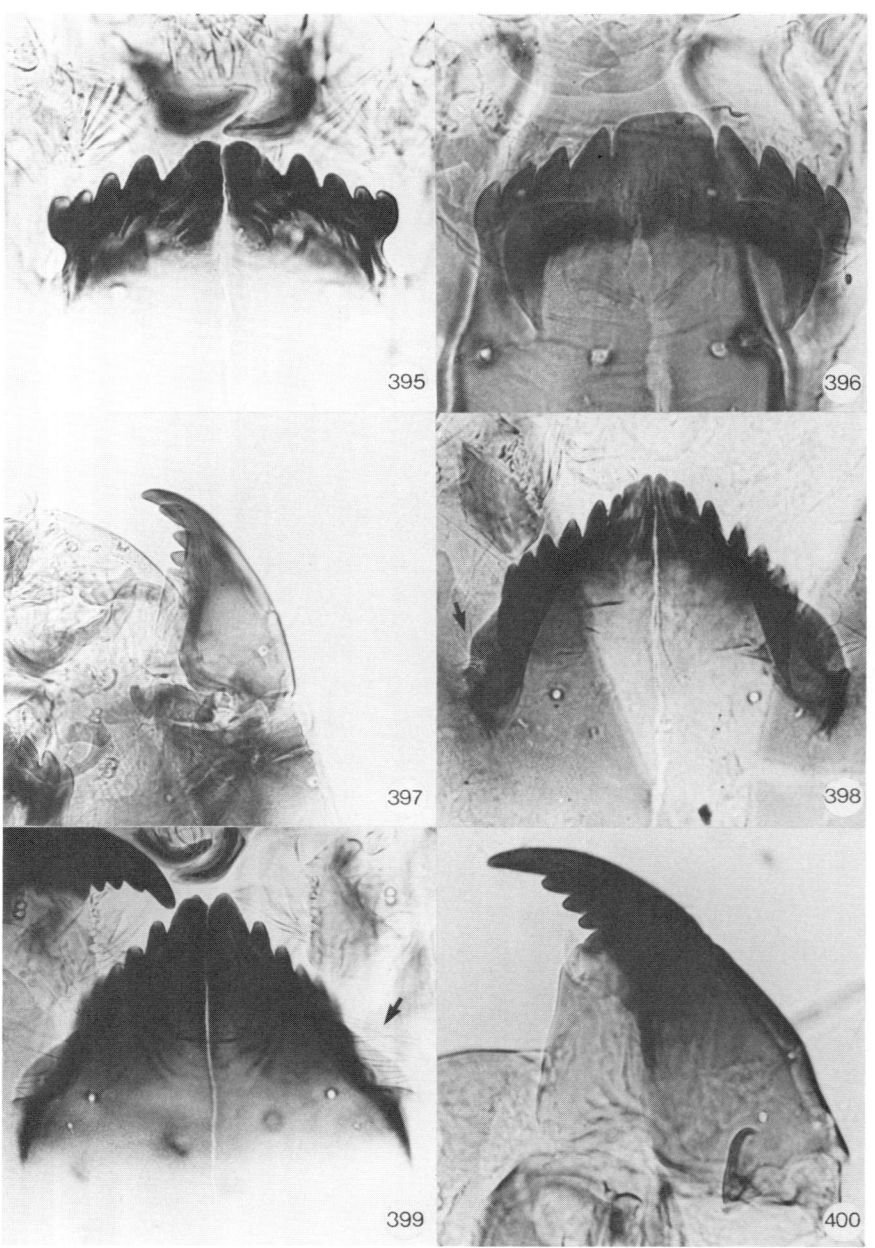

Figs. 395–400. 395, *Pseudorthocladius* – hypostoma; 396, *Pseudosmittia* – hypostoma; 397, *Pseudosmittia* – mandible; 398, 399, *Rheocricotopus* – hypostoma and paralabial plate with setae; 400, *Rheocricotopus* – mandible.

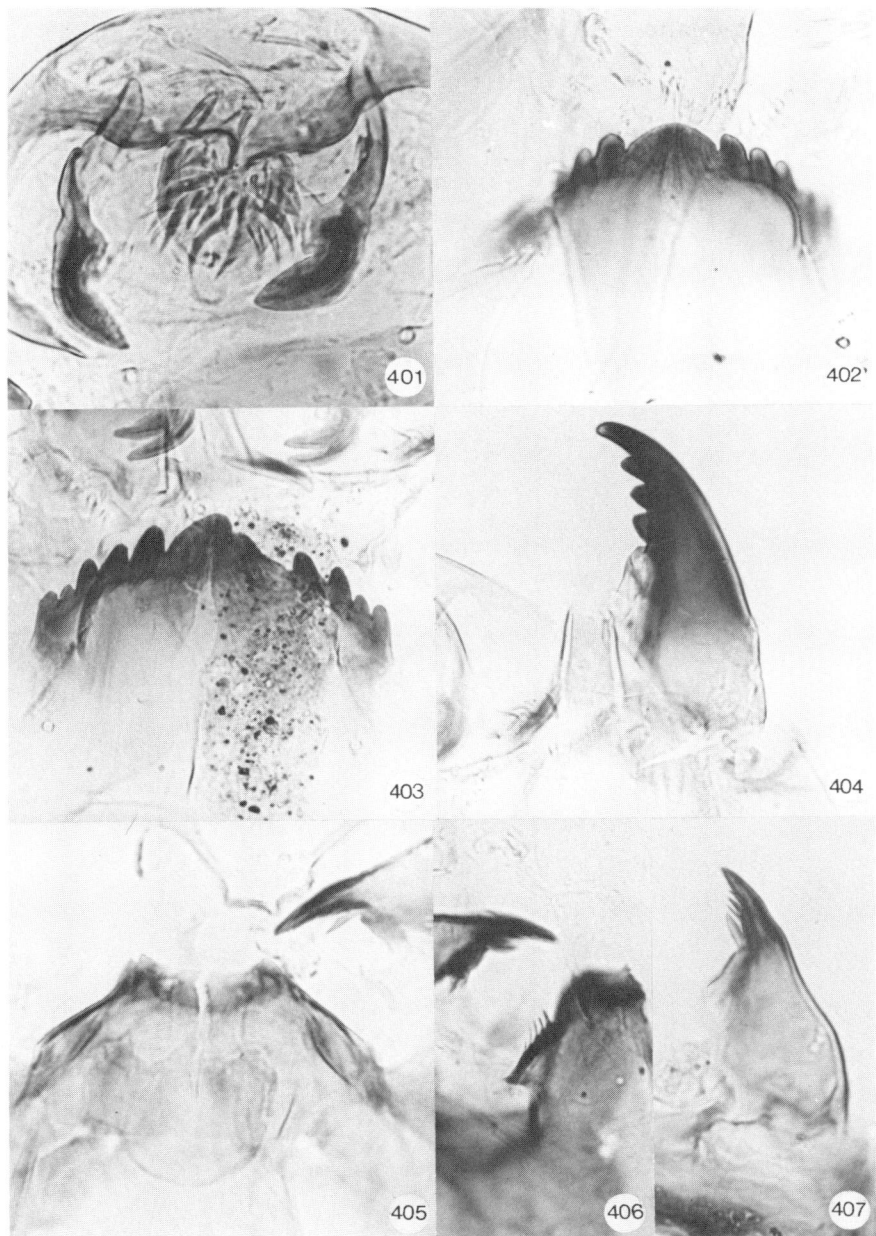

Figs. 401–407. 401, *Rheocricotopus* – SI and premandibles; 402, 403, *Smittia* – hypostoma; 404, *Smittia* – mandible; 405, *Symbiocladius* – hypostoma; 406, *Symbiocladius* – peglike lateral teeth of hypostoma; 407, *Symbiocladius* – mandible.

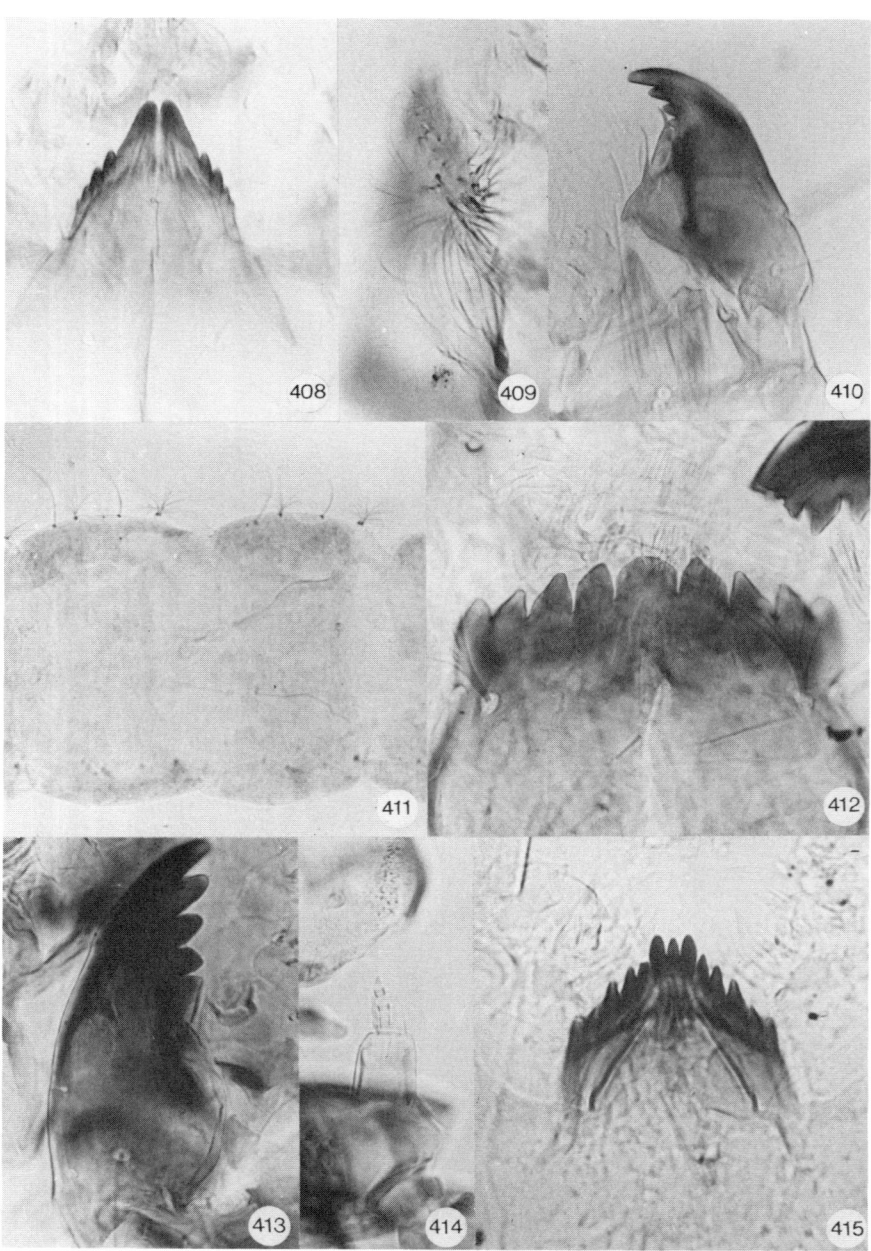

Figs. 408–415. 408, *Synorthocladius* – hypostoma; 409, *Synorthocladius* – paralabial plate with setae; 410, *Synorthocladius* – mandible; 411, *Synorthocladius* – abdominal setae; 412, *Thalassosmittia* – hypostoma; 413, *Thalassosmittia* – mandible; 414, *Thalassosmittia* – antenna; 415, *Thienemanniella* – hypostoma.

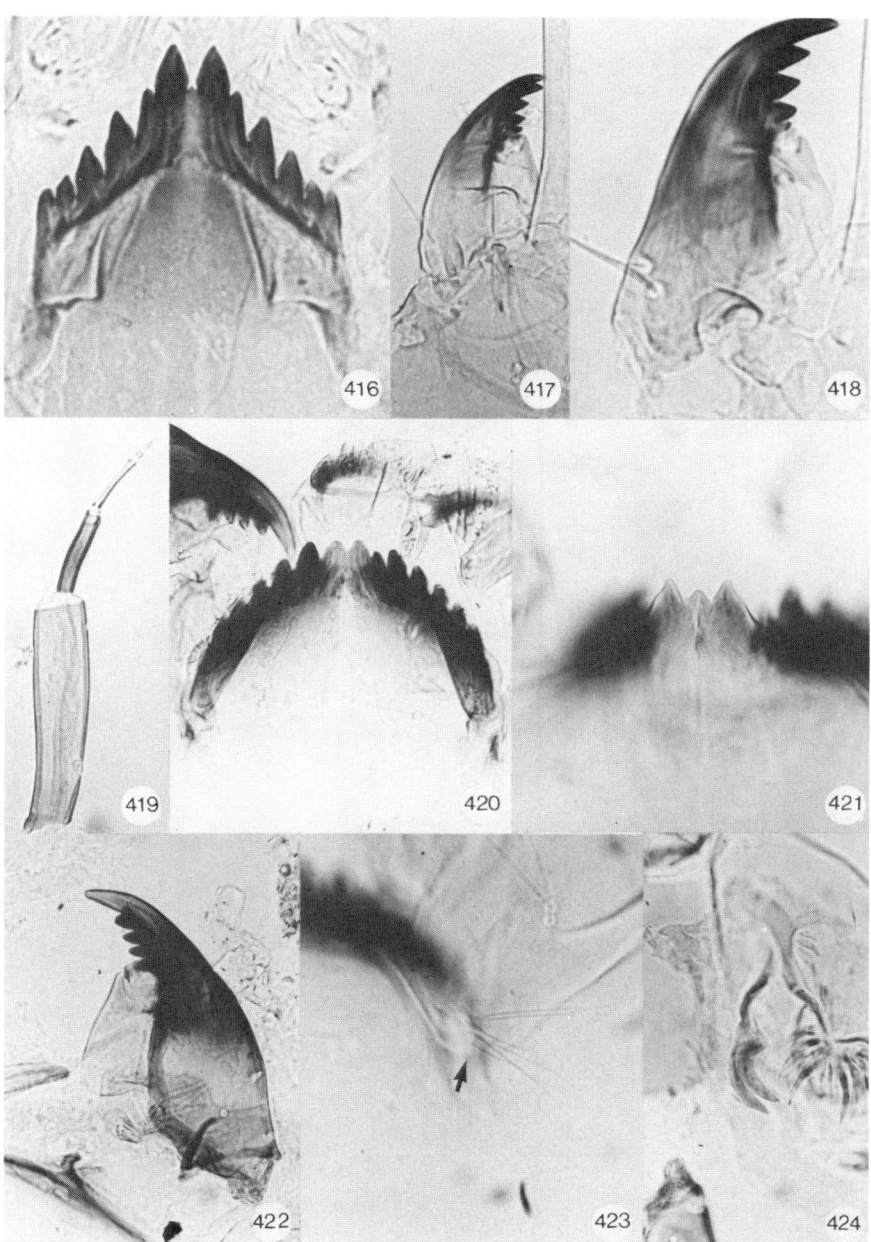

Figs. 416–424. 416, *Thienemanniella* – hypostoma; 417, 418, *Thienemanniella* – mandible; 419, *Thienemanniella* – antenna; 420, *Zalutschia* – hypostoma; 421, *Zalutschia* – median teeth of hypostoma; 422, *Zalutschia* – mandible; 423, *Zalutschia* – paralabial plate with setae; 424, *Zalutschia* – premandible.

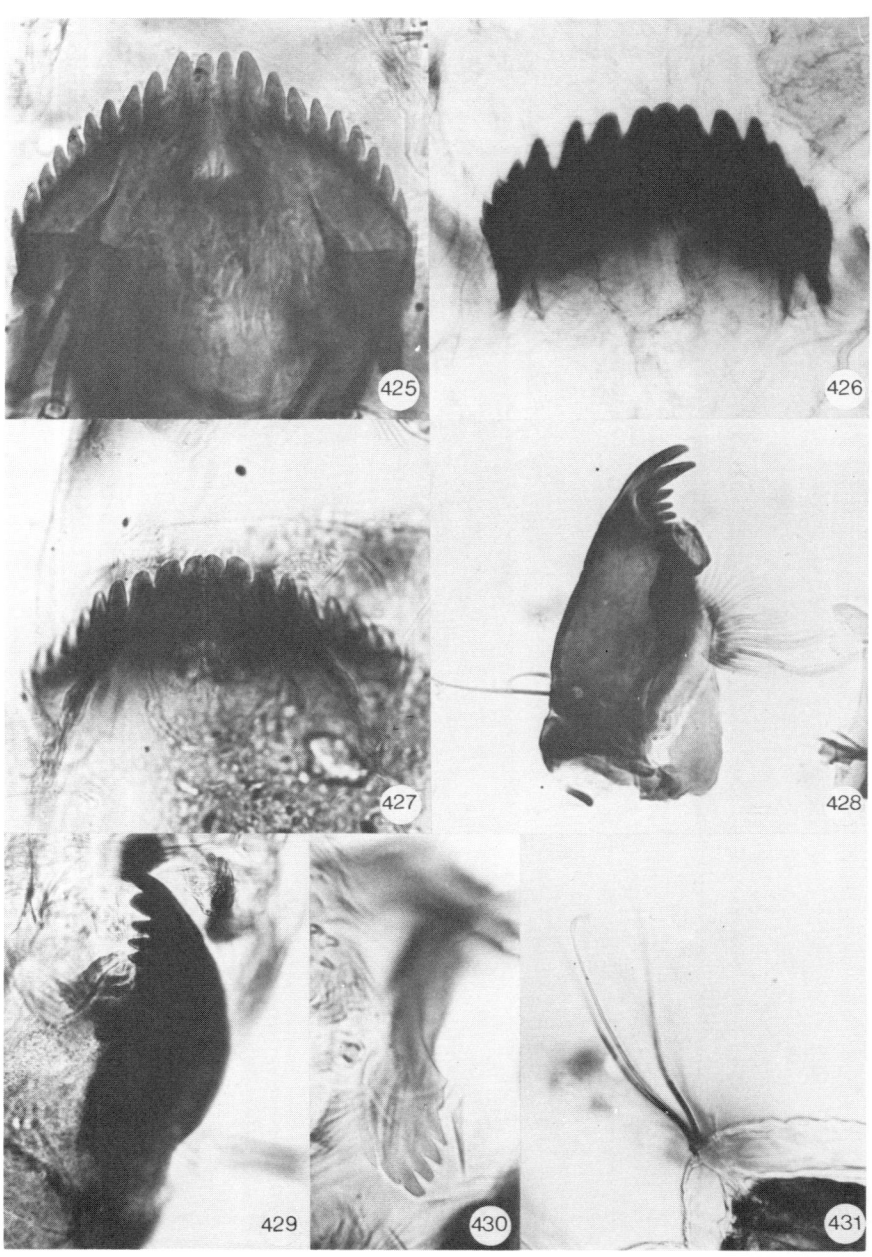

Figs. 425–431. 425–427, *Diamesa* – hypostoma; 428, 429, *Diamesa* – mandible; 430, *Diamesa* – premandible; 431, *Diamesa* – procercus and anal setae.

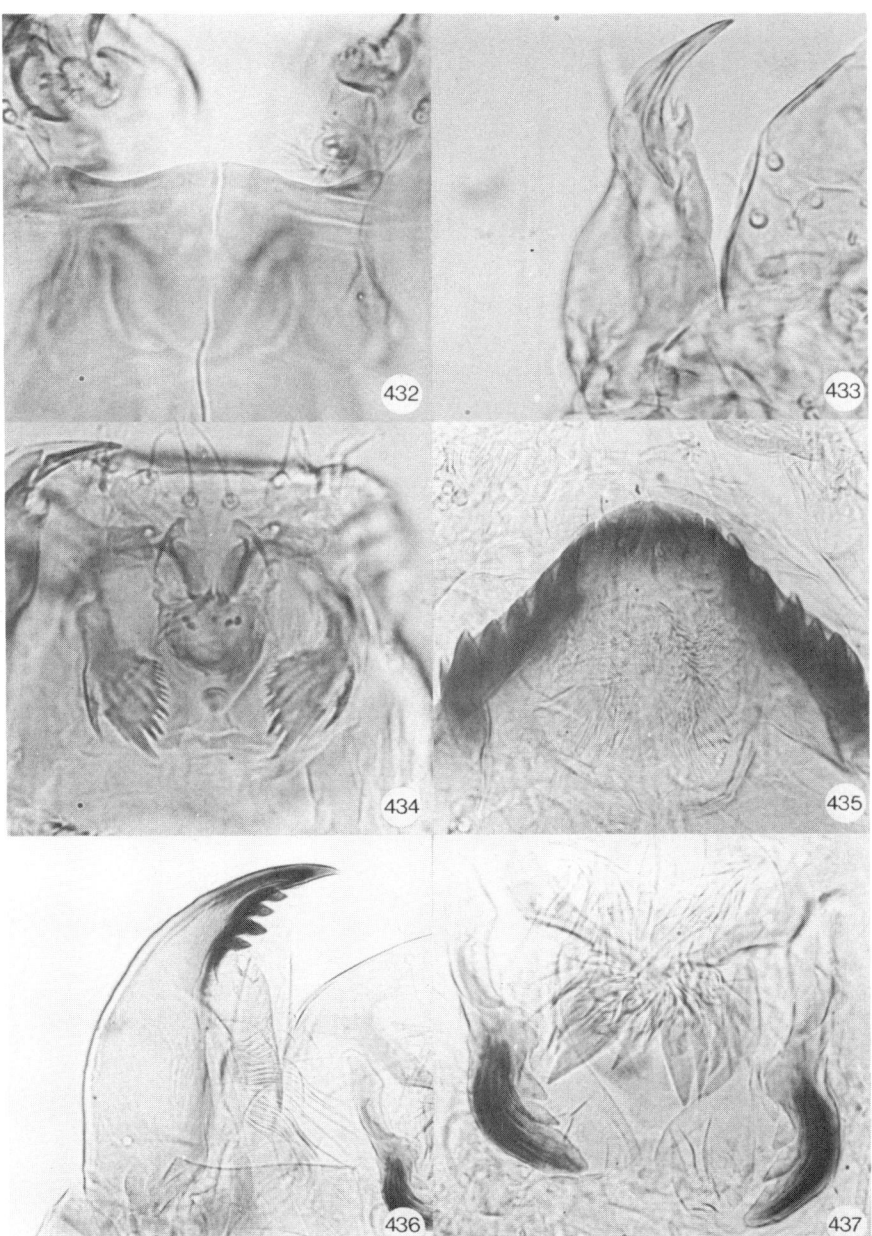

Figs. 432–437. 432, *Potthastia*; *longimanus* type – hypostoma; 433, *Potthastia*; *longimanus* type – mandible; 434, *Potthastia*; *longimanus* type – premandibles; 435, *Potthastia*; *gaedii* type – hypostoma; 436, *Potthastia*; *gaedii* type – mandible; 437, *Potthastia*; *gaedii* type – premandibles.

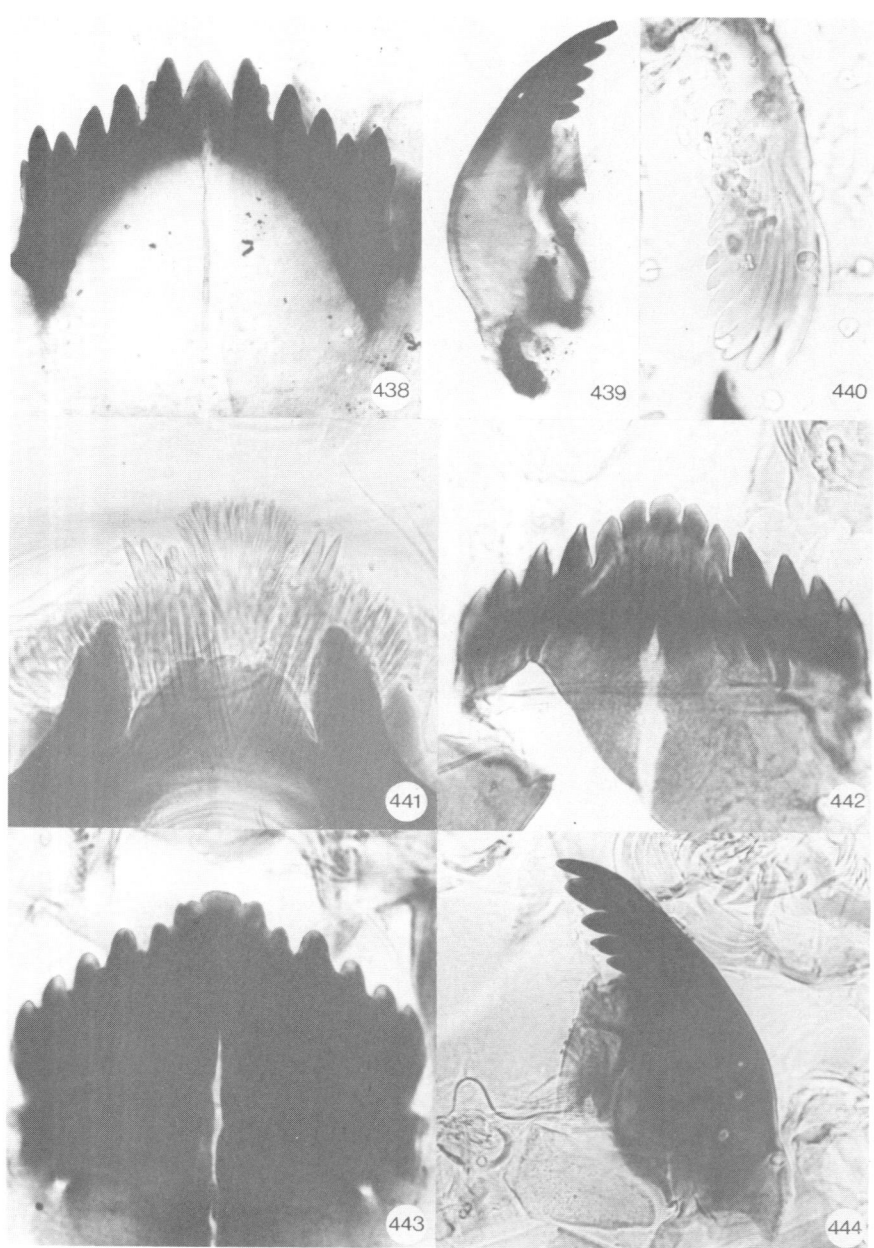

Figs. 438–444. 438, *Pseudodiamesa* – hypostoma; 439, *Pseudodiamesa* – mandible; 440, *Pseudodiamesa* – premandible; 441, *Pseudodiamesa* – prementohypopharyngeal complex; 442, 443, *Pseudokiefferiella* – hypostoma; 444, *Pseudokiefferiella* – mandible.

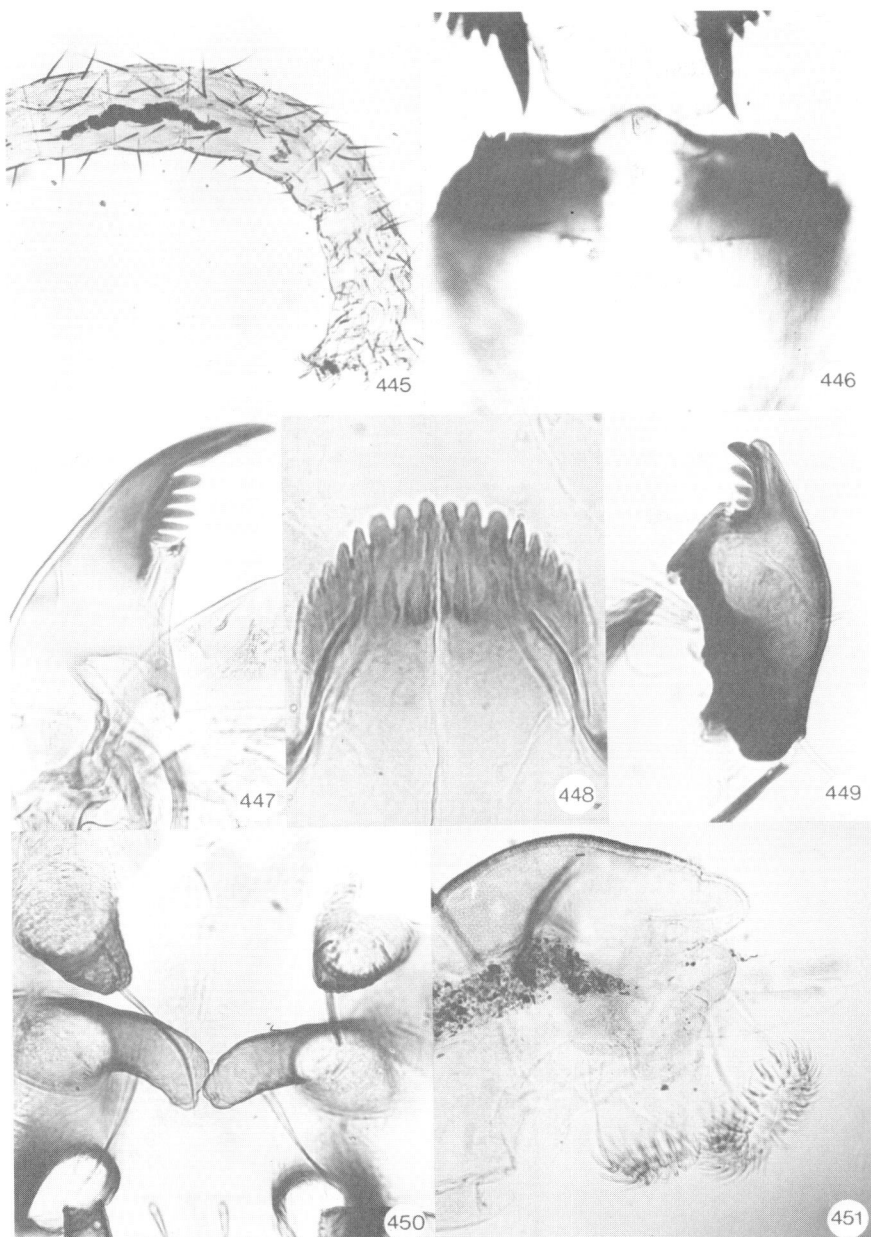

Figs. 445–451. 445, *Pseudokiefferiella* – abdominal segments showing setae; 446, *Protanypus* – hypostoma; 447, *Protanypus* – mandible; 448, *Boreoheptagyia* – hypostoma; 449, *Boreoheptagyia* – mandible; 450, *Boreoheptagyia* – dorsal tubercles on head capsule; 451, *Boreoheptagyia* – anal end.

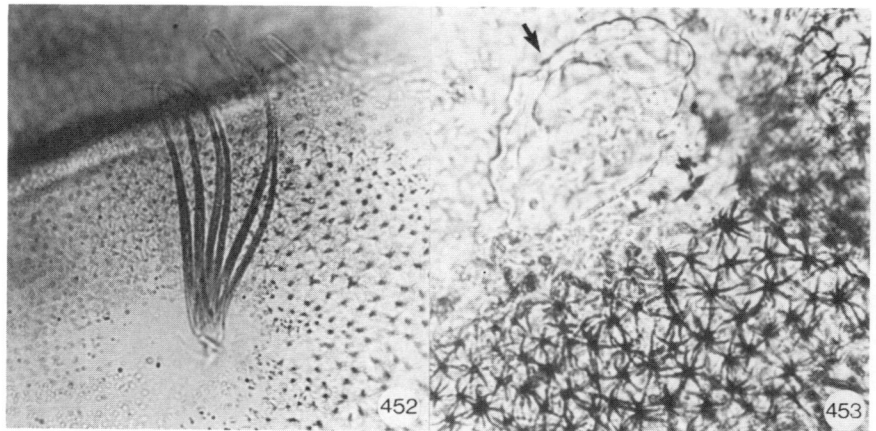

Figs. 452–453. 452, *Boreoheptagyia* – anal setae; 453, *Boreoheptagyia* – multipointed spines and dorsolateral tubercle on abdominal segment.

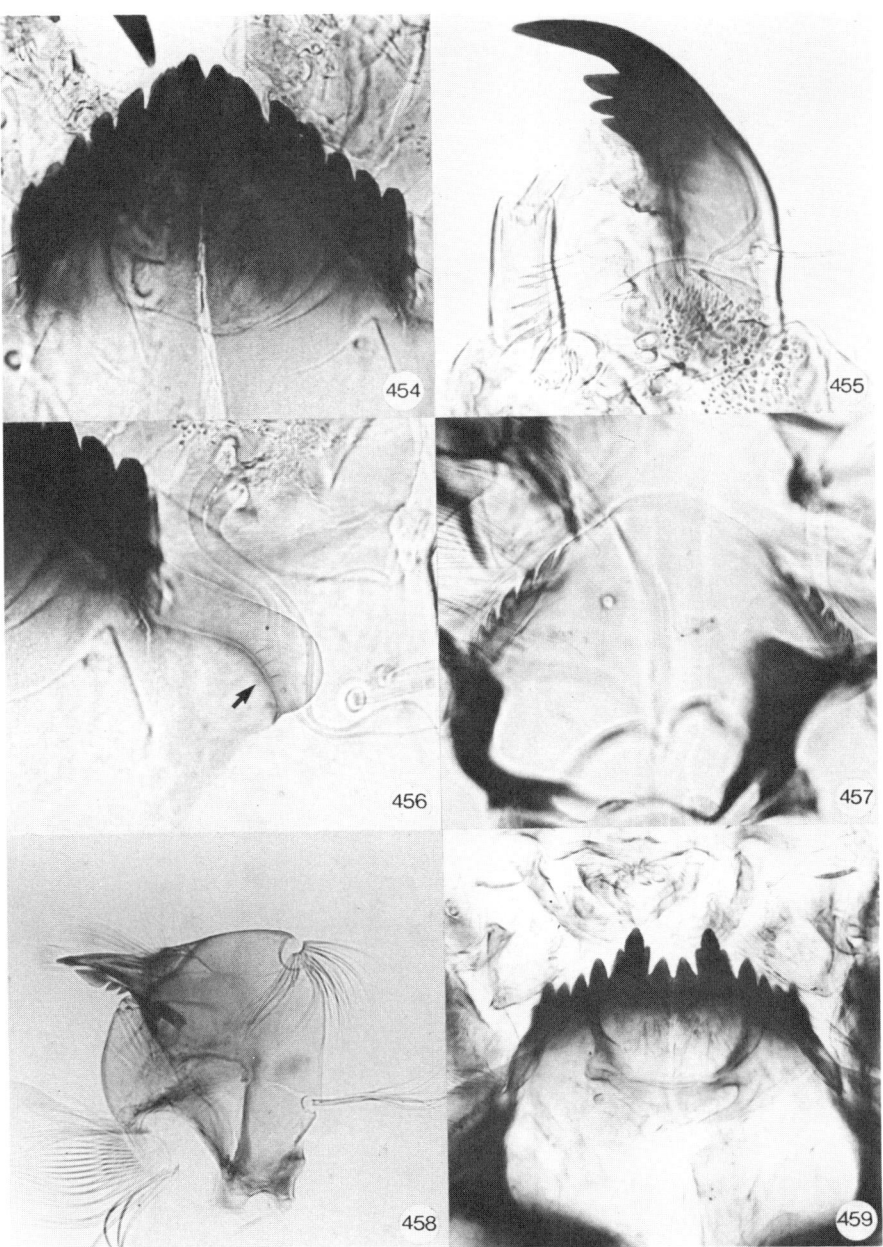

Figs. 454–459. 454, *Monodiamesa* – hypostoma; 455, *Monodiamesa* – mandible; 456, *Monodiamesa* – paralabial plate with setae; 457, *Odontomesa* – hypostoma; 458, *Odontomesa* – mandible; 459, *Prodiamesa* – hypostoma.

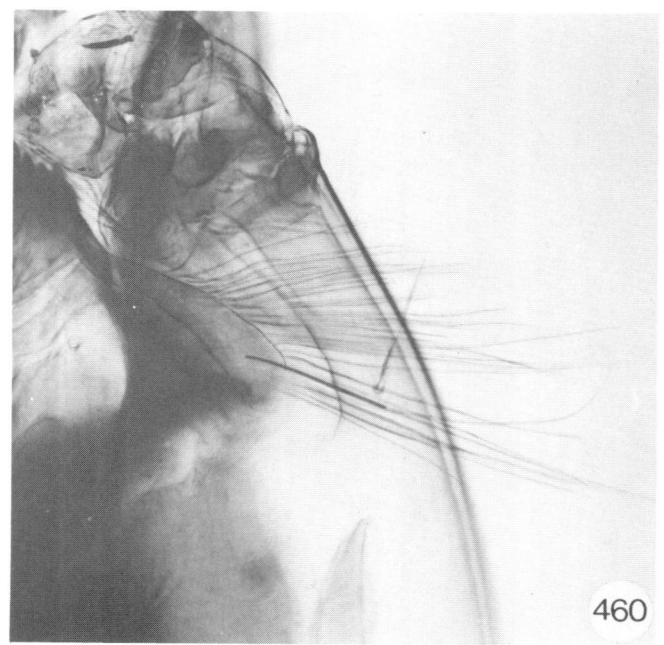

Fig. 460. *Prodiamesa* – paralabial plate with setae.

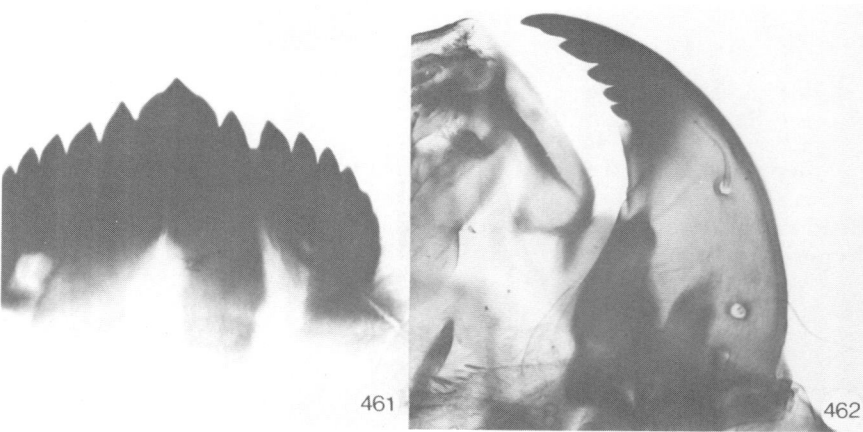

Figs. 461–462. 461, *Paraclunio* – hypostoma; 462, *Paraclunio* – mandible.

# Glossary

**accessory tooth**  In Tanypodinae, tooth situated laterally and usually adjacent to inner tooth of mandible; in other subfamilies; see seta subdentalis.
**anal seta** (pl., **setae**)  Seta borne on apex of procercus, or on dorsolateral areas of preanal abdominal segment.
**anal tubules**  Paired tubular structures, 1–3, usually 2 pairs, borne on the posterior end dorsal to posterior parapods.
**antenna** (pl., **antennae**)  Paired segmented structures, borne one on each side of the head capsule.
**antennal tubercle**  Extension of head capsule bearing antenna.
**anterior parapod** (pl., **parapods**)  Paired, sometimes fused, fleshy appendages arising from anteroventral area of first thoracic segment, bearing claws and/or spines.
**apical tooth**  Tooth on apex of mandible.
**apicodorsal tooth**  Tooth on outer edge of mandible, posterior and usually somewhat dorsal to apical tooth.
**apicomesal projection**  Projection on apex of antennal tubercle.
**A.R.**  Length of first antennal segment/length of remaining segments (note Hirvenoja 1973 reversed this ratio).

**blade**  Modified seta borne on apex of first antennal segment.

**chaetulae laterales**  Generally 6–8 pairs of simple or pectinate scales on each side of pecten epipharyngis.

**egg burster**  Projection on mediodorsal area of head of first instar larva.
**exuvium** (pl., **exuviae**)  Cast skin of larva after molt.
**eye spot**  Pigment areas, 1–3, located on each side of head capsule.

**frontoclypeal apotome**  Dorsomedial subdivision of head capsule extending from occipital margin to labrum.

**head capsule**  Fused sclerites of the head, which form a hard compact case anteriorly bearing mouthparts.
**head ratio**  Length of head/width of head.
**hypostoma**  Sclerotized anteroventral margin of head capsule, usually toothed.

**inner teeth**  Teeth on inner margin of mandible between base of apical tooth and mandibular base.
**instar**  Stage between molts of a larva; first instar is between egg and first molt, second between first and second molts, etc.

**labral lamella** (pl., **lamellae**)  One or more plates borne on labrum between bases of SI or between SI and inner apices of tormae.
**labrum**  Anterodorsal area of head covering roof of mouth.
**Lauterborn's organs**  Paired globular or hemispherical organs borne on second and/or third antennal segments.

**ligula**(in Tanypodinae)  Anteroventral movable toothed part of hypopharyngeal complex.

**mandible**  Jaw, borne one on each side of the head.

**maxillary palpus**(pl., **palpi**)  Anterolateral extension of maxillae.

**occipital margin**  Posterior rim of head capsule.

**ocular sclerite**  Lateral and ventral parts of head capsule, fused ventrally with opposite sclerite and separated dorsally by frontoclypeal apotome.

**palatum**  An area attached to anteroventral margin of labrum and with labrum forming anterior "lip" of mouth.

**paraglossa** (pl., **paraglossae**) (in Tanypodinae)  Paired structures, one on each side of ligula; part of prementohypopharyngeal complex.

**paralabial plate**  Plate lying ventral to posterolateral corner of hypostoma.

**paralabial teeth** (in Tanypodinae)  Row of teeth borne on either side of a membraneous hypostoma.

**pecten epipharyngis**  Scales or toothed bar lying on labrum, anteroventral to tormae.

**pecten hypopharyngis** (in Tanypodinae)  Arched toothed structure lying dorsal to ligula and paraglossa; part of prementohypopharyngeal complex.

**pecten mandibularis**  Row of setae on anterodorsal surface of mandible, about level of inner teeth.

**petiole**  Stalk of a Lauterborn's organ.

**posterior parapod**  Fleshy appendage arising, one on each side, from anal end of larva; apically bearing claws.

**premandible**  Movable appendage attached to ventral surface of labrum, one on either side of ungula.

**prementohypopharyngeal complex**  Multiple structure lying dorsal to hypostoma; in Tanypodinae with three sclerotized parts, ligula, paired paraglossae, and pecten hypopharyngis.

**procercus** (pl., **procerci**)  Paired projections arising one on each side from dorsolateral area of eighth abdominal segment, apically bearing anal setae.

**ring organ**  Circular organ on first antennal segment.

**SI**  Anteroventral setal pair on labrum.

**SII**  Setal pair on labrum dorsal and posterior to SI.

**seta interna**  Setal brush or group of setae arising from inner margin of mandible, distal to inner teeth.

**seta subdentalis**  Modified seta arising from mandibular base behind row of inner teeth.

**socle**  Sclerotized protuberance bearing setae.

**striations**  Longitudinal lines on paralabial plates.

**supra-anal setae**  Pair of setae arising from anal end of larvae between procerci and anal tubules.

**terminal segments**  Distal segments of antennae, excluding first or basal segment.

**torma** (pl., **tormae**)  Transverse sclerotized bars forming anteroventral margin of labrum.

**ungula**  U-shaped or semicircular arch on labrum, anteroventral to pecten epipharyngis.

**ventral tubules**  Tubular projections from ventral or ventrolateral surface of 11th body segment.

# References

Andersen, F. S. 1937. Über die Metamorphose der Ceratopogoniden und Chironomiden Nordost–Grönlands. Meddr. Grönland. 116:1–95.

Anderson, J. B., and W. T. Mason. 1968. A comparison of benthic macroinvertebrates collected by dredge and basket sampler. J. Wat. Pollut. Control Fed. 1:252–259.

Anderson, R. O. 1959. A modified flotation technique for sorting bottom fauna samples. Limnol. Oceanogr. 4:223–225.

Bause, E. 1914. Die metamorphose der gattung *Tanytarsus* und einiger verwandter Tendipedidenarten. Ein Beitrag zur Systematik der Tendipediden. Arch. Hydrobiol. Suppl. 2:1–128.

Beck, E. C., and W. M. Beck. 1969. Chironomidae (Diptera) of Florida. III. The *Harnischia* complex (Chironomidae). Bull. Fla St. Mus. biol. Sci. 13:277–313.

Beck, W. M., and E. C. Beck. 1964. New Chironomidae from Florida (Diptera). Fla Ent. 47:201–207.

Beck, W. M., and E. C. Beck. 1966. Chironomidae (Diptera) of Florida. I. Pentaneurini (Tanypodinae). Bull. Fla St. Mus. biol. Sci. 10:305–379.

Beck, W. M., and E. C. Beck. 1968. The concept of genus in the family Chironomidae. Annls zool. Fenn. 5:14–16.

Beck, W. M., and E. C. Beck. 1970. The immature stages of some Chironomini (Chironomidae). Q.Jl Fla Acad. Sci. 33:29–42.

Beck, W. M. 1977. Environmental requirements and pollution tolerance of common freshwater Chironomidae. U.S. Environmental Protection Agency, Cincinnati, EPA-600/4-77-024:1–261.

Berg, C. O. 1950. Biology of certain Chironomidae reared from *Potamogeton*. Ecol. Monogr. 20:83–101.

Boesel, M. W. 1974. Observations on the Coelotanypodini of the northeastern states, with keys to the known stages. (Diptera: Chironomidae: Tanypodinae). J. Kans, ent. Soc. 47:417–432.

Brinkhurst, R. O. 1974. The benthos of lakes. MacMillan Press Ltd., London. 190 pp.

Brock, E. M. 1960. Mutualism between the midge *Cricotopus* and the alga *Nostoc*. Ecology 41:474–483.

Brundin, L. 1947. Zur Kenntnis der schwedischen Chironomiden. Ark. Zool. 39:1–95.

Brundin, L. 1948. Über die Metamorphose der Sectio Tanytarsariae connectentes (Diptera, Chironomidae). Ark. Zool. 41:1–22.

Brundin, L. 1949. Chironomiden und andere Bodentiere der südschwedischen Urgebirgsseen. Rep. Inst. Freshwat. Res. Drottningholm. 30:1–914.

Brundin, L. 1956. Zur Systematik der Orthocladiinae (Dipt. Chironomidae). Rep. Inst. Freshwat. Res. Drottningholm. 37:5–185.

Brundin, L. 1966. Transantarctic relationships and their significance, as evidenced by chironomid midges. With a monograph of the subfamilies Podonominae and Aphroteniinae and the Austral Heptagyiae. K. svenska Vetensk-Akad. Handl. 11:1472.

Bryce, D. 1960. Studies on the larva of the British Chironomidae (Diptera), with keys to the Chironominae and Tanypodinae. Trans Soc. Br. Ent. 14:19–62.

Buckley, B. R., and J. E. Sublette. 1964. Chironomidae (Diptera) of Louisiana. II. The limnology of the upper part of Cane River Lake, Natchitoches Parish, Louisiana, with particular reference to the emergence of Chironomidae. Tulane Stud. Zool. 11:151–166.

Campbell, J., J. F. Flannigan, M. Friesen, A. Furutani, R. D. Hamilton, S. G. Lawrence, S. L. Leonhard, and B. E. Townsend. 1973. A preliminary compilation of literature pertaining to the culture of aquatic invertebrates or macrophytes. Fish. Res. Bd Can., Tech. Rep. 227:1–25.

Cannings, R. A. 1975. A new species of *Chironomus* (Diptera: Chironomidae) from saline lakes in British Columbia. Can. Ent. 107:447–450.

Chernovskii, A. A. 1949. Identification of larvae of the midge family Tendipedidae [in Russian]. Isd. Akad. Nauk. S.S.S.R. 31:1–186.

Codreanu, R. 1939. Recherches biologiques sur un Chironomidae, *Symbiocladius rhithrogenae* (Zav.), ectopariasite "cancérigène" des Ephémères torrenticoles. Archs. zool. exp. gén. 81:1–283.

Coffman, W. P. 1967. Community structure and trophic relations in a small woodland stream, Linesville Creek, Crawford County, Pennsylvania. Ph.D. Thesis, Univ. Pittsburgh. 376 pp.

Cummings, K. W. 1962. An evaluation of some techniques for the collection and analysis of benthic samples with special emphasis on lotic waters. Am. Midl. Nat. 67:477–504.

Curry, L. L. 1958. Larvae and pupae of the species of *Cryptochironomus* (Diptera) in Michigan. Limnol. Oceanogr. 3:427–442.

Curry, L. L. 1961. A key for the larval forms of aquatic midges. (Tendipedidae: Diptera) found in Michigan. A.E.C. and N.I.H. Rep. 1. 162 pp.

Curry, L. L. 1965. A survey of environmental requirements for the midge (Diptera: Tendipedidae). Pages 127–144 *in* Biological problems in water pollution. Cincinnati U.S. Pub. Health Serv. Publ. 99-WP-25.

Danks, H. V. 1971a. Overwintering of some north temperate and arctic Chironomidae (Dipt.). I. The winter environment. Can. Ent. 103:589–604.

Danks, H. V. 1971*b*. Life history and biology of *Einfeldia synchrona* (Diptera: Chironomidae). Can Ent. 103:1597–1606.

Danks, H. V., and D. R. Oliver. 1972. Seasonal emergence of some high arctic Chironomidae (Diptera). Can. Ent. 104:661–686.

Davies, B. R. 1976. The dispersal of Chironomidae larvae: A review. J. ent. Soc. sth. Afr. 39:39–62.

Eaton, A. E. 1875. Breves dipterarum uniusque lepidopterarum Insulae Kerguelensi indigenarum diagnoses. Entomologist's mon. Mag. 12:58–61.

Edgar, W. D., and P. S. Meadows. 1969. Case construction, movement, special distribution and substrate selection in the larva of *Chironomus riparius* Meigen. J. exp. Biol. 50:247–253.

Edmondson, W. T., and G. G. Winberg. 1971. A manual on methods for the assessment of secondary productivity in fresh water. IBP Handbook 17, Int. Biol. Prog. London. Blackwell Scientific Publications, London. 358 pp.

Edwards, F. W. 1929. British non-biting midges (Diptera, Chironomidae). Trans. R. ent. Soc. Lond. 77:279–430.

Edwards, F. W. 1937. Chironomidae (Diptera) collected by Prof. Aug. Thienemann in Swedish Lappland. Ann. Mag. nat. Hist. Ser. 10, 20:140–148.

Edwards, F. W., and A. Thienemann. 1938. Neuer Beitrag zur Kenntnis der Podonominae (Dipt. Chironomidae). Zool. Anz. 122:152–158.

Enderlein, G. 1912. Die Insekten des Antarkto-Archiplata-Gebietes. K. svenska Vetensk-Aka. Handl. 48:106–110.

Fittkau, E. J. 1954. *Trichocladius nivalis* Goetgh. (Chironomidenstudien III). Ber. limnol. Flussstn. Freudenthal. 6:17–27.

Fittkau, E. J. 1962. Die Tanypodinae (Diptera Chironomidae); (Die Tribus Anatopyniini, Macropelopiini und Pentaneurini). Abh. Larvalsyst. Insekten 6:1–453.

Fittkau, E. J. 1966. *Chironomus*, nicht *Tendipes*. Bemerkungen zu einem Beschluss der I.C.Z.N., der Internationalen Zoologischen Nomenklaturkommission. Arch. Hydrobiol. 62:269–271.

Fittkau, E. J. 1968. Eine neue Tanypodinae-Gattung, Djalmabatista (Chironomidae, Dipt.), aus dem brasilienischen Amazonasgebiet. Amazoniana 1:327–349.

Fittkau, E. J., D. Schlee, and F. Reiss. 1967. Chironomidae. Pages 346–381 *in* J. Illies, ed. Limnofauna Europeae. G. Fischer, Stuttgart.

Fries, B. F. 1830. Beskrifning öfver ett nytt slägte *Hydrobaenus* hörande till Tipulariae. K. svenska Vetensk-Akad. Handl. 1829:176–187.

Garrett, C. B. D. 1925. Seventy new Diptera. Cranbrook, B.C. 16 pp.

Goetghebuer, M. 1922. Nouveaux matériaux pour l'étude de la faune de Chironomides de Belgique. Annls Biol. lacustre. 11:38–62.

Goetghebuer, M. 1932. Diptères (Nématocères). Chironomidae IV. Orthocladiinae, Corynoneurinae, Clunioninae, Diamesinae. Faune Fr. 23:1–204.

Goetghebuer, M. 1937–1954. Tendipedidae (Chironomidae). b) Subfamilie Tendipedinae (Chironomidae). A. Die Imagines. Pages 1–138 *in* E. Lindner, ed. Die Fliegen der palaearktischen Region. 13*c*.

Goetghebuer, M. 1939*a*. Tendipedidae (Chironomidae). e) Subfamilie Corynoneurinae. A. Die Imagines. Pages 1–14 *in* E. Lindner, ed. Die Fliegen der palaearktischen Region. 13*f*.

Goetghebuer, M. 1939*b*. Tendipedidae (Chironomidae). e) Subfamilie Diamesinae. A. Die Imagines. Pages 1–28 *in* E. Lindner, ed. Die Fliegen der palaearktischen Region. 13*d*.

Gowin, F. 1943. Orthocladiinen aus Lunzer Fliessgewässern. II. Arch. Hydrobiol. 40:114–122.

Grandjean, F. 1964. Oribates mexicains (1$^{re}$ serie) *Dampfiella* Selln. et *Beckiella* n.g. Acarologia 6:694–711.

Grodhaus, G. 1967. Identification of chironomid midges commonly associated with waste stabilization lagoons in California. Calif. Vector Views 14:1–11.

Grodhaus, G. 1976. Two species of *Phaenopsectra* with drought-resistant larvae (Diptera: Chironomidae). J. Kans. ent. Soc. 49:405–418.

Hamilton, A. L. 1965. An analysis of a freshwater benthic community with special reference to the Chironomidae. Ph.D. Thesis, Univ. British Columbia, Vancouver. 216 pp.

Hamilton, A. L. 1969. A method of separating invertebrates from sediments using longwave ultraviolet light and fluorescent dyes. J. Fish. Res. Bd Can. 26:1667–1672.

Hamilton, A. L., O. A. Sæther, and D. R. Oliver. 1969. A classification of the Nearctic Chironomidae. Fish. Res. Bd Can. Tech. Rep. 124:1–42.

Hansen, D. C., and E. F. Cook. 1976. The systematics and morphology of the Nearctic species of *Diamesa* Meigen, 1835 (Diptera: Chironomidae). Mem. Am. ent. Soc. 30:1–203.

Harnisch, O. 1924. Metamorphose und System der Gattung *Cryptochironomus* K. s.l. Zool. Jahrb. 47:271–308.

Hauber, U. A. 1947. The Tendipedinae of Iowa (Diptera). Am. Midl. Nat. 38:456–465.

Hilsenhoff, W. L. 1969. An artificial substrate device for sampling benthic stream invertebrates. Limnol. Oceanogr. 14:465–471.

Hinton, H. E. 1960. A fly larva that tolerates dehydration and temperatures from $-270°$ to $102°C$. Nature, Lond. 188:336–337.

Hirvenoja, M. 1961. Description of the larva of *Corynocera ambigua* Zett. (Dipt. Chironomidae) and its relation to the subfossil species *Dryadotanytarsus edentulus* Anders and *D. duffi* Deevey. Suom. hyönt. Aikak. 27:105–110.

Hirvenoja, M. 1973. Revision der Gattung *Cricotopus* van der Wulp und ihrer Verwandten (Diptera, Chironomidae). Annls zool. Fenn. 10:1–363.

Holmgren, A. E. 1869. Bidrag til kännodomen om Beeren Eilands och Spetsbergens Insekt-Fauna. K. svenska Vetensk-Akad. Handl. 8:1–55.

Hynes, H. B. N. 1970. The ecology of running waters. Univ. Liverpool Press. 555 pp.

Jackson, G. A. 1977. Nearctic and Palaearctic *Paracladopelma* Harnisch and *Saetheria* n. gen. (Diptera: Chironomidae). J. Fish. Res. Bd Can. 34:1321–1359.

Johannsen, O. A. 1905. Aquatic nematocerous Diptera II, Chironomidae. Pages 76–327 *in* I. G. Needham, K. I. Morton, and O. A. Johannsen, eds. May flies and midges of New York. Bull. N.Y. St. Mus. 86.

Johannsen, O. A. 1907. Notes on the Chironomidae. Pages 400–401 *in* H. Skinner, ed. Notes and news. Ent. News 18.

Johannsen, O. A. 1937*a*. Aquatic Diptera. III. Chironomidae: Subfamilies Tanypodinae, Diamesinae and Orthocladiinae. Mem. Cornell Univ. agric. Exp. Stn 205:3–84.

Johannsen, O. A. 1937*b*. Aquatic Diptera. IV. Chironomidae: Subfamily Chironominae. Mem. Cornell Univ. agric. Exp. Stn 210:3–56.

Jonasson, P. M. 1958. The mesh factor in sieving techniques. Verh. int. Verein. theor. angew. Limnol. 13:860–866.

Jonasson, P. M. 1965. Factors determining population size of *Chironomus anthracinus* in Lake Esrom. Mitt. int. Verein. theor. angew. Limnol. 13:139–162.

Kajak, Z., K. Dusoge, and A. Prejs. 1968. Application of the flotation technique to assessment of absolute numbers of benthos. Ekol. pol. Ser. A. 16:607–620.

Kalugina, N. S. 1959. Changes in morphology and biology of chironomid larvae in relation to growth (Dipt. Chir.) [in Russian]. Trudȳ vses. gidrobiol. Obshch. 9:85–107.

Kalugina, N. S. 1961. Taxonomy and development of *Endochironomus albipennis* Mg., *E. tendens* F. and *E. impar* Walk. (Diptera, Tendipedidae)[in Russian]. Ent. Obozr. 40:900–919.

Kieffer, J. J. 1906*a*. Description de nouveaux Diptères Nematocères d'Europe. Annls Soc. scient. Brux. 30:311–348.

Kieffer, J. J. 1906*b*. Description d'un genre nouveau et de quelques espèces nouvelles de Diptères de l'Amérique du Sud. Annls Soc. scient. Brux. 30:349–358.

Kieffer, J. J. 1906*c*. Diptera Fam. Chironomidae. Pages 1–78 *in* P. Wytsman, ed. Genera Insect. 42.

Kieffer, J. J. 1908. I. Neue und bekannte Chironomiden. Pages 1–10, 33–39, 78–84 *in* J. J. Kieffer and A. Thienemann, eds. Neue und bekannte Chironomiden und ihre Metamorphose. Z. wiss. Insekt- Biol. 4.

Kieffer, J. J. 1909. Diagnoses de nouveaux Chironomides d'Allemagne. Bull. Soc. Hist. nat. Metz. 26:37–56.

Kieffer, J. J. 1911*a*. Nouvelles descriptions des Chironomides obtenus d'éclosion. Bull. Soc. Hist. nat. Metz. 27:1–60.

Kieffer, J. J. 1911*b*. Nouveaux Tendipédides du groupe *Orthocladius* (Dipt.) Bull. Soc. ent. Fr. 8:181–187; 199–202.

Kieffer, J. J. 1911*c*. Description d'un Chironomide d'Amérique formant un genre nouveau. Bull. Soc. Hist. nat. Metz. 27:103–105.

Kieffer, J. J. 1912. Nouveaux Chironomides (Tendipedidae) de Ceylan. Spolia zeylan. 8:1–24.

Kieffer, J. J. 1913*a*. Nouvelle étude sur les Chironomides de l'Indian Museum de Calcutta. Rec. Indian Mus. 9:119–197.

Kieffer, J. J. 1913*b*. *Dasyhelea halophila* n. sp. eine neue halophile Zuckmücke. Pages 255–256 *in* Zur flora und fauna der Strandtümpel von Rovigno (in Istrien). Biol. Zbl. 33.

Kieffer, J. J. 1913*c*. Nouveaux Chironomides (Tendipedides) d'Allemagne. Bull. Soc. Hist. nat. Metz. 3:7–35.

Kieffer, J. J. 1913*d*. Chironomidae et Cecidomyidae — Voyage de Ch. Alluaud et R. Jeanell en Afrique orientale (1911–1912). Résult. scient. Insectes Diptères: 1–43.

Kieffer, J. J. 1915. Neue chironomiden aus Mitteleuropa. Broteria Ser. zool. 13:65–87.

Kieffer, J. J. 1918*a*. Beschreibung neuer, auf Lazarettschiffen des östlichen Kriegsschauplatzes und bei Ignalino in Litauen von Dr. W. Horn gesammelter Chironomiden, mit Übersichtstabellen einiger Gruppen von paläarktischen Arten (Dipt.). Ent. Mitt. 7:35–53, 94–110, 163–170, 177–188.

Kieffer, J. J. 1918*b*. Chironomides d'Afrique et d'Asie conservés au Musée National Hongrois de Budapest. Annls hist.-nat. Mus. natn. hung. 16:31–139.

Kieffer, J. J. 1919. Chironomiden der nördlichen Polarregion. Pages 40–48, 110–120 *in* J. J. Kieffer and A. Thienemann. Chironomiden gesammelt von Dr. A. Koch (Münster i. W.) auf den Lofoten, der Bäreninsel und Spitzbergen (Dipt.). Ent. Mitt. 8.

Kieffer, J. J. 1920*a*. Un nouveau genre de Chironomide (Dipt.). Bull. Soc. ent. Fr. 25:333–334.

Kieffer, J. J. 1920*b*. Tableau synoptique suivant des Chironomides paléarctiques appartenant aux genres *Polypedilum* et *Limnochironomus*. Annls Soc. scient. Brux. 39:159–167.

Kieffer, J. J. 1921a. Chironomides nouveaux ou peu connus de la région paléarctique. Bull. Soc. Hist. nat. Metz. 29:51–109.

Kieffer, J. J. 1921b. Synopse de la tribu des Chironomariae (Diptères). Annls Soc. scient. Brux. 40:269–276.

Kieffer, J. J. 1921c. Diagnoses de nouveaux genres et espèces de Chironomidae (Dipt.). Bull. Soc. ent. Fr. 26:287–289.

Kieffer, J. J. 1921d. Chironomides de l'Afrique equatoriale. Annls Soc. ent. Fr. 90:1–56.

Kieffer, J. J. 1922a. Chironomides nouveaux ou peu connus de la région paléarctique. Annls Soc. scient. Brux. 42:71–128, 138–180.

Kieffer, J. J. 1922b. Nouveaux Chironomides a larves aquatiques. Annls Soc. scient. Brux. 41:355–366.

Kieffer, J. J. 1923. Chironomides de l'Afrique equatoriale. Annls Soc. ent. Fr. 92:149–204.

Kieffer, J. J. 1924a. Chironomides nouveaux ou rares de l'Europe centrale. Bull. Soc. Hist. nat. Metz. 30:11–110.

Kieffer, J. J. 1924b. Quelques Chironomides nouveaux et remarquables du Nord de l'Europe. Annls Soc. scient. Brux. 43:390–397.

Kieffer, J. J. 1925. Deux genres nouveaux et plusieurs espèces nouvelles du groupe des Orthocladiariae (Diptères, Chironomidae). Annls Soc. scient. Brux. 44:555–566.

Krüger, Fr. 1938. *Tanytarsus*-Studien I. Die Subsectio *Atanytarsus*. Zugleich variations-statistiche Untersuchungen zum Problem der Artbildung bei Chironomiden. Arch. Hydrobiol. 33:208–256.

Krüger, Fr. 1945. Eutanytarsariae der *Gregarius*-Gruppe (Diptera: Chironomidae) aus Schleswig-Holstein (Tanytarsarienstudien IV). Arch. Hydrobiol. 40:1084–1115.

Krüger, Fr., and A. Thienemann. 1941. Terrestrische Chironomiden XI. Die Gattung *Gymnometriocnemus* Geotgh. (Mit einem Beitrag von. M. Goetghebuer, Gent). Zool. Anz. 135:185–195.

Kruseman, G. 1933. Tendipedidae Neerlandicae. I: Genus *Tendipes* cum generibus finitimis. Tijdschr. Ent. 76:119–216.

Kurazhkovskaya, T. N. 1966. Structure of the intestine and salivary glands of chironomid larvae (Diptera) [in Russian]. Trudȳ Inst. Biol. Vodokhran. 12:286–296.

Lehmann, J. 1969. Zur Ökologie und Verbreitung dreier für Schleswig-Holstein neuer Chironomidenarten (Diptera, Nematocera). Faun. ökol. Mitt. 3:262–268.

Lehmann, J. 1972. Revision der europäischen Arten (Puppen ♂♂ und Imagines ♂♂) der Gattung *Eukiefferiella* Thienemann. Beitr. Ent. 22:347-405.

Lellák, J. 1968. Positive Phototaxis der Chironomiden-Larvulae als regulierender Faktor ihrer Verteilung in stehenden Gewässern. Annls zool. Fenn. 5:84–87.

Lenz, F. 1921. Chironomidenpuppen und-larven. Bestimmungstabellen. Dt. ent. Z. 3:148–162.

Lenz, F. 1939. Tendipedidae (Chironomidae). e) Subfamilie Corynoneurinae. B. Die metamorphose der Corynoneurinae. Pages 14–19 *in* E. Lindner, ed. Die Fliegen der palaearktischen Region. 13*f.*

Lenz, F. 1941*a.* Die Metamorphose der Chironomiden-Gattung *Cryptochironomus*. Zool. Anz. 133:29–41.

Lenz, F. 1941*b.* Die Jugendstadien der Sectio Chironomarie (Tendipedini) connectentes (Subf. Chironominae = Tendipedinae). Zusammenfassung und Revision. Arch. Hydrobiol. 38:1–69.

Lenz, F. 1954–1962. Tendipedidae (Chironomidae). b) Subfamilie Tendipedinae (Chironominae). B. Die Metamorphose der Tendipedinae. Pages 139–260 *in* E. Lindner, ed. Die Fliegen der palaearktischen Region. 13*c.*

Lenz, F. 1955. Revision der Gattung *Endochironomus* Kieff. (Diptera, Tendipedidae). Z. angew. Zool. 8:109–121.

Lipina, N. N. 1939. New forms of young stages of Chironomidae in the experimental lakes of sapropel station in Zaluchie [in Russian]. Trudy Lab. sapropel. Otlozh. 1:90–107.

Macan, T. T. 1958. Methods of sampling the bottom fauna in stony streams. Mitt. int. Verein. theor. angew. Limnol. 8:1–21.

Macan, T. T. 1970. Biological studies of the English lakes. American Elsevier Publishing Co., New York. 260 pp.

Malloch, J. R. 1915. The Chironomidae or midges of Illinois, with particular reference to the species occurring in the Illinois River. Bull. Ill. St. Lab. nat. Hist. 10:275–543.

Martin, J. E. H. 1977. The insects and arachnids of Canada. Part 1. Collecting, preparing and preserving insects, mites, and spiders. Agric. Can. Publ. 1643. 182 pp.

Maschwitz, D. E. 1975. Revision of the Nearctic species of the subgenus *Polypedilum* (Chironomidae: Diptera). Ph.D. Thesis, Univ. Minnesota. 324 pp.

McCauley, V. J. E. 1974. Instar differentiation in larval Chironomidae (Diptera). Can. Ent. 106:179–200.

Meigen, J. W. 1800. Nouvelle classificaton des mouches à deux ailes (Diptera L.) d'après un plan tout nouveau. Paris. 40 pp.

Meigen, J. W. 1803. Versuch einer neuen Gattungseinteilung der europäischen zweiflügeligen Insekten. Illigers Mag. Insekten:259–281.

Miller, R. B. 1941. A contribution to the ecology of the Chironomidae of Costello Lake, Algonquin Park, Ontario. Univ. Toronto Stud. 49:1–63.

Morgan, M. J. 1949. The metamorphosis and ecology of some species of Tanypodinae (Diptera, Chironomidae). Entomologist's mon. Mag. 85:119–126.

Mozley, S. C. 1970. Morphology and ecology of the larva of *Trissocladius grandis* (Kieffer) (Diptera, Chironomidae), a common species in the lakes and rivers of northern Europe. Arch. Hydrobiol. 67:433–451.

Mozley, S. C. 1971. Maxillary and premental patterns in Chironominae and Orthocladiinae (Diptera, Chironomidae). Can. Ent. 103:298–305.

Needham, J. G., and P. R. Needham. 1941. A guide to the study of fresh-water biology. Comstock Publishing Co., Inc., Ithaca. 89 pp.

Neff, S. E., and E. F. Benfield. 1970. Notes on the status of the genus *Demeijerea* Krusemann (Diptera: Chironomidae). Proc. ent. Soc. Wash. 72:126–132.

Oldroyd, H. 1966. The future of taxonomic entomology. Syst. Zool. 15:253–260.

Oliver, D. R. 1968. Adaptations of Arctic Chironomidae. Annls zool. Fenn. 5:111–118.

Oliver, D. R. 1971a. Description of *Einfeldia synchrona* n. sp. (Diptera: Chironomidae). Can. Ent. 103:1591–1595.

Oliver, D. R. 1971b. Life history of the Chironomidae. Ann. Rev. Ent. 16:211–230.

Oliver, D. R. 1976. Chironomidae (Diptera) of Char Lake, Cornwallis Island, N.W.T., with descriptions of two new species. Can. Ent. 108:1053–1064.

Oliver, D. R. 1977. *Bicinctus*-group of the genus *Cricotopus* van der Wulp (Diptera: Chironomidae) in the Nearctic with a description of a new species. J. Fish. Res. Bd Can. 34:98–104.

Oliver, D. R. 1981. Description of *Euryhapsis* new genus including three new species (Diptera: Chironomidae). Can. Ent. 113:711–722.

Oliver, D. R., D. McClymont, and M. E. Roussel. 1978. A key to some larvae of Chironomidae (Diptera) from the Mackenzie and Porcupine River watersheds. Can. Fish. Mar. Serv. Tech. Rep. 791. 73 pp.

Pagast, F. 1933. Chironomidenstudien. Stettin. ent. Ztg 94:286–300.

Pagast, F. 1947. Systematik and Verbeitung der um die Gattung *Diamesa* gruppierten Chironomiden. Arch. Hydrobiol. 41:435–596.

Paine, G. H., and A. R. Gaufin. 1956. Aquatic Diptera as indicators of pollution in a mid-western stream. Ohio J. Sci. 56:291–304.

Paloumpis, A. A., and W. C. Starrett. 1960. An ecological study of the benthic organisms in three Illinois River flood plain lakes. Am. Midl. Nat. 64:406–435.

Pankratova, V. Ya. 1970. Larvae and pupae of midges of the subfamily Orthocladiinae (Diptera, Chironomidae = Tendipedidae) of the USSR fauna [in Russian]. Izd. Nauka, Leningrad. 344 pp.

Pennak, R. W. 1953. Fresh water invertebrates of the United States. Ronald Press Co., New York. 769 pp.

Philippi, R. A. 1865. Aufzählung der Chilenischen Dipteren. Verh. zool.-bot. Ges. Wien 15:595–782.

Potthast, A. 1915. Über die Metamorphose der *Orthocladius*-Gruppe. Ein Beitrag zur Kenntnis der Chironomiden. Arch. Hydrobiol. Suppl. 2:243–376.

Reiss, F. 1968. Beitrag zur Taxonomie und Phylogenie palaearktischer *Neozavrelia*-Arten (Diptera, Chironomidae) mit der Beschreibung zweier neurer Arten aus Afghanistan und den Alpen. Gewäss. Abwäss. 47:7–19.

Reiss, F. 1969*a*. Die neue, europäisch verbreitete Chironomidengattung *Parapsectra* mit einem brachypteren Artvertreter aus Mooren (Diptera). Arch. Hydrobiol. 66:192–211.

Reiss, F. 1969*b*. *Krenopsectra fallax* gen. n. sp. n. (Diptera, Chironomidae) aus den Alpen und Pyrenäen. Annls zool. Fenn. 6:435–442.

Reiss, F., and E. J. Fittkau. 1971. Taxonomie und Ökologie europäisch verbreiteter *Tanytarsus*-Arten (Chironomidae, Diptera). Arch. Hydrobiol. Suppl. 40:75–200.

Rempel, J. G. 1936. The life-history and morphology of *Chironomus hyperboreus*. J. biol. Bd Can. 2:209–221.

Roback, S. S. 1953. Tendipedid larvae from the St. Lawrence River (Diptera: Tendipedidae). Notul. Nat. 253:1–4.

Roback, S. S. 1955. The tendipedid fauna of a Massachusetts cold spring (Diptera: Tendipedidae). Notul. Nat. 270:1–8.

Roback, S. S. 1957. The immature tendipedids of the Philadelphia area (Diptera: Tendipedidae). Monogr. Acad. nat. Sci. Philad. 9:1–152.

Roback, S. S. 1963. The genus *Xenochironomus* (Diptera: Tendipedidae) Kieffer, taxonomy and immature stages. Trans. Am. ent. Soc. 88:235–245.

Roback, S. S. 1969. Notes on the food of Tanypodinae larvae. Ent. News 80:13–18.

Roback, S. S. 1971. The adults of the subfamily Tanypodinae (=Pelopiinae) in North America (Diptera: Chironomidae). Mon. Acad. nat. Sci. Philad. 17:1–410.

Roback, S. S. 1972. The immature stages of *Paramerina smithae* (Sublette) (Diptera: Chironomidae: Tanypodinae). Proc. Acad. nat. Sci. Philad. 124:11–15.

Roback, S. S. 1974a. Insects (Arthropoda: Insecta). Pages 313–376 *in* C. W. Hart and S. L. H. Fuller, eds. Pollution ecology of freshwater invertebrates. Academic Press, New York, London.

Roback, S. S. 1974b. The immature stages of the genus *Coelotanypus* (Chironomidae; Tanypodinae; Coelotanypodini) in North America. Proc. Acad. nat. Sci. Philad. 126:9–19.

Roback, S. S. 1976. The immature chironomids of the Eastern United States I. Introduction and Tanypodinae — Coelotanypodini. Proc. Acad. nat. Sci. Philad. 127:147–201.

Roback, S. S. 1977. The immature chironomids of the Eastern United States II. Tanypodinae — Tanypodini. Proc. Acad. nat. Sci. Philad. 128:55–87.

Roback, S. S. 1978. The immature chironomids of the Eastern United States III. Tanypodinae — Anatopyniini, Macropelopiini and Natarsiini. Proc. Acad. nat. Sci. Philad. 129:151–202.

Roback, S. S. 1980. New name for *Anceus* Roback nec *Anceus* Risso. Ent. News 91:32.

Roback, S. S., and W. W. Moss. 1978. Numerical taxonomic studies on the congruence of classifications for the genera and subgenera of Macropelopiini and Anatopyniini (Diptera: Chironomidae: Tanypodinae). Proc. Acad. nat. Sci. Philad. 129:125–150.

Roback, S. S., and K. J. Tennessen. 1978. The immature stages of *Djalmabatista pulcher* [= *Procladius* (*Calotanypus*) *pulcher* (Joh.)]. Proc. Acad. nat. Sci. Philad. 130:11–20.

Rosenberg, D. M., and A. P. Wiens. 1976. Community and species responses of Chironomidae (Diptera) to contamination of fresh waters by crude oil and petroleum products, with special reference to the Trail River, Northwest Territories. J. Fish. Res. Bd Can. 33:1955–1963.

Rosenberg, D. M., A. P. Wiens, and O. A. Sæther. 1977. Life histories of *Cricotopus* (*Cricotopus*) *bicinctus* and *C.* (*C.*) *mackenziensis* in the Fort Simpson area, Northwest Territories. J. Fish. Res. Bd Can. 34:247–253.

Sæther, O. A. 1968. Chironomids of the Finse area, Norway, with special reference to their distribution in a glacier brook. Arch. Hydrobiol. 64:426–483.

Sæther, O. A. 1969. Some Nearctic Podomoninae, Diamesinae, and Orthocladiinae (Diptera: Chironomidae). Bull. Fish. Res. Bd Can. 170:1–154.

Sæther, O. A. 1970. Chironomids and other invertebrates from North Boulder Creek, Colorado. Univ. Colo. Stud. Ser. Biol. 31:57–114.

Sæther, O. A. 1971a. Nomenclature and phylogeny of the genus *Harnischia* (Diptera: Chironomidae). Can. Ent. 103:347–362.

Sæther, O. A. 1971b. Notes on general morphology and terminology of the Chironomidae (Diptera). Can. Ent. 103:1237–1260.

Sæther, O. A. 1971c. Four new and unusual Chironomidae (Diptera). Can. Ent. 103:1799–1827.

Sæther, O. A. 1973a. Four species of *Bryophaenocladius* Thien., with notes on other Orthocladiinae (Diptera: Chironomidae). Can. Ent. 105:51–60.

Sæther, O. A. 1973b. Taxonomy and ecology of three new species of *Monodiamesa* Kieffer, with keys to Nearctic and Palaearctic species of the genus (Diptera: Chironomidae). J. Fish. Res. Bd Can. 30:665–679.

Sæther, O. A. 1974. Morphology and terminology of female genitalia in Chironomidae (Diptera). Ent. Tidskr. 95:216–223.

Sæther, O. A. 1975a. Two new species of *Heterotanytarsus* Spärck, with keys to Nearctic and Palaearctic males and pupae of the genus (Diptera: Chironomidae). J. Fish. Res Bd Can. 32: 259–270.

Sæther, O. A. 1975b. Two new species of *Protanypus* Kieffer, with keys to Nearctic and Palaearctic species of the genus (Diptera: Chironomidae). J. Fish. Res. Bd Can. 32:367–388.

Sæther, O. A. 1975c. Nearctic and Palaearctic *Heterotrissocladius* (Diptera: Chironomidae). Bull. Fish. Res. Bd Can. 193:1–67.

Sæther, O. A. 1976. Revision of *Hydrobaenus, Trissocladius, Zalutschia, Paratrissocladius*, and some related genera (Diptera: Chironomidae). Bull. Fish. Res. Bd Can. 195:1–287.

Sæther, O. A. 1977a. Taxonomic studies on Chironomidae: *Nanocladius, Pseudochironomus* and the *Harnischia* complex. Bull. Fish. Res. Bd Can. 196:1–143.

Sæther, O. A. 1977b. Female genitalia in Chironomidae and other Nematocera: morphology, phylogenies, keys. Bull. Fish. Res. Bd Can. 197:1–209.

Sæther, O. A. 1977c. *Habrobaenus hudson*; n. gen., n. sp. and the immatures of *Baeoctenus bicolor* Sæther (Diptera: Chironomidae). J. Fish. Res. Bd Can. 34:2354–2361.

Sæther, O. A. 1979. New name for *Beckiella* Saether, 1977 (Diptera: Chironomidae) nec *Beckiella* Grandjean, 1964 (Acari: Oribatei). Ent. Scand. 10:315.

Sæther, O. A. 1980a. Glossary of chironomid morphology terminology (Diptera: Chironomidae). Ent. Scand. 14:1–51.

Sæther, O. A. 1980b. New name for *Oliveria* Sæther, 1976 (Diptera: Chironomidae) nec *Oliveria* Sutherland, 1965 (+Cnidaria= Anthozoa), with a first record for the European continent. Ent. Scand. 11:399–400.

Santos Abreu, E. 1918. Ensayo de una monografia de los Tendipedidos de las Islas Canarias. Mems R. Acad. Cienc. Artes Barcelona 14:159–326.

Saunders, L. G. 1924. On the early stages of *Cardiocladius* (Diptera: Chironomidae). Entomologist's mon. Mag. 60:227–231.

Saunders, L. G. 1928a. The early stages of *Diamesa* (*Psilodiamesa*) *lurida* (Diptera: Chironomidae). Can. Ent. 60:261–264.

Saunders, L. G. 1928b. Some marine insects of the Pacific coast of Canada. Ann. ent. Soc. Am. 21:521–545.

Saunders, L. G. 1930. The larvae of the genus *Heptagyia*, with descriptions of a new species (Diptera: Chironomidae). Entomologist's mon. Mag. 66:209–214.

Schlee, D. 1968. Vergleichende Merkmalsanalyse zur Morphologie und Phylogenie der *Corynoneura*-Gruppe (Diptera: Chironomidae). Zugleich eine allgemeine Morphologie der Chironomiden-Imago (♂) Stuttg. Beitr. Naturk. 180:1–150.

Scott, K. M. F. 1967. The larval and pupal stages of the midge *Tanytarsus* (*Rheotanytarsus*) *fuscus* Freeman (Diptera: Chironomidae). J. ent. Soc. sth Afr. 30:174–184.

Serra-Tosio, B. 1971. Contribution à l'étude taxonomique, phylogénétique, biogéographique et écologique des Diamesini (Diptera, Chironomidae) d'Europe. Thèse Univ. scient. Méd. Grenoble T.I:1–303, T. II:304–462.

Serra-Tosio, B. 1972. Ecologie et biogéographie des Diamesini d'Europe (Diptera, Chironomidae). Trav. Lab. Hydrobiol. Piscic. Univ. Grenoble 63:5–175.

Shilova, A. I. 1955. Some abundant tendipedid species of the Amu-Darya drainage basin [in Russian]. Ént. Obozr. 34:313–322.

Shilova, A. I. 1966. The larvae of *Odontomesa fulva* Kieff. (Diptera, Chironomidae, Orthocladiinae [in Russian]. Trudȳ Inst. Biol. Vodokhran. 12:239–250.

Shiozawa, D. K., and J. R. Barnes. 1977. The microdistribution and population trends of larval *Tanypus stellatus* Coquillett and *Chironomus frommeri* Atchley and Martin (Diptera: Chironomidae) in Utah Lake, Utah. Ecology 58:610–618.

Simmons, G. M., and A. Winfield. 1971. A feasibility study using conservation webbing as an artificial substrate in macrobenthic studies. Va J. Sci. 22:52–59.

Skuse, F. A. A. 1889. Diptera of Australia, Pt. VI. The Chironomidae. Proc. Linn. Soc. N.S.W. 2:215–311.

Soponis, A. R. 1977. A revision of the Nearctic species of *Orthocladius* (*Orthocladius*) van der Wulp (Diptera: Chironomidae). Mem. ent. Soc. Can. 102:1–187.

Spärck, R. 1922. Beiträge zur Kenntnis der Chironomiden-Metamorphose I-IV. Ent. Meddr 14:32–109.

Steffan, A. W. 1965. *Plecopteracoluthus downesi* gen. et sp. nov. (Diptera: Chironomidae), a species whose larvae live phoretically on larvae of Plecoptera. Can. Ent. 97:1323–1344.

Steffan, A. W. 1968. Zur Evolution und Bedeutung epizoischer Lebensweise bei Chironomiden-Larven (Diptera). Annls zool. Fenn. 5:144–150.

Stone, A. 1941. The generic names of Meigen 1800 and their proper application. Ann. ent. Soc. Am. 34:404–418.

Strenzke, K. 1940. Terrestrische Chironomiden V. *Camptocladius stercorarius* DeGeer. Zol. Anz. 132:115–123.

Strenzke, K. 1950. Systematik, Morphologie und Ökologie der terrestrischen Chironomiden. Arch. Hydrobiol. Suppl. 18:207–414.

Strenzke, K., and H. Remmert. 1957. Terrestrische Chironomiden. XVII. *Thalassosmittia thalassophila* (Begu. u. Goetgh.). Kieler Meeresforsch. 13:263–273.

Sublette, J. E. 1957. The ecology of the macroscopic bottom fauna in Lake Texoma (Dennison Reservoir), Oklahoma and Texas. Am. Midl. Nat. 57:371–402.

Sublette, J. E. 1960. Chironomid midges of California. I. Chironominae, exclusive of Tanytarsini (=Calopsectrini). Proc. U.S. natn. Mus. 112:197–226.

Sublette, J. E. 1964. Chironomidae (Diptera) of Louisiana. I. Systematics and immature stages of some lentic chironomids of West-Central Louisiana. Tulane Stud. Zool. 11:109–150.

Sublette, J. E. 1967. Type specimens of Chironomidae (Diptera) in the Canadian National Collection, Ottawa. J. Kans. ent. Soc. 40:290–331.

Sublette, J. E., and M. S. Sublette. 1965. Family Chironomidae (Tendipedidae). Pages 142–181 *in* A. Stone et al. eds. A catalog of the Diptera of America, north of Mexico. U.S. Dep. Agric. Handbook 276:1–1696.

Sublette, J. E., and M. F. Sublette. 1974a. A review of the genus *Chironomus* (Diptera, Chironomidae). V. The *maturus*-complex. Stud. nat. Sci. 1:1–41.

Sublette, J. E., and M. F. Sublette. 1974b. A review of the genus *Chironomus* (Diptera, Chironomidae). VII. The morphology of *Chironomus stigmaterus* Say. Stud. nat. Sci. 1:1–64.

Sulc, K., and J. Zavřel. 1924. O epoikickych a parasitickych larvach Chironomidu. Über epoikische und parasitische Chironomidenlarven. Acta Soc. Sci. nat. moravo-siles. 1:353–391.

Sutherland, P. K. 1965. Rugose corals of the Henryhous formation (Silurian) in Oklahoma. Bull. Okla. geol. Surv. 109:1–92.

Thienemann, A. 1921. Die metamorphose der Chironomiden-gattungen *Camptocladius, Dyscamptocladius* und *Phaenocladius*, mit Bemerkungen über die Artdifferenzierung bei den Chironomiden überhaupt. Arch. Hydrobiol. Suppl. 2:809–850.

Thienemann, A. 1926a. Hydrobiologische Untersuchungen an den kalten Quellen und Bächen der Halbinsel Jasmund auf Rügen. Arch. Hydrobiol. 17:221–336.

Thienemann, A. 1926b. Hydrobiologische Untersuchungen an Quellen. VII. Insekten aus norddeutschen Quellen mit besonderer Berücksichtigung der Dipteren. Dt. ent. Z.: 1–50.

Thienemann, A. 1929. Chironomiden-Metamorphosen. II. Die Sectio *Tanytarsus genuinus*. Arch. Hydrobiol. 20:93–123.

Thienemann, A. 1933. Chironomiden-Metamorphosen. III. Zur Metamorphose der Orthocladiariae. (Mit einen Beitrag von O. Harnisch.) Dt. ent. Z.: 1–39.

Thienemann, A. 1934. Chironomiden-Metamorphosen. VIII. *Phaenocladius*. Encycl. ent. 7:29–46.

Thienemann, A. 1935. Chironomiden-Metamorphosen. X. *Orthocladius-Dactylocladius* (Dipt.). Stettin. ent. Ztg 96:201–224.

Thienemann, A. 1936. Alpine Chironomiden. (Ergebnisse von Untersuchungen in der Gegend von Garmisch-Partenkirchen, Oberbayern.) Arch. Hydrobiol. 30:167–262.

Thienemann, A. 1937a. Nachtrag zur Orthocladiinenfauna Niederländisch-Indiens. Arch. Hydrobiol. Suppl. 15:119–120.

Thienemann, A. 1938. Zur Metamorphose der Podonominae. Pages 154–158 *in* F. W. Edwards and A. Thienemann. Neuer Beitrag zur Kenntnis der Podonominae (Dipt. Chironomidae). Zool. Anz. 122.

Thienemann, A. 1939. Dritter Beitrag zur Kenntnis der Podonominae (Dipt., Chironomidae). (Chironomiden aus Lappland IV.) Zool. Anz. 128:161–176.

Thienemann, A. 1941. Lappländische Chironomiden und ihre Wohngewässer. (Ergebnisse von Untersuchungen im Abiskogebiet in Schwedisch-Lappland.) Arch. Hydrobiol. Suppl. 17:1–253.

Thienemann, A. 1942. *Trichocladius*-Arten aus den Lunzer Seen. (Chironomiden aus dem Lunzer Seengebiet. V). Arch. Hydrobiol. 39:294–315.

Thienemann, A. 1944. Bestimmungstabellen für die bis jetzt bekannten Larven und Puppen der Orthocladiinen (Diptera, Chironomiden). Arch. Hydrobiol. 39:551–664.

Thienemann, A. 1951. *Tanytarsus*-Studien II. Die Subsectio *Paratanytarsus*. Auf Grund der nachgelassenen Papiere Friedrich Wilh. Carl Krüger's. Arch. Hydrobiol. Suppl. 18:595–632.

Thienemann, A. 1954. *Chironomus*. Leben, Verbreitung und wirtschaftliche Bedeutung der Chironomiden. Binnengewässer 20:1–834.

Thienemann, A., and O. Harnisch. 1932. Chironomiden-Metamorphosen. IV. Die Gattung *Cricotopus* v.d.w. Zool. Anz. 99:135–143.

Thienemann, A., and J. J. Kieffer. 1916. Schwedische Chironomiden. Arch. Hydrobiol. Planktonk. Suppl. 2:483–554.

Thienemann, A., and F. Krüger. 1939. Terrestrische Chironomiden II. Zool. Anz. 127:246–258.

Townes, H. K., Jr. 1945. The Nearctic species of Tendipedini [Diptera, Tendipedidae (=Chironomidae)]. Am. Midl. Nat. 34:1–206.

Walshe, B. M. 1951. The feeding habits of certain chironomid larvae (subfamily Tendipedinae). Proc. zool. Soc. Lond. 121:63–79.

Waltl, J. 1835. Neue Arten von Dipteren aus der Umgegend von Muechen, benannt und beschreiben von Meigen, aufgefunden von Dr. J. W. Waltl. Faunus 2:66–73.

Webb, D. W. 1969. New species of Chironomids from Costello Lake, Ontario (Diptera: Chironomidae.) J. Kans. ent. Soc. 42:91–108.

Webb, D. W. 1972. The immature stages of *Chironomus aethiops* (Townes) with keys to the species of the known immature stages of the subgenus *Dicrotendipes* (Chironomidae: Diptera). Trans. Ill. St. Acad. Sci. 65:74–76.

Welch, H. E. 1973. Emergence of Chironomidae (Diptera) from Char Lake, Resolute, Northwest Territories. Can J. Zool. 51:1113–1123.

Welch, H. E. 1976. Ecology of Chironomidae (Diptera) in a polar lake. J. Fish. Res. Bd Can. 33:227–247.

Welch, P. S. 1948. Limnological methods. McGraw-Hill Co., New York. 471 pp.

Winnertz, J. 1846. Beschreibung einiger neuer Gattungen aus der Ordnung der Zweiflügler. Stettin. ent. Ztg 7:11–20.

Wirth, W. W. 1949. A revision of the Clunionine midges with descriptions of a new genus and four new species. Univ. Calif. Publs Ent. 8:151–182.

Wirth, W. W. 1957. The species of *Cricotopus* midges living in the blue-green alga *Nostoc* in California (Diptera: Tendipedinae). Pan-Pacif. Ent. 33:121–126.

Wirth, W. W., and J. L. Gressitt. 1967. Diptera: Chironomidae (midges). Antarctic Res. Ser. 10:197–203.

Wirth, W. W., and J. E. Sublette. 1970. A review of the Podonominae of North America with descriptions of three new species of *Trichotanypus* (Dipt.: Chir. J. Kans. ent. Soc. 43:335–354.

Wülker, W. 1957. Über die Chironomiden der *Parakiefferiella*-Gruppe (Diptera: Tendipedidae: Orthocladiinae). Beitr. Ent. 7:411–429.

Wülker, W., J. E. Sublette, M. F. Sublette, and J. Martin. 1971. A review of the genus *Chironomus* (Diptera, Chironomidae) I. The *staegeri* group. Stud. nat. Sci. 1:1–88.

Wulp, F. M. van der. 1874. Dipterologische aanteekneningen. Tijdschr. Ent. 17:109–148.

Wundsch, H. H. 1943. Die Metamorphose von *Demeijerea rufipes* L. (Dipt. Tendip.). Zool. Anz. 141:27–32.

Wurtz, C. B., and S. S. Roback. 1955. The invertebrate fauna of some Gulf Coast rivers. Proc. Acad. nat. Sci. Philad. 107:167–206.

Zavřel, J. 1926. Metamorfosa několika nových Chironomidu. Acta Soc. Sci. nat. moravo-siles. 3:251–282.

Zavřel, J. 1939. Chironomidarum larvae et nymphae II. (Genus *Eukiefferiella* Th.). Acta Soc. Sci. nat. moravo-siles. 11:1–29.

Zavřel, J. 1941a. Chironomidarum larvae et nymphae IV. (Genus *Metriocnemus* v.d. Wulp). Acta Soc. Sci. nat. moravo-siles. 13:1–28.

Zavřel J. 1941b. Chironomidarum larvae et nymphae III. ("*Pseudokiefferiella*"). Ent. Listy 4:9–14.

Zavřel, J. 1942. Polypodie der Chironomiden-Puppen. Acta Soc. Sci. nat. moravo-siles. 14:1–40.

Zavřel, J., and A. Thienemann. 1919. Die Metamorphose der Tanypinen (II Tiel). Arch. Hydrobiol. Suppl. 2:655–784.

Zetterstedt, J. W. 1838. Dipterologis Scandinaviae. Sect. 3:Diptera: 477–868.

# Index

(Page numbers of principal entries are in boldface; synonyms are in italic type.)

Abiskomyia 12, 95, 102, 106, **110**, 111, 126, 209
Ablabesmyia 10, 36, 38, 39, 40, **41**, 46, 160, 161, 164, 165, 180
Acalcarella 10, 58, 61, 70, **71**, 72, 190
Acricotopus 12, 103, 107, **111**, 112, 209, 210
acrocinata 27
adrumbatus 37
Allopsectrocladius 140, 141
Alotanypus 26, 32
Anatopynia 26, 36, 37, 45
Anatopyniini 26
annularius 65, 66
Anodonta 112
anthracinus 65, 66
*Apedilum* 84
Apelopia 9, 31, 177
Aphroteniinae 8, 18
Aponteus 29
Apsectrotanypus 9, 26, 32, 33, **34**, 35, 36, 177, 178
araucana 51
Arctopelopia 10, 27, 38, 47
Atanytarsus 95
Axarus 11, 90, 202, 203
Baeoctenus 12, 103, 108, **112**
Beckidia 10, 58, 61, 70, **72**, 73
Beckiella 72
bellus 37
berneri 44
bicinctus 119, 120, 123, 212
Boreochlini 5, 10, 49, 50, **51**
Boreochlus 10, 51, **52**
Boreoheptagyia 13, **153**, 154, 235, 236
Boreoheptagyini 5, 13, 152, **153**
Brillia 12, 104, 108, **112**, 113, 123, 125, 210
Bryophaenocladius 12, 106, 110, **113**, 114, 115, 124, 138, 142
Buchonomyiinae 8
burmanicus 52
californicus 67
Calopsectra 95, 99
Calopsectrini 8
Calotanypus 35

Camptochironomus 64
Camptocladius 12, 106, 110, **114**, 115, 142, 144, 146
Cantopelopia 10, 27, 38
Cardiocladius 12, 103, 107, **115**, 210
Ceratopogonidae
Chaetocladius 12, 105, 109, **116**, 125, 131, 136
Chaetolabis 11, 64, 65, 66, 186
Chasmatonotus 12
Chernovskiia 11, 57, 61, 70, **73**, 190
Chironomidae 5, 7, 8, 9, 18, 24, 25, 49
Chironominae 5, 8, 10, 16, 18, 21, 23, 24, 25, 27, **54**, 55, 56, 88, 92, 101, 159
Chironomini 5, 10, 54, 55, **56**, 57, 64, 71, 81, 84, 85, 89, 91, 92, 99, 102, 113
Chironomus 8, 11, 57, 60, 63, **64**, 65, 66, 67, 68, 69, 70, 71, 75, 80, 82, 85, 87, 88, 89, 160, 161, 166, 167, 186
Cladopelma 11, 58, 62, 70, 71, **73**, 74, 75, 190, 191
Cladotanytarsus 11, 93, **94**, 95, 97, 98, 204
Clinotanypus 9, **29**, 30, 176
Coelotanypodini 5, 9, 26, 27, **28**, 29, 31
Coelotanypus 9, 29, **30**, 176, 177
Conchapelopia 10, 38, 47
conformis 128
Constempellina 11, 92, 93, 94, **95**, 204
Corynocera 11, 92, 93, 94, **96**, 205
Corynoneura 12, 102, 106, **116**, 117, 147, 211
Corynoneurella 117
Corynoneurinae 117
Cricotopus 12, 103, 104, 107, 108, **117**, 118, 119, 120, 123, 133, 135, 137, 139, 170, 171, 211, 212, 213
Cryptochironomus 11, 57, 61, 65, 70, 71, 73, **74**, 75, 76, 191
*Cryptocladopelma* 73

259

Cryptotendipes   11, 58, 62, 70, 71, 74, **75**, 191, 192
culiciformis   37
cylindraceus   119, 120, 211
Cyphomella   11, 58, 61, 70, **75**, 76, 77, 192
Dactylocladius   145
Demeijerea   69, 70
Demicryptochironomus   11, 57, 61, 70, 71, 74, **76**, 77, 192
Derotanypus   9, 26, 27, 32, 33, **35**, 37, 178
Diamesa   13, 152, 154, **155**, 156, 157, 174, 175, 232
Diamesinae   5, 8, 13, 15, 16, 18, 21, 24, 25, 101, 148, **151**, 152, 153, 154
Diamesini   5, 13, 152, 153, **154**, 155
Dicrotendipes   11, 59, 63, 64, 65, **66**, 67, 69, 80, 187
Diplocladius   12, 101, 104, 108, 113 **121**, 147, 148, 213, 214
*Diplomesa*   157
dispar   69
distylus   128
Ditanytarsus   97
Djalmabatista   9, 33, **35**, 36
downesi   131
Einfeldia   11, 57, 60, 63, 64, 65, **67**, 68, 69, 80, 187, 188
elachista   84, 85
Endochironomus   11, 60, 63, 65, **68**, 69, 86, 87, 99, 188, 189
Ephemeroptera   145
Epoicocladius   12, 104, 109, **121**, 122, 129, 214
*Eucricotopus*   117
Eudactylocladius   12, 133, 134, 220
Eukiefferiella   12, 103, 105, 107, 109, **122**, 123, 136, 214, 215
*Euorthocladius*   12, 133, 134, 221
*Euphaenocladius*   143, 144
Euryhapsis   12, 104, 108, 113, **123**, 215, 216
exigiguus   98
festivellus   119, 120, 212
fulva   152
fuscus   119, 120
gaedii   152, 156, 233
Gillotia   54, 57, 61, 71, 74, 76
Glyptotendipes   11, 57, 59, 60, 63, 64, 67, 68, **69**, 70, 80, 189
Goeldichironomus   18, 54, 64

Graceus   11, 54
Guttipelopia   10, 39, 40, **41**, 42, 180
Gymnometriocnemus   12, 105, 110, 114, 115, **124**, 138, 142, 216
Halocladius   12, 103, 108, **124**, 125, 135, 143
halophilus   65, 66
Harrisonini   152
Harnischia   11, 16, 54, 57, 58, 60, 61, **70**, 71, 72, 73, 74, **77**, 78, 193
Heleniella   12, 105, 109, 113, **125**, 130, 216
Helopelopia   10
Heptagyia   154
Heptagyini   152
Heterotanytarsus   12, 95, 102, 106, 111, 117, **126**, 216
Heterotrissocladius   12, 105, 109, **126**, 127, 128, 137, 216, 217
Hydrobaeninae   8
Hydrobaenus   12, 105, 109, **127**, 128, 129, 133, 137, 217
inculta   46
intersectus   119, 120
*Isocladius*   12, 117, 118, 119, 120, 212, 213
Karelia   10, 41
karelicus   125, 130
kiefferi   50
Kiefferulus   11, 57, 60, 64, 65, **80**, 196
Krenopelopia   10, 38, 39, 40, **42**, 43
Krenosmittia   12, 105, 109, **128**, 129, 142, 217
*Kribioxenus*   82
Labrundinia   10, 39, 40, **43**, 44, 45, 46, 180, 181
lapponicus   128
laricomalis   119, 120, 213
Larsia   10, 27, 39, 40, **43**, 44, 48, 181
Lasiodiamesa   10, 51, **52**, 53, 184
Lauterbornia   11, 93, 94, 97
Lauterborniella   11, 57, 60, 81, 83, 85, 89, 196, 197
Limnochironomus   66, 67
Limnophyes   12, 104, 105, 108, 110, 125, **129**, 130, 217, 218
Linaceus   52
Lobodiamesini   152
longimana   156
longimanus   152, 233
longiseta   129

lugubris 128
Macropelopia 10, 26, 32, 33, 34, 35, 36, 45, 178, 179
Macropelopiina 26, 32
Macropelopiini 5, 9, 26, 27, 28, 31, **32**, 33, 35, 37
macropodus 73
maeaeri 127
marcidus 127
Meropelopia 10
Merotanypus 9, 35
Mesocricotopus 12
Mesopelopia 10
Mesopsectrocladius 140, 141
Metriocnemus 12, 103, 104, 107, 108, 113, 116, **130**, 131, 218, 219
Microchironomus 54, 58, 62, 71, 74
*Microcricotopus* 131
Micropsectra 12, 93, 94, **96**, 97, 98, 99, 168, 169, 205
Microtendipes 11, 56, 57, 59, 62, **81**, 82, 83, 84, 197
minima 42
Monodiamesa 13, 148, **149**, 150, 151, 237
Monopelopia 10, 39, 40, 43, **44**, 181
Monopsectrocladius 140, 141
Nanocladius 12, 101, 102, 103, 107, **131**, 132, 219
Natarsia 10, 26, 32, 38, 39, 40, **44**, 45, 48, 181
Natarsiini 26
nigrohalterale 84
Nilodorum 18, 54, 65
Nilotanypus 10, 38, 39, 40, 43, **45**, 46, 182
Nilothauma 11, 59, 62, **82**, 83, 197, 198
Nostoc 19
nymphoides 69
obnixus 119, 120, 213
Odontomesa 13, 148, 149, **150**, 237
Oliveria 132
Oliveridia 12, 105, 109, 128, **132**, 133, 219
Omisus 11, 59, 62, **83**, 89, 198
Oreadomyia 12
Orthocladiinae 5, 8, 12, 15, 16, 18, 21, 24, 25, 54, **100**, 101, 102, 106, 111, 113, 115, 117, 121, 130, 131, 133, 136, 140, 142, 143, 147, 148, 152

Orthocladius 12, 104, 108, 114, 118, **133**, 134, 137, 139, 212, 220, 221
Pagastia 13, 152
Pagastiella 11, 55, 56, 57, 60, 71, **83**, 84, 198
Paraboreochlus 49, 51
Parachaetocladius 13
Parachironomus 11, 58, 61, 70, 71, **77**, 78, 193
Paracladius 13, 103, 108, **134**, 135, 139, 143, 221, 222
Paracladopelma 11, 58, 61, 62, 70, 76, 77, **78**, 79, 80, 194
Paraclunio 13, **159**, 174, 175, 238
Paracricotopus 13, 103, 107, 112, **135**, 136, 222, 223
Parakiefferiella 13, 105, 109, **136**, 137, 223
Paralauterborniella 11, 59, 62, **84**, 85, 198, 199
Paralimnophyes 129
Paramerina 10, 38, 39, 40, 41, **46**, 48, 182
Parametriocnemus 13, 105, 110, **137**, 138, 224
Parapelopia 27, 32
Paraphaenocladius 13, 105, 109, 114, 137, **138**, 224
Paratanytarsus 12, 50, 52, 93, 94, 95, **97**, 98, 206
Paratendipes 11, 59, 62, **85**, 199
*Paratrichocladius* 13, 104, 108, 134, **138**, 139
Parochlus 10, 49, **50**, 51, 184
Parorthocladius 145
Pedionomus 54, 59, 63, 87
Pelopia 8, 30, 42
Pelopiinae 8
Pentaneura 10, 38, 39, 40, 43, 45, **46**, 47, 182
Pentaneurini 5, 10, 26, 27, 28, 32, **38**, 39, 41, 45, 46
*Pentapedilum* 11, 86, 87, 88, 200
Phaenopsectra 11, 59, 60, 63, 68, 69, **85**, 86, 87, 99, 199, 200
Phragmites 88
Phycoidella 13, 103, 107, **139**, 140, 225
Phytotendipes 11, 69, 70, 189
pilipes 128
Plecoptera 132

261

*Plecopteracoluthus*  12, 131, 132, 219
plumosus  64, 65, 66, 186
Podonominae  5, 8, 10, 15, 16, 17, 18, 21, 24, 25, **49**, 50
Podonomini  5, 10, 49, **50**, 51
Podonomus  52
Pogonocladius  12, 133, 134, 220, 221
Polypedilum  11, 59, 60, 63, **86**, 87, 88, 200, 201
Potthastia  13, 152, 155, **156**, 233
Procladiina  26, 32
Procladius  10, 27, 32, 33, 34, 35, **36**, 37, 162, 163, 179
Prodiamesa  13, 121, 148, 149, **150**, 151, 174, 175, 237, 238
Prodiamesinae  5, 8, 13, 16, 18, 21, 24, 25, 101, 148, 149, 150, 151
Prodiamesini  152
Protanypodini  5, 13, 152, 153, 154, **158**
Protanypus  13, 101, 148, 151, 153, **158**, 172, 173, 235
Protenthes  30
Psammocladius  124, 125
Psectrocladius  13, 101, 104, 108, 131, 136, **140**, 141, 225, 226, 227
Psectrotanypus  10, 26, 32, 33, 34, 35, 36, **37**, 38, 179, 180
Pseudochironomini  5, 11, 54, 55, 56, 57, **91**, 92
Pseudochironomus  11, 55, 56, **91**, 203
Pseudodiamesa  13, 152, 155, **157**, 234
Pseudokiefferiella  13, 155, 156, **157**, 158, 234, 235
Pseudorthocladius  13, 104, 109, 129, **141**, 142, 228
Pseudosmittia  13, 106, 110, 115, **142**, 143, 144, 146, 160, 161, 174, 175, 228
Psilodiamesa  152
Psilotanypus  10, 32, 36, 37
reductus  65, 66
reversus  119, 120
Rheocricotopus  13, 101, 103, 107, **143**, 174, 175, 228, 229
Rheopelopia  10, 38, 47
*Rheorthocladius*  133
Rheotanytarsus  12, 93, 94, **97**, 98, 206
riparius  37

Robackia  11, 58, 61, 70, 72, 73, **79**, 194, 195
Saetheria  11, 58, 61, 70, 78, **79**, 80, 195
salinarius  65, 66
*Saunderia*  146
Scirpus  88
semireductus  65, 66
*Sergentia*  86
signaticornis  69
Smittia  13, 106, 110, 115, 122, 142, **143**, 144, 146, 229
Spaniotoma  112, 114, 121, 122, 123, 140, 144, 145
Sphagnum  53
staegeri  65, 66
Stempellina  12, 92, 93, 94, 95, **98**, 111, 207
Stempellinella  12, 92, 93, 94, 95, **99**, 100, 207
Stenochironomus  11, 24, 25, 54, 59, 62, **88**, 201
Stictochironomus  11, 59, 62, **89**, 99, 202
subpilosus  127
sylvestris  119, 120, 212, 213
Symbiocladius  13, 106, 110, **144**, 145, 229
Sympotthastia  13, 152
*Syncricotopus*  138, 139
Syndiamesa  52, 157
Synorthocladius  13, 103, 107, 112, 121, **145**, 146, 230
Syntanytarsus  97
Tanypodinae  5, 8, 9, 15, 16, 17, 18, 20, 21, 23, 24, 25, **26**, 27, 28, 36, 42, 54, 101, 239, 240
Tanypodini  5, 9, 26, 27, 28, 29, **30**, 31
Tanypus  8, 9, 28, **30**, 31, 177
Tanytarsini  5, 8, 11, 17, 54, 55, 56, 57, 81, 84, 90, 91, **92**, 93, 95, 96, 98, 111
Tanytarsus  12, 68, 85, 89, 93, 94, 97, 98, **99**, 100, 208
Telmatogeton  159
Telmatogetoninae  5, 8, 13, 18, 21, 24, 25, 101, 154, **159**
Telmatopelopia  38
Telopelopia  27, 38
Tendipedidae  8
Tendipedinae  8
Tendipes  8, 64, 66, 67, 80

Thalassomyia 159
Thalassosmittia 13, 106, 110, 142, 144, **146**, 230
thienemanni 52
Thienemanniella 13, 102, 106, 117, **146**, 147, 230, 231
Thienemannimyia 10, 38, 39, 40, **47**, 48, 182
thummi 65, 66
tibialis 119, 120
tremulus 119, 120, 211
Tribelos 85, 86
*Trichocladius* 112, 117, 139, 143
Trichotanypus 10, 51, **53**, 185
trifascia 119, 120, 211

Tripodura 11, 87, 88, 201
triquetra 139
Trissocladius 127, 133, 147
Trissopelopia 10, 39, 40, 42, **47**, 48, 182, 183
Wirthiella 64
Xenochironomus 11, 57, 60, 65, **89**, 90, 202, 203
Xenopelopia 38, 47
Zalutschia 13, 104, 108, 128, **147**, 148, 231
Zavrelia 12, 93, 94, 95, 99, **100**
*Zavreliella* 81, 92
Zavrelimyia 10, 39, 40, 42, 44, 45, **48**, 183

263